Plasma Applications
in Gases, Liquids and Solids
Technology and Methods

Recommended Titles in Related Topics

Plasma Polymer Films
edited by Hynek Biederman
ISBN: 978-1-86094-467-3

Plasma Physics in Active Wave Ionosphere Interaction
by Spencer P Kuo
ISBN: 978-981-323-212-9

Visual and Computational Plasma Physics
by James J Y Hsu
ISBN: 978-981-4619-51-6

Magnetic Helicity, Spheromaks, Solar Corona Loops, and Astrophysical Jets
by Paul M Bellan
ISBN: 978-1-78634-514-1

Fundamentals of Theoretical Plasma Physics:
Mathematical Description of Plasma Waves
by Hee J Lee
ISBN: 978-981-327-675-8

Plasma Applications
in Gases, Liquids and Solids
Technology and Methods

Edited by

Claudia Riccardi
H Eduardo Roman

University of Milano-Bicocca, Italy

World Scientific

NEW JERSEY · LONDON · SINGAPORE · BEIJING · SHANGHAI · HONG KONG · TAIPEI · CHENNAI · TOKYO

Published by

World Scientific Publishing Co. Pte. Ltd.
5 Toh Tuck Link, Singapore 596224
USA office: 27 Warren Street, Suite 401-402, Hackensack, NJ 07601
UK office: 57 Shelton Street, Covent Garden, London WC2H 9HE

Library of Congress Control Number: 2023946325

British Library Cataloguing-in-Publication Data
A catalogue record for this book is available from the British Library.

ISBN 978-981-12-7592-0 (hardcover)
ISBN 978-981-12-7593-7 (ebook for institutions)
ISBN 978-981-12-7594-4 (ebook for individuals)

For any available supplementary material, please visit
https://www.worldscientific.com/worldscibooks/10.1142/13400#t=suppl

Typeset by Stallion Press
Email: enquiries@stallionpress.com

Preface

The application of plasma physics to the manufacturing and processing of materials has become a new challenging frontier. Partially ionized discharges are used in industry, and the performance of plasmas has already had a large commercial and technological impact. However, the science of non-thermal equilibrium plasmas is of great interest and the forms in which plasmas can be used and applied are growing and new phenomena and physics are becoming involved. This book features chapters on challenging areas of plasma science and examples of issues at the forefront. The underlying concept is that electric discharges in plasmas are expanding toward new horizons, assuming new forms and methods of applications. The chapters are concerned with general issues of the plasma state and its interaction with matter surface and bulk, as well as gases and liquids. New applications at the nanoscale and the diagnostic tools and simulations are highlighted.

The book has gained from the contributions of internationally recognized experts in their fields, and as such it represents a useful and up-to-date review of the main achievements in each area of research and applications.

Chapter 1 reviews the behavior of non-equilibrium electrical discharges in gases, at atmospheric pressure, and their fundamental chemical and physical properties. The chapter offers an introductory overview on chemical kinetics and dynamics, addressing the origin of chemical reactions and basic notions on transition state theory. Of particular interest are out-of-equilibrium environments such as chemical processes governed by vibrational excited states, and effects of the electron energy distribution on the coupling between molecular degrees of freedom and plasma constituents. Furthermore, key experimental aspects of plasma related parameters, commonly used in plasma chemistry, are quantitatively discussed in a consistent and unified fashion. Optical diagnostic techniques are examined as probes of space–time evolution of the relevant observables. The chapter concludes with a detailed discussion on plasma CO_2 conversion, with special emphasis on nanosecond repetitively plasma discharge techniques.

Chapter 2 deals with the optimization and development of plasma processing of polymeric materials, which are of importance for a wide range of manufacturing applications. The method, based on material chemistry concepts, is discussed in detail and is focused all the way down from the initial to the final produced material. The chapter reviews different technological solutions regarding the film deposition, surface treatment and etching of polymer materials, obtained by the implementation of both low- and atmospheric pressure plasma techniques.

Chapter 3 discusses the issue of plasma techniques applied to the treatment of liquids, of which water cleaning represents one of the most important applications. Electro-physical methods based on the use of electrical discharges of a specific type and power are discussed in detail. In particular, non-equilibrium low-temperature plasmas in liquids, including thin plasma layers on a liquid surface, are considered. The methods are relevant to a variety of technological and biomedical applications, civil engineering processes and environmental protection issues. Specifically, the chapter discusses the effects of streamers, which are bright, short-living, chaotically spreading and branching thin current filaments, and the methods for their production in/on dielectric and conductive liquids.

Chapter 4 deals with the application of plasma techniques to create new materials with outstanding scaling properties at the nanoscale. The chapter focuses on the plasma-assisted supersonic jet deposition technique, which is becoming very popular due to its precision and controllability in the fabrication of thin films, displaying both a broad range of widths and a variety of morphologies. The method is described in detail, with emphasis on the different processes taking place in the plasma chamber, where the nanoparticles constituting the building blocks of the film are generated, and in the deposition chamber, where the supersonic jet carries the nanoparticles to be deposited on a substrate. Monte Carlo simulations' results are also discussed to illustrate the dynamics and expansion of Ar^+ ions in a jet. The chapter closes with a description of the different morphologies effectively produced in the laboratory using TiO_2-based nanoparticles as the building blocks of the deposited thin films.

The present book is unique as it collects in a unified and organized fashion the broad field of low-energy plasma applications in technology regarding the three states of matter: gases, liquids and solids. It could also be useful as an accompanying textbook in advanced graduate courses related to plasma applications in a broad sense.

Claudia Riccardi and H. Eduardo Roman
Milan, July 2023

Acknowledgments

We would like to thank our long standing collaborators, Prince Alex, Ruggero Barni, Ilaria Biganzoli, Stefano Caldirola, Chiara Carra, Matteo Daghetta, Elisa Dell'Orto, Fabio Di Fonzo, Silvia Freti, Francesco Fumagalli, Simone Magni, Alessandro Mietner, Vittorio Morandi, Cecilia Piferi, Moreno Piselli, Riccardo Siliprandi and Dario Tassetti, for their constant support and for creating a friendly working atmosphere over the years. A special thanks to the researchers and technicians of the Plasma Prometeo Center of the physics department for their precious contributions. The Politecnico di Milano and the Istituto Tecnologico Italiano, IMM-CNR Bologna, are gratefully acknowledged.

UNIVERSITA' DEGLI STUDI DI MILANO BICOCCA

25

1998 | 2023

The publication of this book marks and celebrates the 25th anniversary of the University of Milano-Bicocca.

Contributors

Luca Matteo Martini

Dipartimento di Fisica, Università di Trento,

Via Sommarive 14, 38123 Povo (TN), Italy

luca.martini.1@unitn.it

Giorgio Dilecce

CNR - Istituto per la Scienza e Tecnologia dei Plasmi,

Via Amendola 122/D, 70126 Bari, Italy

giorgio.dilecce@cnr.it

Paolo Tosi

Dipartimento di Fisica, Università di Trento,

Via Sommarive 14, 38123 Povo (TN), Italy

paolo.tosi@unitn.it

Fabio Palumbo

CNR-Istituto di Nanotecnologia, Unità di Bari, Consiglio Nazionale delle Ricerche, c/o Dipartimento di Chimica,

Università degli Studi di Bari Aldo Moro, via Orabona 4, 70125, Bari (Italy)

fabio.palumbo@cnr.it

Fiorenza Fanelli

CNR-Istituto di Nanotecnologia, Unità di Bari, Consiglio Nazionale delle Ricerche, c/o Dipartimento di Chimica,

Università degli Studi di Bari Aldo Moro, via Orabona 4, 70125, Bari (Italy)

fiorenza.fanelli@cnr.it

Antonella Milella
Dipartimento di Chimica, Università degli Studi di Bari Aldo Moro, via Orabona 4, 70125, Bari (Italy)
and
CNR-Istituto di Nanotecnologia, Unità di Bari, Consiglio Nazionale delle Ricerche, c/o Dipartimento di Chimica, Università degli Studi di Bari Aldo Moro, via Orabona 4, 70125, Bari (Italy)
antonella.milella@uniba.it

Pietro Favia
Dipartimento di Chimica, Università degli Studi di Bari Aldo Moro, via Orabona 4, 70125, Bari (Italy)
and
CNR-Istituto di Nanotecnologia, Unità di Bari, Consiglio Nazionale delle Ricerche, c/o Dipartimento di Chimica, Università degli Studi di Bari Aldo Moro, via Orabona 4, 70125, Bari (Italy)
pietro.favia@uniba.it

Francesco Fracassi
Dipartimento di Chimica, Università degli Studi di Bari Aldo Moro, via Orabona 4, 70125, Bari (Italy)
and
CNR-Istituto di Nanotecnologia, Unità di Bari, Consiglio Nazionale delle Ricerche, c/o Dipartimento di Chimica, Università degli Studi di Bari Aldo Moro, via Orabona 4, 70125, Bari (Italy)
francesco.fracassi@uniba.it

Yuri Akishev
SRC RF TRINITI, 108840, Moscow, Troitsk, Pushkovykh Street, Vladenie 12, Russia,
and
NRNU MEPhI, 115409, Moscow, Kashirskoe Shosse, 31, Russia
akishev@triniti.ru

Stefano Caldirola
Dipartimento di Fisica, Università di Milano-Bicocca, Piazza della Scienza 3, 20126 Milano, Italy
stefano.caldirola@unimib.it

Claudia Riccardi
Dipartimento di Fisica, Università di Milano-Bicocca,
Piazza della Scienza 3, 20126 Milano, Italy
claudia.riccardi@unimib.it

H. Eduardo Roman
Dipartimento di Fisica, Università di Milano-Bicocca,
Piazza della Scienza 3, 20126 Milano, Italy
hector.roman@unimib.it

© 2023 World Scientific Publishing Company

https://doi.org/10.1142/9789811275937_fmatter

Contents

Fabio Palumbo, Fiorenza Fanelli, Antonella Milella,
Pietro Favia and Francesco Fracassi

Yuri Akishev

Chapter 1

Non-equilibrium Plasmas in Gases at Atmospheric Pressure

Luca Matteo Martini[*,‡], Giorgio Dilecce[†,§] and Paolo Tosi[*,¶]

Dipartimento di Fisica, Università di Trento
Via Sommarive 14, 38123 Povo (TN), Italy
†*CNR — Istituto per la Scienza e Tecnologia dei Plasmi*
Via Amendola 122/D, 70126 Bari, Italy
‡*luca.martini.1@unitn.it*
§*giorgio.dilecce@cnr.it*
¶*paolo.tosi@unitn.it*

This chapter discusses the non-equilibrium (NEQ) in plasma discharges operating at atmospheric pressure, providing an overview of their fundamental chemical and physical aspects. In the first part of the chapter, we recall the basic notions of chemical kinetics and dynamics. In particular, we discuss why chemical reactions occur and the basis of transition state (TS) theory. We focus on aspects of chemistry peculiar to out-of-the-equilibrium environments, like chemical processes involving vibrationally excited states and ions. After this brief recall of basic notions of chemistry, we discuss the crucial role of the electron energy distribution function (EEDF) in determining the way the discharge power is transferred to the different degrees of freedom of the plasma constituents. It turns out that the input energy can be partitioned among the various quantum states of the gas constituents so that non-Boltzmann distributions occur. This possibility allows boosting the reactivity by exploiting mode-selective chemistry. An exciting opportunity, so much investigated at present, is given by the potential role of the vibrational excitation of CO_2 in favoring the reduction reaction to $CO+O$.

The second part of the chapter discusses key experimental aspects of plasma-mediated processes. Using gas discharges to stimulate the desired chemical reactions requires the accurate assessment of several process parameters, such as conversion, yield and selectivity, plasma power, specific energy input and process efficiency. These quantities, which are regularly employed in plasma chemistry, are often poorly specified. Therefore, we review their definition and the procedure to retrieve them from the experimental observables, discussing how to avoid common missteps. We also present a selection of optical diagnostics techniques, finalized at investigating the space and time evolution of physical and chemical observables. Indeed, revealing the NEQ nature of non-thermal plasmas requires dedicated diagnostics to detect the local value of critical quantities. Finally, since the plasma activation of carbon dioxide is a hot field nowadays, we discuss a selection of the most recent research works on plasma CO_2 conversion, with particular attention to the case of the nanosecond repetitive discharges that represent one of the most challenging environments for the microscopic investigation of plasma discharges.

Contents

Dedication

This work is dedicated to our dear friend and colleague Nicola Gatti, who passed away prematurely. He was a brilliant researcher and a kind-hearted person. He will be greatly missed by everyone who had the pleasure of knowing him.

1. Introduction

The successful and widespread use of gas discharge technologies in many applications relies on their being systems in highly non-equilibrium (NEQ) conditions. The majority of applications can be classified as 'Plasma Chemistry' [1]. The role of NEQ is to initiate non-thermal chemistry, which is characterized by a significant production of reactive species compared to heating the gas. This peculiarity leads to a *high-T* chemistry in a relatively *cold* environment. The players of this 'miracle' are the charged species and the internal excitation of molecules, both vibrational and electronic. Consolidated industrial applications mainly pertain to surface treatment requiring spatial uniformity over large areas, which can be achieved by low-pressure discharges (≤ 1 mbar). Gas-phase chemistry, instead, requires processing large mass fluxes that atmospheric pressure discharges can fulfill. The target for industrial applications, in this case, is to reach competitive energy efficiencies by stressing and tailoring the NEQ conditions in such a way as to channel as much energy as possible into the process of interest.

The NEQ locution itself reveals an underlying philosophy. Thermodynamic equilibrium (TE) is the starting point, NEQ is a variety of deviations from it. In the modern approach, TE is seen as *an active collection of elementary balances, all in equilibrium. In contrast, a departure from equilibrium can be specified as the non-equilibrium state of some of these balances* [2]. The consequence of the practical description of a NEQ plasma system is, again quoting from [2], *any situation contains some equilibrium so that the statistical laws are practically applicable.*

In other words, conditions are looked for within which one or more balances are still in equilibrium and can be described by simple statistical laws, i.e., the Maxwell distribution for free particles, the Boltzmann distribution for bound states, the Saha equation for the ionization balance, the Planck law for radiation. Technically, this approach can be pursued by analytical calculus using a few parameters, namely two or more temperatures.

Conditions for a correct application of this approach are rarely found in gas discharges for plasma chemistry. As mentioned, extreme NEQ is looked for to enhance the efficiency and applicability of plasma technology. The description and optimization of gas discharge processes are better approached from an opposite point of view, in which the concept itself of temperature is forgotten. Each plasma component is treated with its own identity. Each quantum level has its kinetic equation; all are coupled by the detailed cross-sections of elementary processes. We shall refer to this as the *microscopic* approach. It has been made possible by the impressive progress in numerical computation, the progressive accumulation of data on elementary processes, and the advancement of selective diagnostic techniques.

In this chapter, we outline the main features of NEQ in gas discharges, with a particular focus on vibrational kinetics and chemistry, and illustrate these concepts in a few applications of moderate and atmospheric pressure discharges. For this purpose and to establish a common language, we first recall the basics of chemical kinetics, including an introduction to the chemistry of ions and vibrationally excited molecules.

2. Equilibrium Chemistry: A Primer

2.1. *Why do chemical reactions occur?*

A process proceeds spontaneously if the total entropy $S + S'$ of the system and its environment increases, i.e.,

$$dS + dS' \geq 0. \tag{1}$$

Since $dS' = -dq/T$, where dq is the heat *supplied* to the system, it follows that for any change

$$dS - \frac{dq}{T} \geq 0. \tag{2}$$

If the pressure P is constant and there is no work other than expansion work, then $dq_p = dH$ (enthalpy) and one obtains

$$T\,dS \geq dH. \tag{3}$$

It is worth noting that the latter expression allows us to express the condition for a spontaneous process only in terms of the system's state functions, thus neglecting the environment. Since the Gibbs function is $G = H - TS$, if the system changes at a constant T, the criterion for spontaneous change is

$$dG_{T,P} = dH - T\,dS \leq 0. \tag{4}$$

Therefore, the direction of spontaneous chemical processes is toward the minimum Gibbs function, an alternative way to say that the total entropy (system plus environment) increases.

In the case of a chemical reaction, $G_{T,P}(n_1, n_2, \ldots)$ changes as a function of the mixture composition. If ξ measures the progress of the reaction, the chemical equilibrium is reached when

$$\left(\frac{\partial G}{\partial \xi}\right)_{T,P} = 0. \tag{5}$$

The equilibrium constant K_p is the ratio between the partial pressures of the products and the reactants (for gas-phase reactions and treating the gases as perfect). If $K_p > 1$, the partial pressure of the products exceeds that of the reactants and vice versa for $K_p < 1$. K_p can be expressed as a function of $\Delta G^{\ominus}$, the standard reaction Gibbs function, that is, the variation in G when the reactants in their standard states ($P^{\ominus} = 10^5$ Pa) change to products in their standard states,

$$K_p = \exp\left(-\frac{\Delta G^{\ominus}}{RT}\right) = \exp\left(-\frac{\Delta H^{\ominus}}{RT}\right) \exp\left(\frac{\Delta S^{\ominus}}{R}\right). \tag{6}$$

At constant pressure, the equilibrium composition is controlled by T. If $\Delta H^{\ominus} < 0$, the reaction is *exothermic*, and increased T favors the reactants. If $\Delta H^{\ominus} > 0$, the reaction is *endothermic* and increased T favors the products. If the entropy rises so much that $T\Delta S > \Delta H$, an endothermic reaction is driven by the increase of entropy of the system.

At the equilibrium, the number of reactants (products) is the sum of the populations over the reactant (product) states, proportional to the partition function. Since the equilibrium constant is the ratio of the number of products and reactants, it can be expressed as the ratio between the respective partition functions. Therefore, the physical basis of the equilibrium constant is that the equilibrium composition reflects the overall Boltzmann distribution of populations on the reactant and product states. In summary, the equilibrium composition depends on T and the relative free energies $G^{\ominus}$ of reactants and products. Time is not explicitly considered here. Thermodynamics only tells us that by waiting long enough, the equilibrium is eventually reached.

2.2. How quickly does a reaction mixture approach equilibrium?

The reaction rate, the speed at which a reaction occurs, is proportional to the reactants' concentrations raised to some power. The constant of proportionality $k(T)$ is called rate constant. For an elementary (single step) process, the equilibrium constant K turns out to be the ratio between the forward k_f and reverse k_r rate constants

$$K = \frac{k_f}{k_r}. \tag{7}$$

The rate constant $k(E)$ or $k(T)$ can be calculated by integrating the reactive cross-section over the appropriate distribution.

$$k(v) = v\sigma(v), \tag{8}$$

$$k(T) = \left(\frac{\mu}{2\pi k_B T}\right)^{\frac{3}{2}} 4\pi \int \sigma v \exp\left(-\frac{\mu v^2}{k_B T}\right) v^2 \, dv$$

$$= \sqrt{\frac{8k_B T}{\pi \mu}} \int \sigma \frac{E_t}{k_B T} \exp\left(-\frac{E_t}{k_B T}\right) d\left(\frac{E_t}{k_B T}\right). \tag{9}$$

The last integral has the dimension of a cross-section, while the first term is the thermal relative-velocity. However, the integral is not a thermally averaged cross-section, but rather an average of the cross-section over molecules' thermal flux. The average energy of the flux molecules is higher than the average energy since faster molecules contribute more to the flux.

The above expressions are valid for reactants in a particular state and assuming that the kinetic energy follows a thermal distribution. In general, one has to sum over all possible reactants' states

$$k(T) = \sum_i p_i k_i, \tag{10}$$

where the probability of each state i is

$$p_i = \frac{1}{Q(T)} \exp\left(-\frac{E_i}{k_B T}\right). \tag{11}$$

The partition function Q can be written as an integral over the density of states

$$Q = \int_0^\infty \exp\left(-\frac{E}{k_B T}\right) \rho(E) \, dE. \tag{12}$$

Then the general expression for the thermal rate constant is

$$k(T) = Q^{-1} \int_0^\infty k(E) \rho(E) \exp\left(-\frac{E}{k_B T}\right) dE. \tag{13}$$

2.3. *Energy threshold and barrier: Transition state theory*

For some reactions, the cross-section vanishes below a threshold value that represents the minimum energy needed for the reaction to take place. Endothermic reactions have a natural thermodynamic threshold. However, the reaction threshold can often be higher than the thermochemical threshold. In this case, one speaks of an *activation barrier* (both for endothermic and exothermic reactions). The barrier prevents many exothermic reactions from occurring at room temperature, with the notable consequence, as an example, that it is possible to handle fuels.

One can depict a chemical reaction as a collisional event starting from the reactant geometry and ending in the products. The so-called potential energy surface (PES) is the potential energy as a function of the system geometry throughout the rearrangement from reactants to products. On the PES, the minimum energy path leading from reactants to products identifies the reaction coordinate. At a given point of the reaction coordinate, chemistry occurs: Reactants' bonds are broken while new bonds form in a continuous transformation from reactants to products. Being a local maximum in the PES due to the partial bonds' breaking, this region is where the barrier is located along the reaction coordinate. Because the latter passes through the local minima of the surface, the potential energy increases sideways from the path. Thus, near the barrier, the PES shows a saddle that separates the reactants from the products. The system's configuration at the saddle point corresponds to a molecular complex, the transition state (TS), that owns one imaginary frequency corresponding to the unbound motion that carries the system across the barrier from the reactants to the products, see Fig. 1.

The reaction rate is then determined by the rate at which the colliding molecules pass through the TS, which, being unbound along the reaction coordinate, naturally evolves into the products. The idea at the base of the TS theory (Eyring, Wigner, Evans, Polanyi) is that it must be possible to calculate the reaction rate by the methods of statistical mechanics if the PES is known [3]. The procedure assumes a large number of systems in thermal equilibrium and calculates the probability of being at the TS on the PES. The next step calculates the rate at which the systems at the TS decompose into the products. Eventually, the rate of barrier crossing turns out to be proportional to $N^{\ddagger}(E - E_0)$, the number of internal (bound) states of the TS whose internal energy ε_i is in the range $0 \leq \varepsilon_i \leq E - E_0$, where E is the total energy and E_0 the energy of the ground state of the

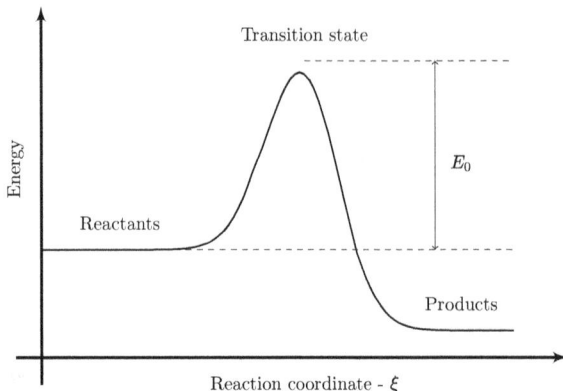

Fig. 1. Schematic energy profile for an exothermic reaction with a barrier. E_0 also accounts for the difference in zero-point energy between the TS and the reactants.

TS with respect to the reactants' ground state. The rate per unit energy of crossing the barrier is

$$\gamma(E) = \frac{1}{h} N^{\ddagger}(E - E_0). \tag{14}$$

The factor $1/h$ arises from the one-dimensional number density of transition states along the reaction coordinate per unit translational energy. The rate constant is obtained dividing the rate by the reactant concentration $\rho(E)$

$$k(E) = \frac{N^{\ddagger}(E - E_0)}{h\rho(E)}. \tag{15}$$

Computing the reaction rate is thus reduced to counting how many internal states of the TS are in the range $0 \leq \varepsilon_i \leq E - E_0$. The thermal rate constant turns out to be

$$k(T) = \frac{k_{\mathrm{B}} T}{h} \frac{Q^{\ddagger}}{Q} \exp\left(-\beta E_0\right), \tag{16}$$

where $\beta = (k_{\mathrm{B}} T)^{-1}$, $Q^{\ddagger}$ is the partition function for the internal states of the TS , and Q, the partition function of the reactants. In this formula, we recognize three factors with a clear physical meaning. The first one is related to the reactants flux along the reaction coordinate, and the last one gives the fraction of molecules with enough energy to pass the barrier. The middle factor, that is the ratio between partition functions, corresponds to the ratio between the number of populated states of TS and reactants. In this way, a dynamic quantity as the reaction rate is led back to the TS structure [4].

2.4. Thermodynamics vs kinetic control of chemical reactions

Thermodynamics tells us that if we wait long enough, then

$$\frac{[\text{products}]}{[\text{reactants}]} = \exp\left(-\frac{\Delta G^{\ominus}}{RT}\right). \tag{17}$$

TS theory tells us that the rate constant is

$$k(T) = \frac{k_{\mathrm{B}}T}{h}\frac{Q^{\ddagger}}{Q}\exp\left(-\beta E_0\right). \tag{18}$$

A reaction may follow different paths with different products starting from the same reactants. A *thermodynamically controlled* reaction depends on the difference between $G^{\ominus}$ of the products and the reactants and yields the lowest energy products. A *kinetically controlled reaction* depends on the energy difference between the TS and the reactants and follows the lowest energy barrier pathway. By varying the reaction conditions, one can affect the product distribution.

2.5. Chemistry of vibrationally excited molecules

For endoergic reactions and exoergic reactions with an energy barrier, the essential requirement is that the reactants own total energy above the barrier height. Since the reactants' total energy can be partitioned between translational and vibrational energy, the question arises about the most efficient energy's form to surmount the barrier. The first attempt to tackle the problem was made by Polanyi and Wong [5] for the case of an atom–diatom reaction. Polanyi's rules identify the barrier's position, in the path from reagents to products, as the key to answering the question. Suppose the particle configuration at the barrier is more similar to that of the reactants. In this case, one speaks of an *early barrier*, and the translational energy is most beneficial to surmount it. On the contrary, vibrational energy is more efficient for surmounting a *late barrier*, corresponding to a configuration that looks more like that of the products. The crucial concept behind Polanyi's rules is that kinetic energy must be available along the reaction coordinate to promote the process.

In 2013, Jiang and Guo [6] generalized Polanyi's rules to polyatomic systems by considering the coupling of the reactants' vibrational modes with the reaction coordinate at the TS (sudden vector projection, SVP, model). In the sudden limit (collision time much shorter than intramolecular vibrational energy redistribution, IVR), the coupling between a reactant mode and the reaction coordinate at the TS is approximated by the projection of the corresponding normal mode vectors. If the projection is large, the energy deposited into this mode flows readily into the reaction coordinate, thus enhancing the reactivity. If the projection is small, the coupling is low, and the reactivity is not enhanced [7]. Thus, concerning the vibrational excitation, the possible increase of the chemical reactivity is not naively due to a mere increase in the total reactant energy. Rather, to be effective, the excited vibrational mode must overlap with the TS vibration characterized by the imaginary frequency, i.e., the motion along the reaction path. Some coupling always happens to some extent in a NEQ plasma since electrons excite all the vibrational modes.

2.6. *Ionic chemistry*

The relative importance of ionic reactions depends on the particular plasma considered. In highly ionized plasmas at high pressure, ion reactions contribute to the reactivity by various processes, as charge-transfer, production of excited states, and electron recombination. On the other extreme, ions also play a crucial role in the chemistry of low-density and low-temperature plasmas, as the interstellar space or planetary ionospheres. In these environments, neutral chemistry is hindered by the low collision probability, whereas ion–molecule reactions take advantage of long-range electrostatic interaction. The latter can be understood at an elementary level by considering the ion as a point charge that polarizes a neutral sphere. The induced dipole $\overrightarrow{P}$ is proportional to the electric field $\overrightarrow{E}$ generated by the charge q

$$\overrightarrow{P}(R) = \alpha\,\overrightarrow{E}(R), \tag{19}$$

where α is the (scalar) polarizability ($\mathrm{C\,m^2\,V^{-1}}$).

The radially symmetric potential energy of the dipole in the electric field is

$$V(R) = -\int_0^E \alpha\,E\,dE = -\alpha\,\frac{E^2}{2} = -\alpha\,\frac{q^2}{2(4\pi\varepsilon_0)^2 R^4}. \tag{20}$$

Therefore, the long-range interaction between the point charge q and the induced dipole is proportional to $1/R^4$.

In the centre-of-mass reference frame, the scattering of two particles in a central potential, with collision energy E_{CM} and impact parameter

b, is conveniently reduced to the one-dimensional motion of a particle of reduced mass μ. For this purpose, the kinetic energy of the angular motion is expressed as the function $V_{cen}(R)$, the centrifugal potential, that depends on the (conserved) angular momentum L.

$$V_{cen}(R) = \frac{L^2}{2\mu R^2} = \frac{E_{CM} b^2}{R^2}. \tag{21}$$

Therefore, the one-dimensional relative motion of two particles occurs in an effective potential V_{eff} obtained by adding the centrifugal potential to the potential energy

$$V_{eff}(R) = V(R) + V_{cen}(R). \tag{22}$$

The Langevin–Gioumousis–Stevenson model assumes that an ion–molecule reaction occurs any time the collision energy is sufficient to overcome the centrifugal barrier, namely

$$E_{CM} \geq V_{eff}(R_{max}). \tag{23}$$

The largest impact parameter that satisfies the former inequality is given by

$$b_{max}^2 = \frac{q}{4\pi\varepsilon_0} \sqrt{\frac{2\alpha}{E_{CM}}}. \tag{24}$$

Thus, the Langevin cross-section is

$$\sigma_L = \pi b_{max}^2 = \frac{q}{4\varepsilon_0} \sqrt{\frac{2\alpha}{E_{CM}}}, \tag{25}$$

and the Langevin rate constant K_L turns out to be independent of the collision energy,

$$K_L = \frac{q}{2\varepsilon_0} \sqrt{\frac{\alpha}{\mu}}. \tag{26}$$

So far, K_L is based on an oversimplified model. However, it is useful for a first-order approximation for the upper limit of the reaction rate constant, typically around $10^{-9}\,cm^3 s^{-1}$. Improvements of the Langevin model are possible using a more realistic PES, but maintaining the original assumption that the reaction occurs with unitary probability whenever reactants can approach at short range, namely when the centrifugal barrier is surmounted. This generalization of the Langevin approach defines the so-called capture models. They are based on the assumption that long-range forces mainly govern the reaction dynamics. Therefore, the problem can be simplified using a single coordinate, the distance between the colliding particles. Capture models are unreliable in charge-transfer reactions, typified as

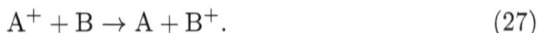

$$A^+ + B \rightarrow A + B^+. \tag{27}$$

Here, the process is controlled by the interaction between the two charge-transfer states and non-adiabatic effects must be taken into account. For the electron jump to occur, there must be a crossing between the potential curves correlating asymptotically with $A^+ + B$ and $A + B^+$, respectively. The two charge states mix at the crossing, whose location and coupling terms determine the reaction probability [8]. In the context of plasma-chemistry, charge-transfer reactions that convert atomic into molecular ions can have important consequences for electron–ion recombination processes since the latter are much faster in the case of molecular ions.

3. Non-equilibrium Features of Gas Discharges

Without much oversimplification, we can state that the acceleration of electrons is the primary energy input from the electromagnetic field to the plasma. The validity of this statement is based on the small mass ratio of the electron over heavy particles, $m_e/m_M \ll 1$. Due to this small ratio, which is easily less than 10^{-4} for molecular gas, the prevalent elastic collisions do not cause significant energy losses allowing electrons to reach thresholds of inelastic processes: Ionization, excitation of bound (either vibrational or electronic) states and direct dissociation of molecules. It is worth noticing that ionization sustains, to a large extent, elevated charge densities into the plasma. Producing a sufficient population of excited states leads to further energy transfer and chemical processes between heavy species. The degree of NEQ depends on the hierarchy of the rate coefficients of the energy transfers between the free and bound degrees of freedom of the plasma constituents. A schematic picture of the energy distribution in a molecular plasma is given in Fig. 2. The thickness of the arrows is proportinal to the amount of energy transfer, and the inelastic processes are in ascending order of required electron energy. Much of the gas heating is due to the de-excitation of bound states. The target of a competitive plasma process is to maximize the energy channeled into the desired processes over the energy lost in gas heating.

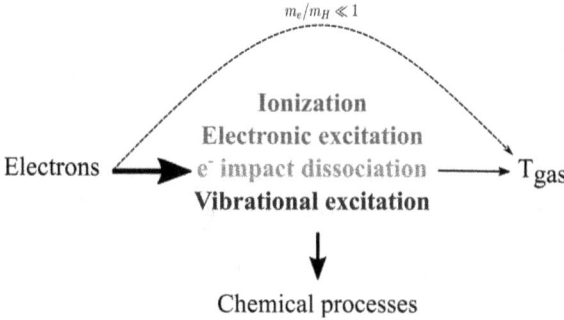

$m_e/m_H \ll 1$

Ionization
Electronic excitation
Electrons ⟷ ▶ e⁻ impact dissociation ⟶ T_{gas}
Vibrational excitation

↓

Chemical processes

Fig. 2. Representation of the energy transfers in a strongly non-equilibrium molecular gas discharge plasma.

A common way to visualize the hierarchy of average energies in electrical discharges still makes use of the temperature parameter,

$$T_e > T_{vib} > T_{ion} \sim T_{rot} = T_{gas}. \tag{28}$$

Typical orders of magnitude are: $T_e \sim (10^4 - 10^5)\,$K, $T_{vib} \sim (10^3 - 10^4)\,$K, $T_{gas} \sim (300 - 10^3)\,$K. Actually, only rotational and kinetic distributions of heavy particles can be described by a temperature. The electron gas and the vibrational excitation usually show non-Maxwell and non-Boltzmann distribution functions, that will be treated in detail in Section 3.1. The kinetic temperature of ions is close to the gas temperature except for the boundary regions with walls and electrodes, where, at low pressure, the ions can gain directional kinetic energy before reaching the wall/electrode. This is an important feature in the plasma etching technology for microelectronics processing.

NEQ plasmas have long been labeled in the Plasma Chemistry literature as *Cold* or *Non-Thermal* Plasmas. A complementary statement was that, as a rule of thumb, discharges below 10 Torr of gas pressure are cold plasmas, above 10 Torr they are thermal plasmas. Such definitions were popular during the 1980s–1990s, a period in which much of the research on plasma technology was devoted to low-pressure discharges for surface modification, while powerful plasma torches were employed in 'hot' technologies like plasma spraying or metal cutting. NEQ is more complicated than being a plasma in a translationally cold gas, and it can be easily found also in atmospheric pressure (ATP) discharge, as it has been clarified in the last three decades, when ATP devices like Corona, dielectric barrier discharges

(DBDs), nanosecond repetitively-pulsed (NRP) discharges have been more intensively studied.

3.1. *The electron component*

The energy transfer from electrons to the gas occurs through collision processes, whose kinetics is quantified by the rate coefficient,

$$k_{\text{coll}} = \int \sigma^{\text{coll}}(v)\, v\, f_e(v)\, dv, \tag{29}$$

where $\sigma^{\text{coll}}(v)$ is the cross-section of the process, $f_e(v)$ is the velocity distribution function and v is the relative velocity of the colliders, that, in case of electron-neutral collision, can be well approximated by the electron velocity. The cross-section is a property of the collider that reflects its quantum-mechanical nature. The electron velocity distribution function is the key property to induce modifications of the rate coefficients. The electron velocity distribution function (EVDF), $f(\vec{r}, \vec{v}, t)$, is the probability distribution of particle velocity $\vec{v}$, at spatial position $\vec{r}$ at time t, such that $f(\vec{r}, \vec{v}, t)\, d\vec{r}\, d\vec{v}$ is the number of particles in the six-dimensional phase space volume, $d\vec{r}\, d\vec{v}$, around the point $(\vec{r}, \vec{v})$ at time t.

The distribution function $f(\vec{r}, \vec{v}, t)$ obeys the Boltzmann equation,

$$\frac{\partial f}{\partial t} + \vec{v} \cdot \nabla_{\vec{r}} f + \frac{\vec{F}}{m} \cdot \nabla_{\vec{v}} f = \left(\frac{\partial f}{\partial t}\right)_{\text{coll}}, \tag{30}$$

where $\vec{F}$ is the Lorentz force, $\vec{F} = q(\vec{E} + \vec{v} \times \vec{B})$, and the term $(\partial f/\partial t)_{\text{coll}}$, the so-called collision integral, accounts for the effects of collisions. At equilibrium of the electron component, the solution to the stationary Boltzmann equation is the Maxwell distribution,

$$f^{\text{M}}(v) = \left(\frac{m}{2\pi k_{\text{B}} T_e}\right)^{3/2} 4\pi v^2 \exp\left(-\frac{mv^2}{2k_{\text{B}} T_e}\right), \tag{31}$$

in which $v = |\vec{v}|$, and the isotropy of the ensemble of velocity vectors is assumed. Expressed in terms of the kinetic energy, one gets the Maxwell–Boltzmann (MB) expression for the electron energy distribution function (EEDF),

$$f^{\text{MB}}(\varepsilon) = 2\sqrt{\frac{\varepsilon}{\pi (k_{\text{B}} T_e)^3}} \exp\left(-\frac{\varepsilon}{k_{\text{B}} T_e}\right), \tag{32}$$

which has dimension of $[\varepsilon^{-1}]$. A convenient representation of the energy distribution is given by the electron energy probability function (EEPF), defined as the EEDF divided by the square root of the kinetic energy, i.e., $f_p(\varepsilon) = f(\varepsilon)/\varepsilon^{1/2}$, having dimension of $[\varepsilon^{-3/2}]$. The EEPF of an MB distribution is a straight line on a semi-log plot. It allows one to quickly visualize if the distribution deviates from an MB one. Indeed, the EEDF can become a non-MB one in the cases of: rapid spatial and/or temporal variations in the electrostatic or electromagnetic fields; the presence of boundaries (electrodes, walls or catalytic surfaces); high anisotropy; inelastic and super-elastic collisions, where the electron kinetic energy is exchanged with internal degrees of freedom of atoms/molecules. Here we want to focus on the last item. A discussion of the other cases can be found in [9]. We make use of the Boltzmann solver BOLSIG+, which solves the Boltzmann equation with a polynomial expansion approximation as a function of the reduced electric field, E/N, i.e., the magnitude of the electric field, E, divided by the gas concentration, N, and a cross-section database for various atomic and molecular gases [10].

EEPFs, calculated using BOLSIG+ for a nitrogen plasma, are shown in Fig. 3. E/N values are expressed in Townsend units, $1\,T_d = 10^{-17}\,\mathrm{Vcm^2}$. Figure 3(a) shows the EEPFs calculated by considering three sets of collision processes in addition to ionization: elastic only, elastic plus vibrational excitation, elastic plus all the inelastic processes (i.e., adding the excitation of electronic states). The influence on the EEPF of these collisions can be appreciated by looking at Fig. 3(b), where representative cross-sections are plotted in the same energy range. In Fig. 3(c) calculations at three E/N values are plotted. It is clear that molecular processes impress a deep footprint on the EEPF shape, and the E/N value acts mainly on the electron average energy.

Finally, electron–electron collisions tend to restore the electron energy micro-balance, and therefore a Maxwellian distribution, when the long-range interaction begins to dominate over the inelastic processes at sufficiently high ionization degrees (Fig. 3(d). This kind of calculation describes the somewhat idealized situation of an electron gas in a stationary state with a local electric field. To be precise, the positive column of a DC discharge is a classic example where such a description fits well. Real discharges have, however, a more complex structure. Apart from the wall and electrodes boundaries, spatial structures are present, in which the EEDF contains non-local contributions. The negative glow, which is of fundamental importance for the self-sustainment of the DC discharge, contains an electron beam-like contribution due to the secondary electrons emitted by the cathode and accelerated in the cathode fall. These

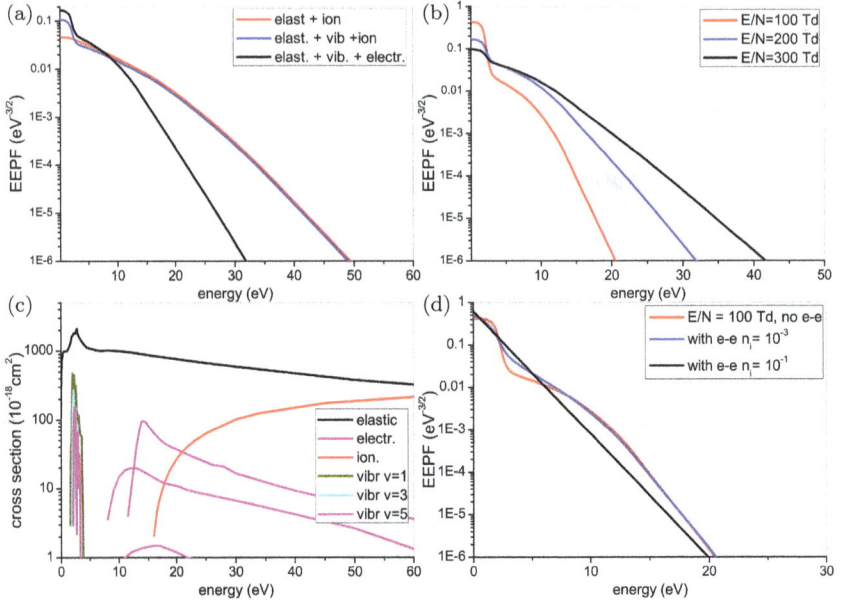

Fig. 3. BOLSIG+ calculations of the EEPF in nitrogen: (a) With progressive inclusion of inelastic collisions for vibrational and electronic nitrogen bound states excitation, $E/N = 100\,\mathrm{Td}$. (b) Cross-sections from the BOLSIG+ database. Just three out of the eight included in the database for vibrational excitation, and three for electronic excitation (of $A^3\Sigma$, $B^3\Pi$ and $C^3\Pi$), plotted with the same line color, to display the energy thresholds and range. (c) EEPF at three values of E/N. (d) Addition of e-e collisions with ionization degrees $n_i = 10^{-3}$ (middle curve) and $n_i = 10^{-1}$ (lower curve).

electrons are responsible for the Townsend breakdown mechanism, and a DC discharge cannot survive without a negative glow. On the contrary, the positive column is not necessary.

In RF (13.56 MHz) discharges, used for surface modification, the stochastic heating, by the spatial expansion and contraction of the sheath boundary close to the RF electrode, provides a non-local group of electrons that sustains the ionization balance in the whole discharge. This results in an approximately two-temperature EEDF (see [11, 12] for in-depth information on non-locality in electron kinetics).

NEQ in the electron kinetics is also found in atmospheric pressure discharges, especially in the streamer breakdown phase of micro-discharges that are the building blocks of dielectric barrier discharges, and the initial phase of NRP discharges that eventually evolve into sparks. The recent approach to their modeling is a fully kinetic particle-in-cell/Monte Carlo

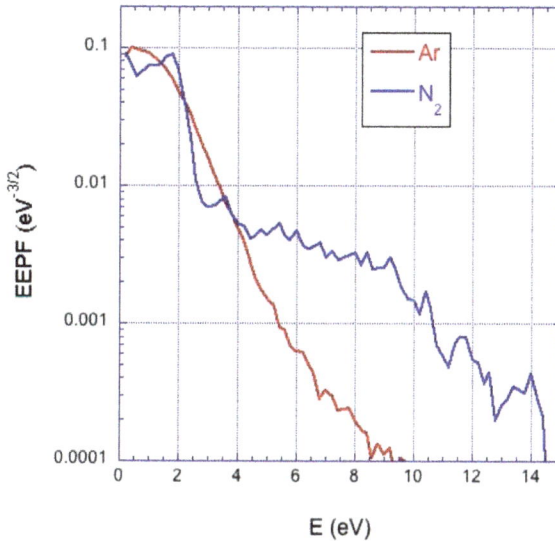

Fig. 4. EEPF in ns DBD at various positions in the discharge gap, in Ar (red curve) and N_2 (blue curve). Unpublished data kindly provided by F. Taccogna and calculated as described in [13].

collision (PIC-MCC). In [13], a high voltage ns pulse dielectric barrier discharge was modeled with a one-dimensional spatial approximation. EEPF calculations in N_2 and Ar are shown in Fig. 4. Although in a completely different physical condition, the quantum footprint of inelastic collisions on the EEPF is still clearly impressed, as well as its dependence on the gas nature, with its own set of bound states and collision processes.

The manipulation of the EEDF allows driving the electron energy into the preferred processes. By *driving* we do not mean to turn knobs such as to tailor perfectly the EEDF to the cross-sections but to look for conditions that maximize as much as possible the relevant rate coefficients. The design of H^- volume sources (for neutral beam injection in nuclear fusion [14,15]) is discussed as an example. The main volume production of H^- occurs by dissociative electron attachment (DEA),

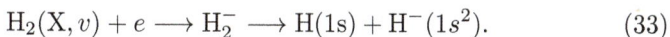

$$H_2(X, v) + e \longrightarrow H_2^- \longrightarrow H(1s) + H^-(1s^2). \tag{33}$$

A dramatic dependence on the vibrational level v[a] of H_2 is found for the DEA cross-section, see Fig. 5(a). The rate constants increase with

---

[a]From now on we use the letters v and w to refer to the vibrational quantum number.

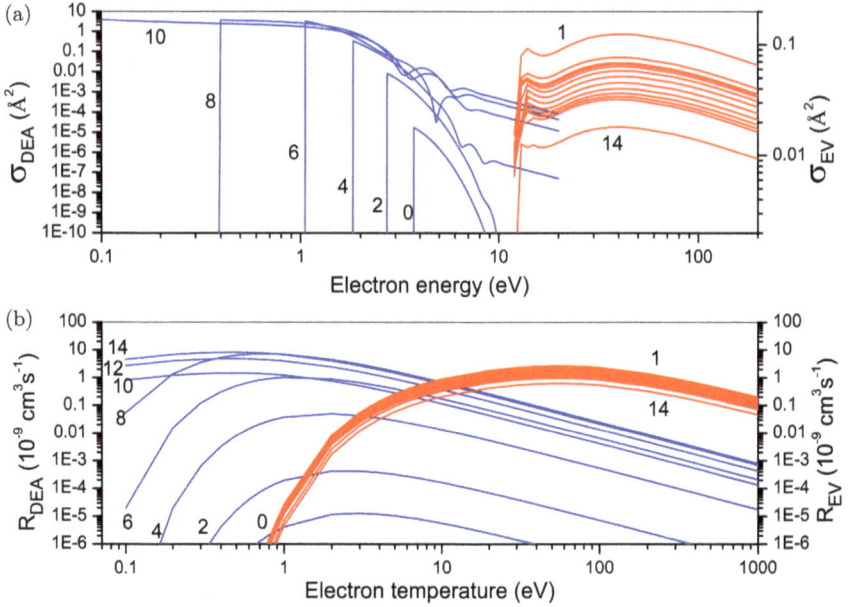

Fig. 5. (Left curves) (a) Cross-sections for dissociative attachment (DEA), Eq. (33), $0 \leq v \leq 10$. (b) Corresponding rate constants calculated for a Maxwellian EEDF as a function of the electron temperature, for $0 \leq v \leq 14$. (Right curves) (a) Cross-sections for radiative cascade (EV), Eq. (34), for $v = 0$ and $1 \leq w \leq 14$. (b) Corresponding rate constants. Data from [16]. Figure reproduced from [9] with permission.

v and have a maximum for very low electron temperatures, as can be appreciated in Fig. 5(b). A NEQ vibrational population of high v-levels can be promoted taking advantage of the radiative cascade (EV) process,

$$e + H_2(X, v) \longrightarrow e + H_2^*(B, C) \longrightarrow e + H_2(X, w) + h\nu, \qquad (34)$$

which populates mainly high v levels [14]. A look at Fig. 5(b) tells us that the best EV performance requires a high electron temperature.

A synergy of DEA and EV, therefore, requires two incompatible EEDFs. This issue is solved in the tandem H^- source [14] employing a magnetic filter [11] that separates two spatial regions, one with high T_e for high vibrational levels population by EV, the other with low T_e for the maximization of DEA. The EEDF is thus tailored for the best performance of H^- negative ions production.

3.2. *Molecular vibrations*

The previous example gives a flavor of the great importance of vibrational states in reaction dynamics and kinetics. Vibrational excitations can have a decisive role in plasma chemistry, given the ability of gas discharges to generate over-thermal vibrational populations. To see how this can happen, let us start with writing down the general form of the vibrational kinetic equation for the occupation number of vibrational level v,

$$\frac{dN_v}{dt} = \left(\frac{dN_v}{dt}\right)_S + \left(\frac{dN_v}{dt}\right)_{VV} - \left(\frac{dN_v}{dt}\right)_{VT} - \left(\frac{dN_v}{dt}\right)_R, \tag{35}$$

where four terms occur: The source (S) that adds quanta to the vibrational manifold, the redistribution without loss of vibrational quanta (VV), the loss by transfer of quanta either to kinetic energy (VT) or to chemical reactions (R). Source terms can add quanta from the bottom or from the top of the vibrational ladder. In the former case, bottom electron collisions yield

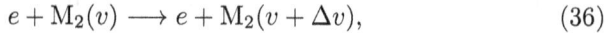

$$e + M_2(v) \longrightarrow e + M_2(v + \Delta v), \tag{36}$$

where for simplicity we consider here a diatomic molecule (M_2). From the top of the ladder, a variety of processes can be envisaged. In addition to radiative cascade, which we have already seen for H_2, Eq. (34), high-v levels can be populated by recombination, as for nitrogen atoms on a silica surface,

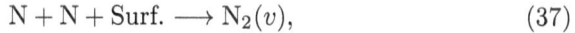

$$N + N + \text{Surf.} \longrightarrow N_2(v), \tag{37}$$

with a nascent vibrational distribution peaked around $v = 25$ [17]. Another possibility is by charge exchange, as for hydrogen [18]:

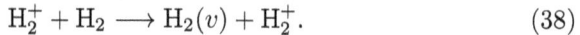

$$H_2^+ + H_2 \longrightarrow H_2(v) + H_2^+. \tag{38}$$

Vibration-to-translation (VT) energy transfer is the most important loss process, in which one or more vibrational quanta are lost in favor of kinetic energy, after collision with a molecule (VTm) or an atom (VTa),

$$M_2(v) + M_2 \longrightarrow M_2(w) + M_2 + \Delta\varepsilon_T, \tag{VTm}$$
$$M_2(v) + A \longrightarrow M_2(w) + A + \Delta\varepsilon_T. \tag{VTa}$$

Equation (VTa) is faster than Eq. (VTm), and in a molecular plasma it comes into play provided the dissociation degree reaches a sufficient value. VT loss acts on all v-levels, but its rate coefficient strongly increases with v and with the gas temperature (see, for example, for N_2, [19] and some relevant plots reported in [9]).

VV transfers deserve a broader discussion. Schematically,

$$M_2(v) + M_2(w) \rightleftarrows M_2(v') + M_2(w') + \Delta\varepsilon. \tag{39}$$

One-quantum processes are faster than multi-quantum ones since quasi-resonance is a preferred route. We restrict to this case to show in a simple way the principles of VV kinetics and its ability in determining strong NEQ in the vibrational distribution function (VDF). Complete numerical calculations of vibrational kinetics also include multi-quantum transfers.

The vibrational energy can be expressed as an expansion as a function of v,

$$E_v(v) = \hbar\omega \left[\left(v + \frac{1}{2} \right) - \chi_e \left(v + \frac{1}{2} \right)^2 + y_e \left(v + \frac{1}{2} \right)^3 + \cdots \right]. \tag{40}$$

Since molecular vibrations are regulated by an anharmonic potential, quasi-resonant VV processes have a positive energy balance,

$$M_2(v) + M_2(v) \rightleftarrows M_2(v - 1) + M_2(v + 1) + 2\hbar\omega\chi_e. \tag{41}$$

In Eq. (41), we stop the approximation at the quadratic term (with coefficient χ_e). The principle of detailed balance for the rate coefficients for the direct, $k_{v,v}^{v-1,v+1}$, and reverse process, $k_{v-1,v+1}^{v,v}$, implies,

$$k_{v,v}^{v-1,v+1} = k_{v-1,v+1}^{v,v} \exp\left(\frac{2\hbar\omega\chi_e}{T_0} \right), \tag{42}$$

where T_0 is the translational gas temperature.

It can be shown that the VDF that makes the VV flux equal to zero for each v is the Treanor distribution [20],

$$N_v = B \exp\left(-\frac{\hbar\omega v}{\Theta_1} + \frac{\hbar\omega\chi_e v^2}{T_0} \right), \tag{43}$$

where B is a normalization factor. This VDF distribution has a minimum at $v_{min} = T_0/2\chi_e\Theta_1$. For $v > v_{min}$, the population of the vibrational states increases. If $\chi_e \to 0$, the second term in the exponential of Eq. (43) is equal to zero and N_v tends to a Boltzmann VDF with $T_v = \Theta_1$. The same results can be obtained if $T_0 \to \Theta_1$.

VV exchanges are faster than VT losses at low v values, and show a milder dependence on the vibrational level and on gas temperature (again for N_2, see [22] and plots in [9]). At low v, the VDF looks like a Treanor distribution, while, on increasing v, VT losses start to compete with VV

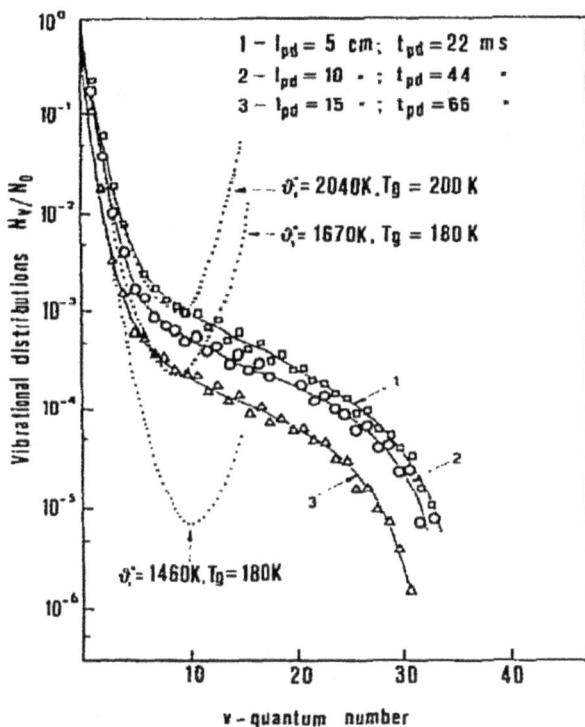

Fig. 6. Vibrational distributions of CO ground state interacting with active nitrogen in the afterglow of a nitrogen discharge. Reproduced from [21] with permission.

exchanges until they begin to dominate, forcing the VDF to a smooth decay (often called plateau), followed by a steep decrease. The plateau's extension depends on the gas temperature since the VT dependence on T_0 is sharper than that of VV. The typical VDF shape due to this VV vs VT competition is shown in Fig. 6, in which measurements of the VDF of CO in the flowing afterglow of a nitrogen discharge [21] are plotted. In that experiment, CO was injected in the post-discharge at various distances from the observation point, yielding different contact times between N_2 vibrationally excited in the discharge and CO. Vibrational quanta in CO were then transferred from N_2 by

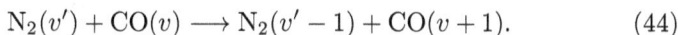

$$N_2(v') + CO(v) \longrightarrow N_2(v'-1) + CO(v+1). \tag{44}$$

In Fig. 6, dotted curves reproduce Treanor distributions that fit the VDF at low v. The effect of VV and VT kinetics in the VDFs is self-explaining.

The data refer to an afterglow cooled down to about 200 K. A comparison with VDFs in a non-cooled afterglow, at about 400 K, can be found in Fig. 3 of Ref. [21], which shows the importance of the gas temperature in enhancing VT losses. It is worth underlining the many orders of magnitude difference between the population of high v-levels and what it would be if the VDF were a Boltzmann distribution at $T_v = \Theta_1$. Nothing better than this figure can give an idea of how much NEQ in the vibrational manifold can impact on the chemistry in gas discharges.

An example of the interplay of electron and vibrational kinetics in NEQ conditions is the destruction and valorization of CO_2 by atmospheric pressure discharges. The first step is the dissociation of carbon dioxide that, in a discharge, can take place by direct electron impact dissociation or through the collision of two vibrationally excited CO_2 molecules,

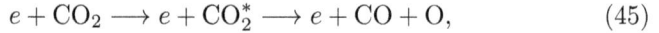

$$e + CO_2 \longrightarrow e + CO_2^* \longrightarrow e + CO + O, \tag{45}$$

and

$$CO_2(v) + M \longrightarrow CO + O + M, \tag{46}$$

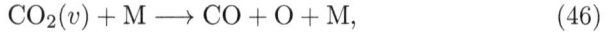

where M can be $CO_2(w)$ or another molecule. The process described by Eq. (45) proceeds via excitation to electronic states with energies $\gtrsim 7.5$ eV, which then lead to dissociation by three routes: The state is pre-dissociative; the state is bound but crosses a dissociative state; the state is bound and makes a cascade to one of the previous kind. The cross-sections reported in Fig. 7 are those measured in [23] and validated in [24]. In Eq. (45), there is a waste of electron energy since more than 7.5 eV are required to activate the process, which exceeds the 5.5 eV bond energy of CO_2 substantially. The vibrational mechanism Eq. (46), which involves the excitation of the anti-symmetric stretch of ν_3 of CO_2, can be, in principle, more efficient since energy is transferred from the bottom of the potential well, with small energy quanta addition. The process starts with the electron impact excitation of v levels,

$$e + CO_2(v) \rightarrow e + CO_2(v + \Delta v), \tag{47}$$

and proceeds with populating the high-v levels via the *climbing* of the vibrational ladder up to the necessary dissociation energy [25]. Since the dissociation of ground state $CO_2(^1\Sigma_g^+)$ on the $CO(^1\Sigma^+) + O(^3P)$ surface is spin-forbidden, one should also consider the barrier due to the inter-crossing system [26, 27]. A complete set of cross-sections for Eq. (47) and all the vibrational modes can be found in [28]. As an example, those

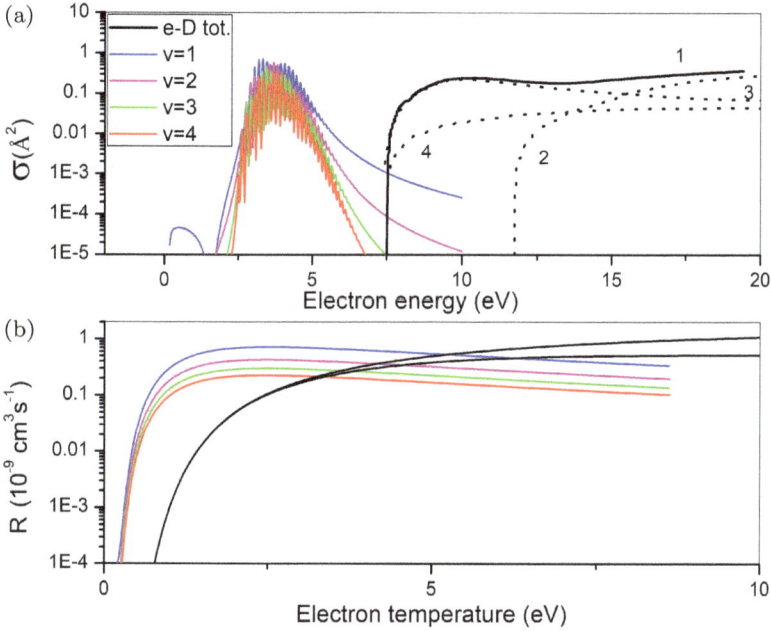

Fig. 7. (a) Cross-sections (colored curves) for electron impact resonant vibrational excitation of low v states, starting from $v = 0$, of the symmetric stretching mode of CO_2 [28]. Electron impact dissociation cross-sections (black curves) taken from Fig. 8 of [23] (1) sum of the partial cross-sections 2-4; (2) with formation of $CO(a^3\Pi)$; (3) by excitation of allowed transitions; (4) by excitation of forbidden transitions). (b) Corresponding rate coefficients, calculated for a Maxwellian EEDF, as a function of electron temperature. The electron impact dissociation is calculated with Polak's total cross-section in the two cases: (1) cross-section equals to zero at the end of the available data, and (2) extrapolated up to 100 eV.

relevant to the symmetric stretching $0 \rightarrow v = 1, 2, 3, 4$ are plotted in Fig. 7(a).

In Fig. 7(b), the rate coefficients for a Maxwellian EEDF as a function of the electron temperature are reported. The relative weight of the two dissociation mechanisms can be tuned by choosing conditions of different electron average energies. A successful strategy for an enhanced contribution of the vibrational mechanism should select conditions with electron temperatures between 1 and 2 eV and a gas temperature as low as possible, to lower VT losses. VT losses, by the way, are a waste of energy that, if too much, would invalidate the better energy efficiency of the vibrational dissociation mechanism.

3.3. *Transient reaction intermediates*

The role of electrons in plasma chemistry is either primary, in that they both generate and sustain the discharge and can directly induce a modification of the gas composition, or secondary, since they produce species which in turn promote further reactions. The vibrational manifold is one such example, but several transient reaction intermediates are also available in NEQ discharges whose concentration is orders of magnitude larger than that achievable at thermal equilibrium.

For example, the plethora of cases in materials processing [1, 29], in devices in use in plasma medicine [30], in plasma-assisted combustion [31], pollutants removal [32], large amounts of reactive oxygen and nitrogen species (ROSs and RONs) are found, like O_3, OH, O, N, and the molecular oxygen singlet metastable $O_2(a^1\Delta_g)$. To get an idea, in Atmospheric Pressure Plasma Jet (APPJ) devices for Plasma Medicine, concentrations are found to be of the order of $10^{15}\,\mathrm{cm}^{-3}$ for OH [33] and $O_2(a^1\Delta_g)$ [30], or up to $(10^{16}-10^{17})\,\mathrm{cm}^{-3}$ for O atoms [30].

3.4. *High pressure*

A naive idea of the relationship between NEQ and gas pressure would conclude that equilibrium should be reached by increasing the collision frequency. It is hopefully clear, from the brief account given up to now, that instead NEQ is rather an intensive property of a gas discharge plasma since it depends on the mutual relationship between the frequencies of microscopic processes. In more quantitative terms, to a first approximation, it is determined by ratios between the rate constants, not between collision frequencies. In other words, it is pressure independent. Of course, this is a first approximation that can be broken by conditions that induce a change in the gas composition, due to large dissociations and chemical reactions, or to the achievement of a huge ionization degree that tends to force to partial equilibrium by charged particles collisions. The same EEDF depends on an intensive property, the E/N reduced electric field. From a practical standpoint, increasing the pressure, one has to either enlarge E or reduce the inter-electrodes gap to resume the same E/N. At high pressure, the E/N value is not only just the applied voltage divided by the gap since local fields come into play in the streamer breakdown mechanism [1]. In the local field approximation, the *local* E/N determines the EEDF [10]. As a whole, at high pressure, NEQ is still there, but just on a shorter time/space scale compared to the case at low pressure.

4. Non-equilibrium Plasma for CO_2 Conversion

As a consequence of the anthropogenic activities, the CO_2 level in the atmosphere is steadily rising. Before the industrial age, the CO_2 concentration was approximately 280 ppm. At the present date, CO_2 level has reached about 418 ppm [34], a net increase of 50%. Apart from CO_2, other greenhouse gases (GHGs), such as CH_4 and N_2O, are emitted in the atmosphere due to anthropogenic activities and massively contribute to global warming [35]. CO_2 and many other GHG (e.g., CH_4) are hard-to-activate stable molecules. NEQ discharges are an attractive tool for the activation of such molecules and, in particular, CO_2 [36–42]. Compared to conventional chemical processes for CO_2 conversion, in NEQ plasmas it is possible to obtain higher energy efficiencies [43]. Besides this crucial property, NEQ plasmas are turnkey systems with excellent scalability. Renewable electricity can easily power NEQ plasma systems that can be used as 'peak-shaver', contributing to the grid stabilization and serving as a way to store renewable electrical energy in a chemical form [44].

In the past decades, the NEQ properties of moderate (larger than 1 mbar) and atmospheric pressure discharges have been widely exploited in many chemical and industrial applications [45–48].

In this section, we focus on an emerging application of NEQ discharges: The activation of stable molecules, and, in particular, the conversion of CO_2. We review how the relevant parameters used to characterize the processes can be obtained. We also present a selection of dedicated diagnostics, useful to reveal the role of NEQ.

4.1. *Macroscopic evaluation of plasma mediated conversion processes*

The characterization of a plasma process relies on the evaluation of two parameters: the power dissipated by the plasma and the average composition of the gas entering and exiting the reactor (or before and after the treatment for batch-reactors). The performance of a plasma-based process can be derived from the parameters as mentioned above.

4.1.1. *The power dissipated by the discharge*

The energy dissipated by the plasma and the steady-state gas composition at the exit of the reactor are quantities that are relatively easy to access. Experimentalists can use commercially available voltage and current probes and fast oscilloscopes for the direct electrical characterization of most discharge types. Voltage $V(t)$ and current $I(t)$ traces collected at the reactor can be used to compute the average power dissipated by the dischage,

$$P\,[\mathrm{W}] = \frac{1}{t_1 - t_0} \int_{t_0}^{t_1} V(t)\,[\mathrm{V}] \cdot I(t + \tau)\,[\mathrm{A}]\,dt, \tag{48}$$

where τ is a delay due to the spurious phase shift introduced by probes, cables and the digitalization system [49]. The integration interval is equal to one or more periods of the excitation waveform for a sinusoidal excitation. For excitation frequencies f up to the kHz range, τ can be neglected ($\tau \ll f^{-1}$). In the MHz range, and at higher frequencies, the spurious phase shift introduced by the probes and the acquisition system becomes appreciable and should be taken into account [50–52]. For microwave (MW) excitation sources, which operate at a narrow frequency bandwidth, directional power meters are available. These devices can directly measure the incident and reflected power from the plasma discharge [53].

For pulsed (broadband) excitation schemes, the integration interval of Eq. (48) has to be chosen equal to the time duration of one or more pulses. An example of a pulsed excitation scheme is the nanosecond repetitively pulsed (NRP), characterized by short and repetitive high-voltage pulses [54]. The high slew rate and short time-duration of the excitation pulses (in the ns or ps time scale) make the measurement of the I/V characteristic of the discharge particularly challenging. NRP I/V waveforms have a broadband frequency spectrum that can extend above 100 MHz. High-frequency signals correspond to excitation wavelengths short enough to become comparable with the physical size of the plasma system (power supply, reactor, cables, probes, etc.). Consequently, it is no more possible to represent the plasma setup with an electrical lumped-element model. Indeed, distributed-element models, like those used for transmission lines, are needed to represent the setup.

Also, the output impedance of the power supply is hardly matched by the plasma reactor one. Consequently, a complex pattern of incoming and reflected energy pulses is generated in the transmission line connecting the power supply to the plasma reactor (see Fig. 8). Thus, fast high-voltage and current probes are needed. Ideal high-voltage probes should have high impedance and low parasitic capacitance to maintain a high bandwidth without affecting the voltage drop on the electrodes. Most of

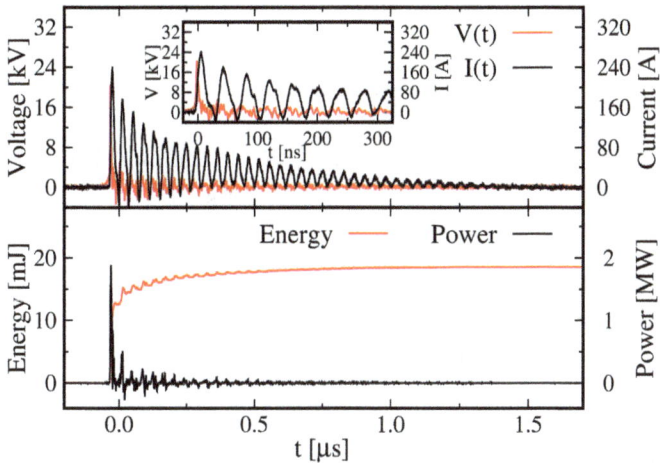

Fig. 8. Current and voltage (upper panel), instantaneous power and dissipated energy (lower panel) of a pin-to-sphere nanosecond pulse discharge ignited in CO_2 at atmospheric pressure. The power supply delivers to the transmission line a 10 ns-long high voltage pulse (see inset in the upper panel).

the time, commercially available high-voltage probes have a consistent stray capacitance and a finite input impedance. Such probes can attenuate the voltage pulse on the electrodes, and don't have a suitable bandwidth to record the actual voltage profile of the discharge.

Khomenko reported that high-parasitic capacitance probes, such as the P6015 by Tektronix, can overestimate the energy dissipated by the plasma by 25% with respect to low-parasitic capacitance probes (e.g., PP066, LeCroy) [55]. A valuable alternative to measuring current and voltage traces at the reactor with commercial probes is to monitor current and voltage signals in the transmission line connecting the power supply to the plasma reactor. Back current shunts (BCS) can be employed to probe signals traveling in a coaxial cable, and many research groups have reported their use [56–58].

Starting from the energy dissipated into the plasma and the reactant mole number, it is possible to define the specific energy input (SEI). The SEI is an intensive quantity helpful in comparing results from different discharge types [43]. For a plug-flow reactor, it is defined as

$$\text{SEI} \left[\text{kJ dm}^{-3}\right] = \frac{P\,[\text{W}]}{\phi\,[\text{cm}^3\text{s}^{-1}]}, \tag{49}$$

or alternatively,

$$\text{SEI} \left[\text{kJ mol}^{-1}\right] = \frac{22.4 \left[\text{dm}^3 \text{ mol}^{-1}\right] \cdot P \left[\text{W}\right]}{\phi \left[\text{cm}^3 \text{ s}^{-1}\right]}, \tag{50}$$

$$\text{SEI} \left[\text{eV mol}^{-1}\right] = \frac{6.24 \cdot 10^{21} \left[\text{eV kJ}^{-1}\right] \cdot 22.4 \left[\text{dm}^3 \text{ mol}^{-1}\right] \cdot P \left[\text{W}\right]}{N_A \cdot \phi \left[\text{cm}^3 \text{ s}^{-1}\right]}, \tag{51}$$

where $\phi \left[\text{cm}^3 \text{ s}^{-1}\right]$ is the flow of reactants entering the reactor and $N_A = 6.022 \cdot 10^{23}$, the Avogadro number. It is worth noting that according to Eqs. (50) and (51), the flow entering the reactor $\phi \left[\text{cm}^3 \text{ s}^{-1}\right]$ must be referred to temperature of 273.15 K and an absolute pressure of 101.325 kPa.

4.1.2. *Conversion, selectivity and energy efficiency*

The characterization of plasma mediated processes requires quantities representative of the progression of the chemical reactions promoted by the discharge. These parameters are functions of the gas composition before and after the reaction and the energy input in the system under investigation.

Let us consider a chemical reaction involving p reactants A_x and q products B_y,

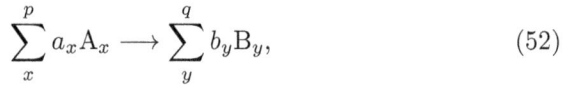

$$\sum_x^p a_x A_x \longrightarrow \sum_y^q b_y B_y, \tag{52}$$

where a_x and b_y are the stoichiometric coefficients of reactants and products, respectively. Stoichiometric coefficients indicate how many molecules B_y are formed during the chemical reaction, Eq. (52), when a given number of molecules A_x is consumed [59]. We define n_x^{in} and n_x^{out} as the moles of reactant x entering and exiting the reactor, respectively.[b] The conversion can be defined as

$$C_x = \frac{n_x^{\text{in}} - n_x^{\text{out}}}{n_x^{\text{in}}} = \frac{n_x^{\text{conv}}}{n_x^{\text{in}}}, \tag{53}$$

where $n_x^{\text{conv}} = n_x^{\text{in}} - n_x^{\text{out}}$ are the moles of reactant x converted in the reactor. If more reactants participate in the reaction, it is possible to define the total conversion C_{tot} as the sum of the individual conversion C_x weighted for the initial molar fraction of the reactant x [60]. Product selectivity S_y^k is another useful quantity to highlight differences in the products formation.

---

[b]Molar fluxes $\dot{n}_x^{\text{in}}$ and $\dot{n}_x^{\text{out}}$ can be used as well for the following discussion.

S_y^k indicates, for a given elemental constituent k made available by the conversion of reactants, how much of it is used for the creation of the product y,

$$S_y^k = \frac{b_y^k \cdot n_y^{\text{out}}}{\sum_x^{\text{cont.} k} a_x^k \cdot n_x^{\text{conv}}}, \tag{54}$$

where the sum runs over all the molar quantity of converted reactants n_x^{conv} containing the elemental constituent k. b_y^k and a_x^k are the stoichiometric weight of the elemental constituent k in the product y, and the reagent x, respectively. It is worth noting that for a given elemental constituent k, the sum of all the S_y^k is equal to 1.

The Y_y^k is another helpful quantity to understand which products are favored and to which extent. Y_y^k indicates, for a given elemental constituent k, how much of it is consumed for the creation of the product y with respect to the total amount of k available from reagents,

$$Y_y^k = \frac{b_y^k \cdot n_y^{\text{out}}}{\sum_x^{\text{cont.} k} a_x^k \cdot n_x^{\text{in}}}. \tag{55}$$

Starting from conversion and SEI, it is possible to derive other useful quantities helpful for the characterization of the process under study. One of the most common choices adopted is to compare the energy spent to convert the reactants to the standard reaction enthalpy ($\Delta H_{298\,K}^{\ominus}$) of the chemical reaction under investigation,

$$\eta = \frac{C_{\text{tot}} \cdot \Delta H_{298\,K}^{\ominus}\,[\text{kJ mol}^{-1}]}{\text{SEI}\,[\text{kJ mol}^{-1}]}. \tag{56}$$

η is called energy efficiency. This metric works well if a single chemical reaction can reasonably describe the process under investigation. For example, in the CO_2 splitting, the overall process can be described by,

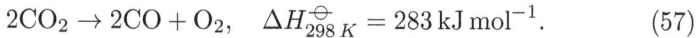

$$2CO_2 \rightarrow 2CO + O_2, \quad \Delta H_{298\,K}^{\ominus} = 283\,\text{kJ mol}^{-1}. \tag{57}$$

Literature data confirm that the reaction's stoichiometry is well reproduced by experiments (see [61, 62] and references therein). A different example is the dry reforming (DR) reaction,

$$CO_2 + CH_4 \rightarrow 2CO + 2H_2, \quad \Delta H_{298\,K}^{\ominus} = 247\,\text{kJ mol}^{-1}. \tag{58}$$

Although Eq. (56) is widely used to estimate the energy efficiency of the DR process [60], the validity of this approach is questionable. Different products are formed in the plasma mediated DR process besides CO and

H_2. The production of light hydrocarbons, oxygenates, carbon powder and water have been reported in addition to syngas [63, 64]. The metric of Eq. (56) is not appropriate for the DR process because Eq. (58) does not represent the process comprehensively.

In selecting the proper metric, other parameters can be considered. If the primary purpose is to evaluate the conversion efficiency, the latter can be defined as the ratio between the reactant moles converted and the plasma energy [65, 66]. If, instead, the production of a single species (e.g., H_2) or a products' family (e.g., syngas [67, 68]) is of interest, the process can be evaluated in terms of energy conversion efficiency (ECE), defined as the lower heating values (LHVs) of products divided by the sum of energy input and the LHV of the converted reactants. The LHV of a compound (or net calorific value) is a measure of the heat release by combustion when the water is in the vapor state (i.e., it does not include the latent heat of condensation of the water)[69]. A more general and universal approach can be pursued by considering in the ECE all the products that have an LHV different from zero.

In other words, the ECE becomes an indicator of the amount of energy, dissipated by the plasma and contained into the converted reactants, which is recovered into the products. A general definition of ECE is

$$\text{ECE} = \frac{\sum_y^q \text{LHV}_y \left[\text{kJ mol}^{-1}\right] \cdot n_y \left[\text{mol}\right]}{E \left[\text{kJ}\right] + \sum_x^p \text{LHV}_x \left[\text{kJ mol}^{-1}\right] \cdot n_x^{\text{conv.}} \left[\text{mol}\right]}, \tag{59}$$

where E is the energy used by the process and the sum is the energy contained in the fraction of reactants that have been converted. Energy storage efficiency (ESE) can also be defined. It is the ability of storing the energy dissipated by the plasma in the products,

$$\text{ESE} = \frac{\left(\sum_y^q \text{LHV}_y \left[\text{kJ mol}^{-1}\right] \cdot n_y \left[\text{mol}\right]\right)}{E \left[\text{kJ}\right]} -$$

$$+ \frac{\left(\sum_x^p \text{LHV}_x \left[\text{kJ mol}^{-1}\right] \cdot n_x^{\text{conv.}} \left[\text{mol}\right]\right)}{E \left[\text{kJ}\right]}. \tag{60}$$

4.1.3. *Gas composition*

The macroscopic characterization of a plug-flow reactor is based on the determination of the number of moles (or the molar flow rate) for each species that enters and exits the reactor (see Fig. 9).

Experimentalists generally take advantage of analytical equipment to monitor the composition of the reactor's input/output. Gascromathographs

(a)

Power supply

P [W]

$\sum_x^p \dot{n}_x^{in}$ → Plasma reactor → $\sum_x^p \dot{n}_x^{out} + \sum_y^q \dot{n}_y^{out}$

Pressure, Temperature

(b)

Flow $[\text{mol s}^{-1}]$

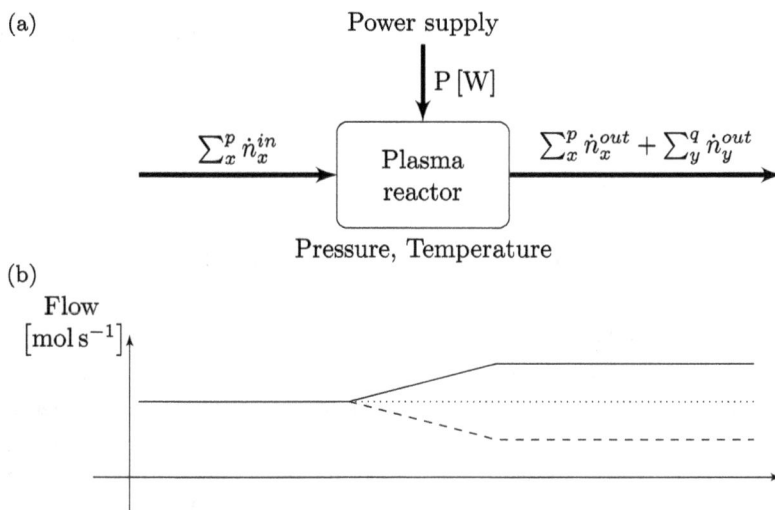

Fig. 9. (a) Typical representation of a plug-flow reactor for plasma processing. Reactants enter the reactor, and unconverted reactants and products are present in the exhaust flow. The other parameters useful for the characterization of the process are the power dissipated by the plasma, temperature and pressure. (b) The value of the molar flows entering and exiting the reactor can be different. The solid line represents a reaction leading to an increase of the molar flow (e.g., CO_2 splitting $2CO_2 \rightarrow 2CO + O_2$). The dotted line represents a reaction where no change in the flow is expected (e.g., reverse water gas shift $CO_2 + H_2 \rightarrow CO + H_2O$). The dashed line represents a reaction where the molar flow decreases (e.g., Sabatier's reaction $CO_2 + 4H_2 \rightarrow CH_4 + 2H_2O$).

(GC), Fourier-transform infrared spectrometers (FTIR), mass spectrometers (MS) or single species detectors can be used. They can sample/analyze a fraction or the whole of the gas exiting the reactor, and provide the molar fraction (m_y^{out}) or the absolute number density of the constituents of the gas mixture. In plug flow reactors, the determination of the species molar fraction is commonly carried out *in line* at the exit of the reactor.[c] The molar fraction of a certain species is defined as

$$m_j^{\text{in/out}} = \frac{n_j^{\text{in/out}}}{\sum_j n_j^{\text{in/out}}} = \frac{\dot{n}_j^{\text{in/out}}}{\sum_j \dot{n}_j^{\text{in/out}}} = \frac{\dot{n}_j^{\text{in/out}}}{\dot{n}_{\text{tot}}^{\text{in/out}}}, \tag{61}$$

[c]In this particular case, the terms flow *entering* and *exiting* the reactor refer to the conditions when the discharge is *off* and *on*, respectively.

where $\dot{n}_{\text{tot}}^{\text{in/out}}$ is the total molar flow entering/exiting the reactor. If we rewrite Eq. (53) as a function of molar fractions, we obtain

$$C_x = \frac{n_x^{\text{in}} - n_x^{\text{out}}}{n_x^{\text{in}}} = \frac{\dot{n}_{\text{tot}}^{\text{in}} \cdot m_x^{\text{in}} - \dot{n}_{\text{tot}}^{\text{out}} \cdot m_x^{\text{out}}}{\dot{n}_{\text{tot}}^{\text{in}} \cdot m_x^{\text{in}}} = \frac{m_x^{\text{in}} - m_x^{\text{out}} \cdot \dot{n}_{\text{tot}}^{\text{out}} / \dot{n}_{\text{tot}}^{\text{in}}}{m_x^{\text{in}}}$$

$$= \frac{m_x^{\text{in}} - \xi \cdot m_x^{\text{out}}}{m_x^{\text{in}}} = 1 - \xi \frac{m_x^{\text{out}}}{m_x^{\text{in}}}. \tag{62}$$

The quantity $\xi = \dot{n}_{\text{tot}}^{\text{out}} / \dot{n}_{\text{tot}}^{\text{in}}$ is the volumetric (molar) flow variation factor. From Eq. (62), we conclude that molar fractions (or the absolute number densities) cannot be used directly to compute the quantities presented in Section 4.1.2. The volumetric (molar) flow variation factor is representative of the variation of the molar flow inside the reactor due to chemical reactions (see Fig. 9(b)).

Pinhão *et al.* [70] proved that neglecting the change in the volumetric (molar) flow can introduce relevant systematic errors. The exact determination of the number of moles entering the reactor is generally straightforward given the availability of mass flow controllers/meters used to feed reactants into the plasma chamber. The determination of the number of moles exiting the reactor is less straightforward. Sometimes, it is possible to avoid the direct measurement of the volumetric flow by estimating it. A good example is the CO_2 splitting reaction, where the chemical process is unambiguously defined, and a balanced chemical reaction can be written. In this case, the molar fractions can be used to calculate relevant quantities with the precaution of applying a correction based on the reaction's stoichiometry [71–77]. A similar approach has been used by Klarennar *et al.* [78] who derived an analytical form for the conversion from the CO_2 and CO absolute number densities [79–83].

A more general and rigorous approach involves the direct estimation of the volumetric flow expansion rate. Different experimental techniques can be employed. Examples are the direct measurement of the flow with a mass flow controller or a bubble flow meter [84], and the addition of a constant flow of an internal standard (IS) before [85, 86] or after the plasma reactor [87–92]. The latter choice appears preferable to avoid the presence of the IS in the discharge zone (see Fig. 10). The analytical equipment will thus sample the species exiting the reactor plus a known amount of IS. If the molar flow changes, the recorded intensity of the signal of the IS will change accordingly. The IS addition after the plasma reactor has the advantage of being easy to implement and does not affect the plasma process. If the quantity of IS added is modest $\left(\dot{n}_{\text{IS}} \ll \dot{n}_{\text{tot}}^{\text{in}}, \dot{n}_{\text{tot}}^{\text{out}}\right)$, the determination of

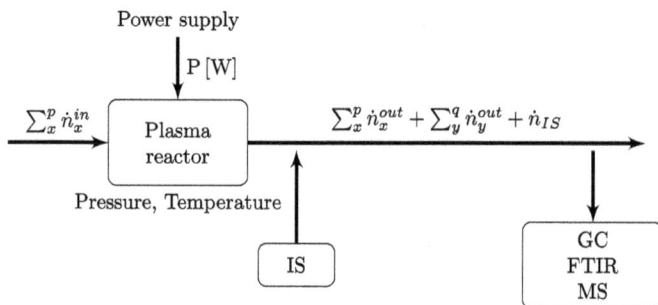

Fig. 10. The internal standard addition methodology can accurately detect the variation of the molar flow. The m_{IS} of the internal standard is inversely proportional to the change in molar flow.

the detector's response factor for the IS is not needed, and the signal of the IS is inversely proportional to the variation of the volumetric (molar) flow [70].[d] This methodology represents the most accurate and reliable way to measure the change in the gas flow at the exit of the reactor.

4.2. Local characterization of plasma-mediated processes

Macroscopic observables (e.g., gas composition, energy input) are insufficient to obtain a complete picture of an NEQ discharge. From macroscopic observables, it is difficult to establish if NEQ has a role or is even present in the process. The degree of NEQ can be, in fact, quite different depending on the type of discharge and experimental conditions employed. The local value of physical observables (e.g., pressure, temperature, etc.) and quantities relevant for understanding the kinetics (e.g., species concentration) can play a decisive role in revealing the NEQ. Nevertheless, NEQ systems at moderate- to high-pressure are characterized by spatial and temporal non-homogeneities. In most cases, the adoption of fast pulsing schemes for the energy injection [87,89,90,93] and the discharge morphology [94–96] require space- and time-resolved dedicated diagnostic techniques.

Diagnostic techniques can be distinguished as intrusive and non-intrusive. A Langmuir probe (LP) is an example of a device that must

---

[d]If A_{IS}^{off} and A_{IS}^{on} are the signals of the IS ($\dot{n}_{IS}$) when the plasma is off (the reactants exit the reactor untreated) and on (reactions take place), respectively, we have that $\xi = A_{IS}^{off}/A_{IS}^{on}$.

be inserted inside the plasma. LPs are used to extract information about electron density and electron temperature [97]. For low-pressure plasmas, retarding field energy analyzers (RFEAs) are used to measure ion energy and flux [98]. These devices are particularly suitable for homogeneous plasmas and can hardly be applied to moderate- to atmospheric-pressure discharges. For the latter, non-invasive optical techniques are recommended. A variety of optical diagnostic techniques have been developed in the field of plasma physics/chemistry. Optical diagnostics can be distinguished into two main categories: (a) passive and (b) active diagnostic techniques [99]. In the following section, some selected diagnostics will be presented to show how it is possible to reveal the NEQ properties of electrical discharges.

4.2.1. *Imaging*

Imaging is probably the most employed passive optical diagnostic technique. It relies on the light emitted by the plasma. Imaging can be done by directly collecting the light from the discharge [100], with the help of spectral filters to limit the bandwidth of the radiation impinging in the sensor or using fast cameras to reveal the time evolution of the discharge [101–103]. Schlieren and shadowgraph photography are also employed to reveal inhomogeneities (i.e., a difference in the refractive index) in optical media [104, 105]. Both are useful to reveal the pressure and temperature gradients produced by pulsed discharges and visualize fluid dynamic phenomena[106, 107].

In the particular case of NEQ discharges generated by repetitive pulses of the duration of around 10 ns, imaging can be used to gain insights on the effect of changing the inter-pulse time. Montesano *et al.* [91, 92] performed imaging of a pin-to-pin NRP discharge to understand the effect of the shortening of inter-pulse time on the spatial features of the discharge. An example of images collected in a CO_2-CH_4 mixture at atmospheric pressure is presented in Fig. 11.

For delays shorter than $40\,\mu s$, successive discharges propagate in the same channel. Also, they are characterized by a lower breakdown voltage and larger current compared to pulses separated by more than $40\,\mu s$. Shortening the inter-pulse interval from $833\,\mu s$ to $40\,\mu s$ increases the ECE from 50% to 65% at constant SEI. Higher densities of lower-energy electrons characterize discharges with lower breakdown voltages and higher discharge currents. This condition (high density of low energy electrons) favors the vibrationally enhanced dissociation mechanism of CO_2 and might explain the improved performance [108]. However, thermal effects due to the

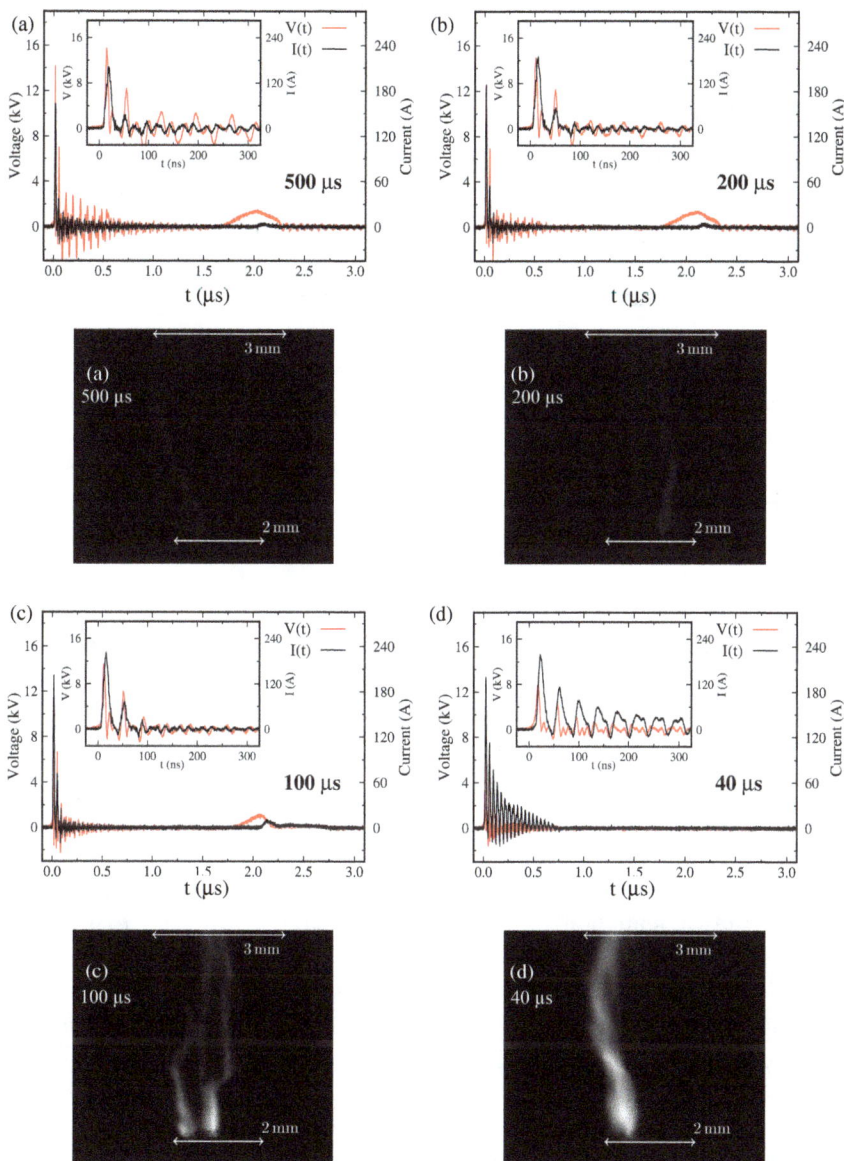

Fig. 11. I/V characteristics (first and third row) of the second pulse in a burst of nanosecond pulses, and commutative images (second and fourth row) of a sequence of three pulses. Images and I/V characteristics are collected at different inter-pulse times: (a) $500\,\mu s$, (b) $200\,\mu s$, (c) $100\,\mu s$ and (d) $40\,\mu s$. Anode (top arrow) and cathode (bottom arrow) sizes and positions are reported in the images. Figure adapted with permission from [92].

shortening of the delay between successive pulses and the persistence of the discharge in the same volume cannot be excluded [92].

4.2.2. *Emission spectroscopy*

Optical emission spectroscopy (OES) is another passive technique that relies on recording the radiation emitted by the plasma after separating its spectral components. OES can be used either for analytical purposes or to extract information about the discharge. OES is widely employed due to its simplicity and the low cost of the devices needed to record the plasma emission (e.g., a spectrometer and a light detector). Nevertheless, extracting quantitative information from OES spectra is challenging. Linking the density of excited species, responsible for the emission of light, to the ground state density is not straightforward. It requires detailed knowledge of the different excitation mechanisms (e.g., electron impact, chemiluminescence, etc.), and of the collisional processes that compete with the radiative de-excitation. The development of a collisional radiative model is thus required, together with the measurement of the absolute emission from the discharge. At moderate- to atmospheric-pressures, absolute emission measurements can be challenging due to the presence of gradients in the plasma and the re-absorption of the emitted radiation. Actinometry can be employed to circumvent all these difficulties. It can be applied to determine the ground state density of a species populated via electron impact in a discharge [109, 110]. It is based on admixing a known amount of a gas (the actinometer, most of the time an atom) to the plasma and assuming the validity of the so-called corona model [111].

The optical emission from the actinometer can be used to monitor variations in electron energy and density. These fluctuations can alter the rate of excitation of the species from the ground state to the excited light-emitting state. It is worth noting that the knowledge of electron impact rate constants for both the actinometer and the species under investigation is required. The accuracy in the determination of the EEDF is thus one of the primary sources of systematic error for this technique [99].

Other parameters are accessible via the analysis of OES spectra and are helpful to reveal the NEQ properties of atmospheric pressure discharges. In particular cases, one can infer the (translational) gas temperature from the rotational distribution function of the emitting molecule [112]. However, one should carefully verify that the rotational temperature of the excited vibro-electronic state reproduces or is in equilibrium with the gas temperature. The latter condition occurs only if the rate of the rotational energy transfer is (much) larger than the overall rate of de-excitation (radiative plus non-radiative) of the light-emitting state [113]. This case, however, is hardly satisfied for collisions with molecular gases, which, differently from atomic gases, are often characterized by high quenching rate constants. Unless rotational thermalization of the excited state is guaranteed, knowing the excitation mechanisms that populate the light-emitting state is mandatory. Electron impact excitation from the ground state projects the rotational distribution function of the ground on the excited state. On the contrary, this is not guaranteed when the latter is populated by other processes (e.g., chemical reactions, dissociation processes, etc.).

Figure 12 reports an example of emission spectrum collected from a nanosecond repetitively pulsed discharge in CO_2, with a small addition of N_2 [114]. The emission spectra of different species are recorded with a gate of 9 ns at the beginning of a discharge pulse (10 ns duration). Emissions from nitrogen's second positive system (SPS) ($N_2(C^3\Pi_u\text{-}B^3\Pi_g)$)

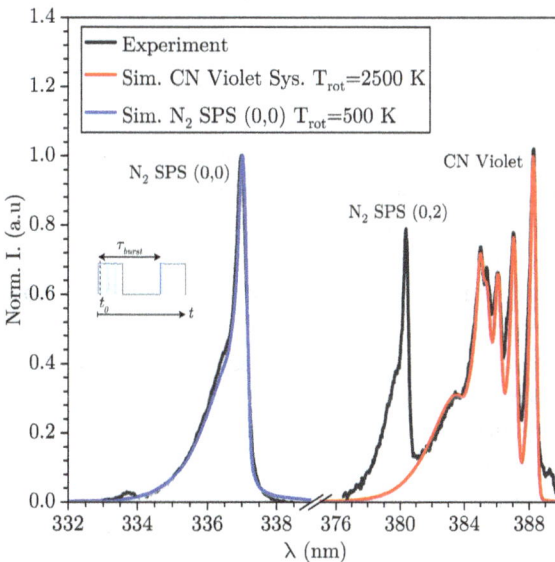

Fig. 12. Emission spectra recorded in the first 9 ns of a nanosecond pulse in CO_2 admixed with 5 % of N_2. Simulation of the SPS(0,0) of N_2 and $\Delta\nu = 0$ sequence of CN Violet System are reported. Figure reproduced from [114] with permission.

and the violet system of CN $(CN(B^2\Sigma^+\text{-}X^2\Sigma^+))$ are shown together with two best-fit simulations. The rotational temperatures that characterize the two emissions are pretty far apart, therefore demonstrating that rotational thermalization is not guaranteed. Understanding the mechanisms that populate the two molecular states is thus critical. The $N_2(C)$ can be populated in the discharge by electron impact excitation from the ground state $N_2(X)$ or by the pooling reaction from the metastable $N_2(A)$,

$$N_2(A) + N_2(A) \rightarrow N_2(C) + N_2(X). \tag{63}$$

However, the latter does not occur in the present case because just before the discharge $N_2(C)$ emission is not observed. Also, if the $N_2(C)$ were produced by the pooling reaction, the vibrational distribution of the $N_2(C)$ state would show an overpopulation of the $v = 1$ state with respect to the $v = 0$ one, which is not observed. Thus, it is safe to assume that the $N_2(C)$ is produced by electron impact from the ground state, ensuring that $N_2(C)$ rotational distribution mirrors that of the $N_2(X)$ ground state. The latter conclusion is not valid for the CN(B) state, which is known to be populated by the recombination reaction,

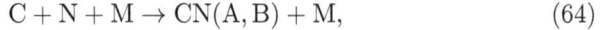

$$C + N + M \rightarrow CN(A, B) + M, \tag{64}$$

which generates supra-thermal vibrational and rotational population [115]. In this case, the rate of rotational energy transfer (RET) is lower than the rate of depopulation and does not guarantee the thermalization of the CN(B) state. Therefore, the correct estimation of the gas temperature is that obtained from the N_2 SPS emission. Measured values of gas temperature reported in [114] ranges from $2000\,\text{K}$ at $10\,\mu\text{s}$ after the discharge pulse to $500\,\text{K}$ at $1\,\text{ms}$. From these data, it is possible to extrapolate a maximum temperature of around $2500\,\text{K}$ closer in time to the discharge event (around $1\,\mu\text{s}$).

The NEQ features of the NRP discharge are evident when comparing the gas temperature with the electron temperature (T_e) measured during the discharge phase. T_e and electron density (n_e) can be estimated from the emission spectrum. The line shape and linewidth of a transition are affected by several broadening processes (e.g., natural, doppler, resonant Stark, etc.). The Stark broadening is due to the interaction of the light-emitting state with charged particles present in the plasma.

The electron density can be derived from the line profile of atomic transitions [116, 117]. In a CO_2 plasma, both O and C lines are present and can be used to this purpose. In Fig. 13, electron densities are presented

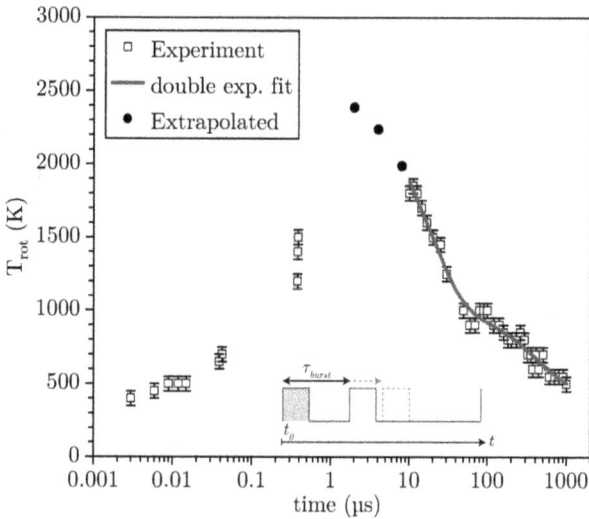

Fig. 13. Gas temperature evolution in a nanosecond pulse (first 10 ns) and its after-pulse in CO_2 admixed with 5 % of N_2. The temperature between 1 μs and 10 μs is extrapolated from a double exponential fit of the after-pulse temperature. Figure adapted from [114] with permission.

together with electron temperatures determined from the fit of C^+ and C^{++} lines simulated with the NIST-LIBS simulation tool assuming that the local thermodynamic equilibrium (LTE) condition is fulfilled in the discharge [118]. When comparing the results presented in Fig. 13 with the gas temperatures in the discharge, NEQ properties are evident, since the electron temperature is several orders of magnitude larger than the gas temperature.

From the observation of emission spectra, other plasma properties can be derived. Both line and continuum (Bremsstrahlung and recombination) radiation can be emitted by the plasma, revealing information about charged particles. From line radiation, it is possible to extract, besides the rotational temperature, the vibrational temperature of an excited state or ground state of a molecule [109, 119, 120]. More complicated is to infer valuable parameters from the radiation emitted by excited states produced via chemical reactions (chemiluminescence). That requires detailed knowledge of the kinetics involved. An interesting example is the use of the flame bands originated by the reaction,

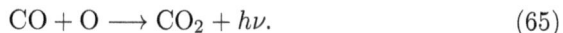

$$CO + O \longrightarrow CO_2 + h\nu. \tag{65}$$

Raposo *et al.* [121] developed an empirical model for simulating flame bands as a function of gas temperature in the spectral range between 260 nm and 700 nm.

4.2.3. *Absorption spectroscopy*

Absorption spectroscopy is a line-of-sight spectroscopic technique mainly employed to get quantitative information about species concentration. It is an active technique in which a light beam is injected into a medium (the sample). The difference between the light intensity before and after the interaction with the sample is recorded and translated in absolute number densities. Many experiments take advantage of absorption spectroscopy to provide an insight into the NEQ nature of electrical discharges. Besides absolute number densities, rotational and vibrational temperatures of molecules in the ground state are accessible. Time-resolved *in situ* Fourier transform infrared (FTIR) spectroscopy has been used in the past years for this purpose. Klarenaar *et al.* [78] demonstrated a method to determine vibrational temperatures of CO_2 and CO in a glow discharge with a time-resolution of $10\,\mu s$. This methodology has been used to measure the time evolution of the temperature of the anti-symmetric stretch (T_3), the Fermi coupled symmetric stretch and bending mode ($T_{1,2}$) together with the rotational temperature (T_{rot}), in a low pressure (1.6 mbar and 6.7 mbar) pulsed glow discharge in CO_2 [80, 122–124]. However, the plasma emission and the large instrumental broadening of the FTIR spectrometers, when working at limited pressure (around 10 mbar) and temperature (lower than 2000 K), can be limiting factors (in terms of signal-to-noise ratio) for the sensitivity of the technique [125].

Narrowband tunable diode lasers can be used to overcome this limitation [126]. Damen *et al.* [125] demonstrated by *in situ* narrow linewidth quantum cascade laser (QCL) absorption spectroscopy the possibility of measuring the gas temperature from the width of the absorption lines in addition to rotational and vibrational temperatures of CO and CO_2 in a pulsed glow discharge in different mixtures of CO_2 and N_2. Rotational temperature estimations of CO and CO_2 were in good agreement with the gas temperature.

In Fig. 14, the effect of N_2 admixing on rotational and vibrational temperatures is presented. During the discharge phase (*plasma on*), the vibrational temperature of the Fermi coupled symmetric stretch and bending mode of CO_2 ($T_{1,2}$) is slightly larger than the rotational temperature; the vibrational temperature of the anti-symmetric stretch mode of CO_2

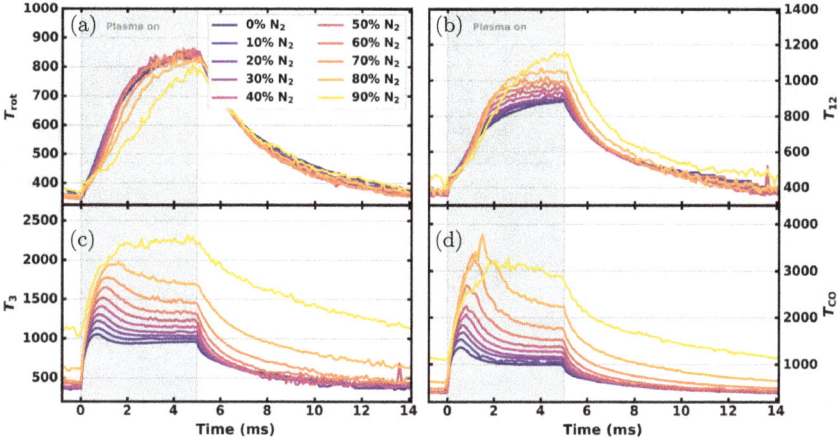

Fig. 14. Time evolution of the rotational and vibrational temperatures in different CO_2/N_2 mixtures in a pulsed (5 ms on-time; 10 ms off-time) glow discharge operating at 6.67 mbar. (a) Rotational temperature. (b) Symmetric stretch and bending mode vibrational temperature of CO_2. (c) Asymmetric stretch vibrational temperature of CO_2. (d) Vibrational temperature of CO. Figure reproduced from [125] with permission.

(T_3) and the vibrational temperature of CO (T_{CO}) are much larger than the rotational temperature. Interestingly, the addition of more N_2 causes the increase of both T_{CO} and T_3 and makes unlikely the thermalization of these two modes with the gas temperature in the afterglow. The increase in T_{CO} and T_3 indicates that vibrationally excited N_2 can serve as an energy reservoir for the asymmetric stretch of CO_2 and the vibrational excitation of CO (Eq. (44)).

The NEQ properties of the glow discharge in CO_2 are evident in the difference between vibrational and rotational temperatures. Figure 15 shows the vibrational temperatures for different amounts of water addition, as measured with QCL absorption spectroscopy [83]. The difference of vibrational temperatures, in particular T_{CO} and T_3, with respect to rotational temperature, tends to decrease when the water fraction increases, suggesting that water induces quenching of the first vibrational level of CO_2 and CO, thus reducing the degree of NEQ.

4.2.4. *Laser-induced fluorescence*

Laser-induced fluorescence (LIF) is an active optical technique similar to the absorption spectroscopy. Instead of collecting the energy lost by the light beam interacting with the sample, LIF determines the absorbed

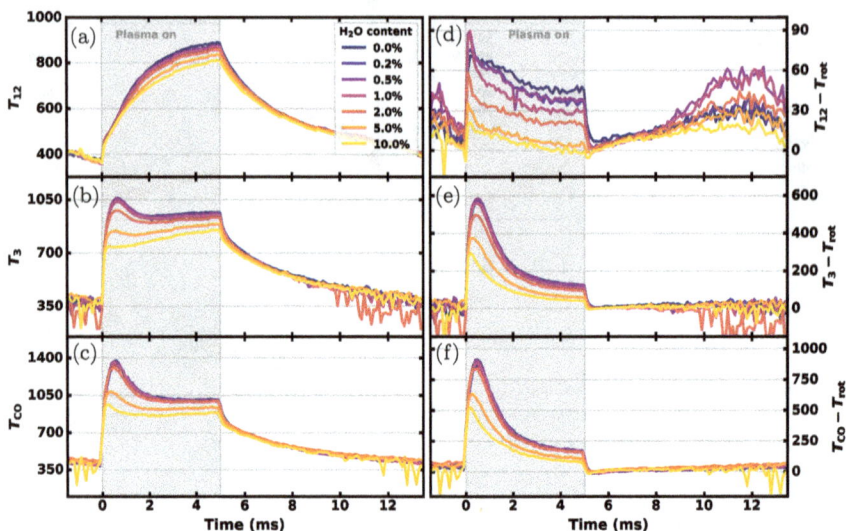

Fig. 15. Time evolution of the rotational and vibrational temperatures of a CO_2 in a pulsed (5 ms on-time; 10 ms off-time) glow discharge operating at 6.67 mbar with H_2O additions (0 to 10 %). (a) Symmetric stretch and bending mode vibrational temperature of CO_2. (b) Asymmetric stretch vibrational temperature of CO_2. (c) Vibrational temperature of CO. (d), (e) and (f) Elevation of the vibrational temperatures above the rotational temperature. Figure reproduced from [83] with permission.

photons [126]. The review by Zare [127] provides an historical overview of the LIF development. A light source (a laser) promotes a species to an (electronic) excited state. The spontaneous emission (fluorescence) of the excited state is then recorded. If pulsed lasers (pulse duration < 10 ns) are used in conjunction with the adoption of the scattering arrangement as the excitation-detection scheme (with the optical axis of the fluorescence detection optics perpendicular to the laser beam), excellent temporal (the exact time duration of the laser pulse) and spatial resolution (much lower than $1\,mm^3$) can be achieved [112].

LIF can be used to overcome the temporal and spatial limitations of quantitative line-of-sight optical techniques (e.g., absorption spectroscopy) or standard analytical techniques for gas composition analysis (e.g., GC, FTIR, MS, etc.). Although LIF is a well-established technique, its application to atmospheric pressure discharges is challenging [128]. Due to the many collisions, energy transfer processes can strongly affect the fluorescence outcome. Knowledge of collisional rate constants for the laser-excited state is then needed [33, 129]. Dilecce *et al.* [128] discussed in detail the case of the hydroxyl radical in different collisional environments. Recently, a new method has been proposed to measure the gas composition from LIF spectra, called collisional energy transfer LIF (CET-LIF) [130, 131]. CET-LIF uses the fluorescence of the laser-prepared state of a suitable molecule M^* to probe the gas composition. Since collisions with the surrounding molecules depopulate M^* by non-radiative energy-transfer processes, changes in the fluorescence spectrum can be related to the variation of the gas composition. In such a way, the excited M^* is actually a quantum sensor of the gas composition. The CET-LIF methodology has been successfully employed to measure the time evolution of the CO_2 conversion in an NRP discharge with high spatial and temporal resolution, by using OH as the probe molecule. OH is produced directly in the discharge adding trace H_2O to the CO_2 flow.

Figure 16 shows a pictorial representation of the discharge vessel and CET-LIF arrangement, a detailed description of the experimental setup can be found in [131]. The gas composition is probed at different delays with respect to the discharge pulse by exciting OH(X, $v = 0$) to the OH(A, $v = 1$) state and recording the fluorescence signal. The gas composition is deduced from the fluorescence spectra as described in [128] assuming that CO_2 dissociates according to the stoichiometry of Eq. (57). The variation

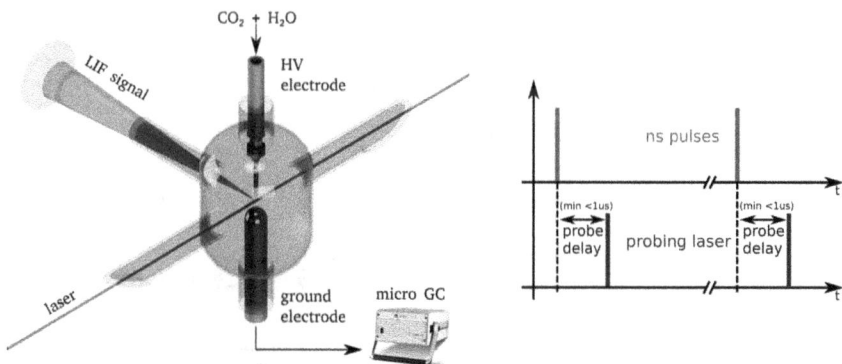

Fig. 16. Left: Schematic representation of the experimental setup for the determination of the CO_2 conversion in a pin-to-sphere NRP discharge; the laser beam used to excite the probing molecule (OH) is injected perpendicularly with respect to the collection of the fluorescence; the effluents of the reaction are analyzed with a micro-gas chromatograph. Right: Triggering schematic representation for the discharge and probing laser; the CET-LIF measurement can be carried out at variable delays with respect to the nanosecond pulse.

of the gas composition in the after-pulse is reflected in the time evolution of the population of the laser excited-state $OH(A, v = 1)$ and the $OH(A, v = 0)$ state that is populated by energy transfer processes, thus affecting the fluorescence spectrum. The availability of accurate rate coefficients for the energy transfer processes involving the $OH(A, v = 0, 1)$ is mandatory for the applicability of the methodology [33, 129].

Results obtained with CET-LIF in an atmospheric pressure NRP discharge operating in CO_2 with 1.35% of H_2O are presented in Fig. 17. The experimental conversion shows a good agreement with the prediction of a 0D computational mode [108]. The combination of the time evolution of the conversion and the insight gained with the computational studies gave the possibility to understand the complex physical and chemical processes taking place. In particular, a single discharge event is characterized by a high conversion, around 70%, that then is gradually lost due to recombination processes and re-circulation of the gas inside the discharge reactor. The kinetic model revealed that most of the CO_2 dissociation occurs from excited vibrational states, as further evidence of the relevant role of NEQ in the CO_2 dissociation mechanism.

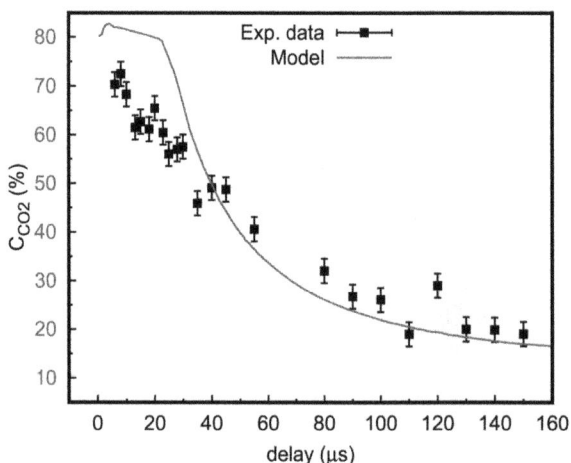

Fig. 17. As a function of the time after one nanosecond pulse measured using CET-LIF methodology. Experimental data are compared with the results from the computational model of [108]. The dissociation in the effluent gas measured by GC is 12% [131]. Figure adapted from [129] with permission.

5. Conclusions

In our, perhaps too quick, ride across the principles and applications of NEQ gas discharges, we have touched upon how the insights into these complicated systems have evolved toward the *microscopic* point of view, progressively abandoning the attempts to describe them by few macroscopic parameters. An evolution that required many decades to be carried out. The example of CO_2 conversion in NRP discharges is higly representative of how complicated a discharge system can be and how much we are in need of a multidisciplinary approach, ranging from electrical engineering to plasma physics and chemistry. All the ingredients for a strong and persistent headache are present: nanosecond-scale transient regimes, spark discharge, fast gas dynamics, complex chemistry involving vibrational kinetics, excited states and ions. It is clear, through this example, that experimental techniques can reveal only a small part of this complexity, and that discharge modeling is mandatory to explore the processes at a microscopic level. To be useful, a model must provide a quantitative, not simply a qualitative, demonstration of phenomena. It must then be realistic and rely on a good set of rate constants for all the essential processes. In the specific case of our example, a model, to be practical, must couple the chemical and discharge kinetics, like in [108], with a modeling of the streamer and its transition into a spark regime, following the methods traced in [13]; and introduce in someway the gas-dynamics. The availability of rate constants is also a work in progress. For example, VV and VT processes in the CO_2 vibrational modes have still to be well characterized quantitatively, especially for high vibrational levels. Constant comparison with good diagnostics is essential in the arduous route toward a decent model. A good diagnostic is not only a well-done experiment, but it must address fundamental quantities related to the kinetics of the discharge processes. Again for the CO_2 case, the macroscopic discharge parameters or the analysis of final products are not enough. The experiments must reach the microscopic details or transient behaviors to be useful for models' validation. The absorption and CET-LIF measurements reported in this chapter are examples of good diagnostics, just like the time-resolved emission spectroscopy measurements of gas temperature, electron density, and temperature reported in [114].

Finally, optical diagnostics also show how much electronically excited states exhibit strong NEQ conditions. This is easily understandable since these states generally have a short lifetime, resulting in insufficient time for reaching equilibrium, for example, in the rotational and vibrational manifolds. But also, as in the CET-LIF development, we have learned how much the detailed knowledge of collision processes of electronic states of molecules and atoms can be helpful or necessary for diagnostic purposes. See also, for example, the case of $N_2(C,v)$ collisional characterization made in [132] and its use for diagnostic purposes in [119]. Research efforts are required in this direction.

References

[1] A. Fridman, *Plasma Chemistry*. Cambridge University Press, New York (2008).

[2] J. A. M. van der Mullen, Excitation equilibria in plasmas; a classification, *Phys. Rep.* **191**(2–3), 109–220 (1990).

[3] H. Eyring, The activated complex in chemical reactions, *J. Chem. Phys.* **3**(2), 107–115 (1935).

[4] R. D. Levine, *Molecular Reaction Dynamics*. Cambridge University Press, Cambridge, UK (2005).

[5] J. C. Polanyi and W. H. Wong, Location of energy barriers. I. effect on the Dynamics of Reactions A + BC, *J. Chem. Phys.* **51**(4), 1439–1450 (1969).

[6] B. Jiang and H. Guo, Relative efficacy of vibrational vs. translational excitation in promoting atom-diatom reactivity: Rigorous examination of polanyi's rules and proposition of sudden vector projection (SVP) model, *J. Chem. Phys.* **138**(23), 234104 (2013).

[7] M. Stei, E. Carrascosa, A. Dörfler, J. Meyer, B. Olasz, G. Czakó, A. Li, H. Guo, and R. Wester, Stretching vibration is a spectator in nucleophilic substitution, *Sci. Adv.* **4**(7), eaas9544 (2018).

[8] P. Tosi, High-resolution measurements of integral cross-sections in ion molecule reactions from low-energy-guided crossed-beam experiments, *Chem. Rev.* **92**(7), 1667–1684 (1992).

[9] F. Taccogna and G. Dilecce, Non-equilibrium in low-temperature plasmas, *Eur. Phys. J. D.* **70**(11), 251 (2016).

[10] G. J. M. Hagelaar and L. C. Pitchford, Solving the Boltzmann equation to obtain electron transport coefficients and rate coefficients for fluid models, *Plasma Sources Sci. Technol.* **14**(4), 722–733 (2005).

[11] V. A. Godyak, Electron energy distribution function control in gas discharge plasmas, *Phys. Plasmas.* **20**(10), 101611 (2013).

[12] V. Kolobov and V. Godyak, Electron kinetics in low-temperature plasmas, *Phys. Plasmas.* **26**(6), 060601 (2019).

[13] F. Taccogna and F. Pellegrini, Kinetics of a plasma streamer ionization front, *J. Phys. D: Appl. Phys.* **51**(6), 064001 (2018).

[14] M. Bacal, Physics aspects of negative ion sources, *Nucl. Fusion.* **46**(6), S250–S259 (2006).

[15] M. Bacal, Negative hydrogen ion production in fusion dedicated ion sources, *Chem. Phys.* **398**, 3–6 (2012).

[16] R. Celiberto, R. K. Janev, A. Laricchiuta, M. Capitelli, J. M. Wadehra, and D. E. Atems, Cross section data for electron-impact inelastic processes of vibrationally excited molecules of hydrogen and its isotopes, *At. Data Nucl. Data Tables.* **77**(2), 161–213 (2001).

[17] M. Rutigliano, A. Pieretti, M. Cacciatore, N. Sanna, and V. Barone, N atoms recombination on a silica surface: A global theoretical approach, *Surf. Sci.* **600**(18), 4239–4246 (2006).

[18] R. K. Janev, W. D. Langer, E. Douglass Jr, and D. E. Post Jr, *Elementary Processes in Hydrogen-Helium Plasmas: Cross Sections and Reaction Rate Coefficients*. Springer, Berlin Heildelberg, Germany (1987).

[19] G. D. Billing and E. R. Fisher, VV and VT rate coefficients in N_2 by a quantum-classical model, *Chem. Phys.* **43**(3), 395–401 (1979).

[20] C. E. Treanor, J. W. Rich, and R. G. Rehm, Vibrational relaxation of anharmonic oscillators with exchange-dominated collisions, *J. Chem. Phys.* **48**(4), 1798–1807 (1968).

[21] S. De Benedictis, M. Capitelli, F. Cramarossa, R. d'Agostino, and C. Gorse, Vibrational distributions of CO in N_2 cooled radiofrequency post discharges, *Chem. Phys. Lett.* **112**(1), 54–58 (1984).

[22] Q. Hong, Q. Sun, M. Bartolomei, F. Pirani, and C. Coletti, Inelastic rate coefficients based on an improved potential energy surface for $N_2 + N_2$ collisions in a wide temperature range, *Phys. Chem. Chem. Phys.* **22**(17), 9375–9387 (2020).

[23] L. S. Polak and D. I. Slovetsky, Electron impact induced electronic excitation and molecular dissociation, *Int. J. Radiat. Phys. Chem.* **8**((1–2)), 257–282 (1976).

[24] A. S. Morillo-Candas, T. Silva, B. L. M. Klarenaar, M. Grofulović, V. Guerra, and O. Guaitella, Electron impact dissociation of CO_2, *Plasma Sources Sci. Technol.* **29**(1), 01LT01 (2020).

[25] I. Armenise and E. Kustova, Mechanisms of coupled vibrational relaxation and dissociation in carbon dioxide, *J. Phys. Chem. A.* **122**(23), 5107–5120 (2018).

[26] D. Y. Hwang and A. M. Mebel, Ab initio study of spin-forbidden unimolecular decomposition of carbon dioxide, *Chem. Phys.* **256**(2), 169–176 (2000).

[27] J. C. S. V. Veliz, D. Koner, M. Schwilk, R. J. Bemish, and M. Meuwly, The $C(^3P) + O_2(^3\Sigma_g^-) \rightarrow CO_2 \leftrightarrow CO(^1\Sigma^+) + O(^1D)/O(^3P)$ reaction: Thermal and vibrational relaxation rates from $15\,K$ to $20000\,K$, *Phys. Chem. Chem. Phys.* **23**(19), 11251–11263 (2021).

[28] V. Laporta, J. Tennyson, and R. Celiberto, Calculated low-energy electron-impact vibrational excitation cross sections for CO_2 molecule, *Plasma Sources Sci. Technol.* **25**(6), 06LT02 (2016).

[29] M. A. Lieberman and A. J. Lichtenberg, *Principles of Plasma Discharges and Materials Processing*. Wiley-Interscience, Hoboken, New Jersey (2005).

[30] X. Lu, G. V. Naidis, M. Laroussi, S. Reuter, D. B. Graves, and K. Ostrikov, Reactive species in non-equilibrium atmospheric-pressure plasmas: Generation, transport, and biological effects, *Phys. Rep.* **630**, 1–84 (2016).

[31] A. Starikovskiy and N. Aleksandrov, Plasma-assisted ignition and combustion, *Prog. Energy Combust. Sci.* **39**(1), 61–110 (2013).

[32] J. Van Durme, J. Dewulf, C. Leys, and H. Van Langenhove, Combining non-thermal plasma with heterogeneous catalysis in waste gas treatment: A review, *Appl. Catal. B: Environ.* **78**(3–4), 324–333 (2008).

[33] L. M. Martini, N. Gatti, G. Dilecce, M. Scotoni, and P. Tosi, Rate constants of quenching and vibrational relaxation in the $OH(A^2\Sigma^+, v = 0, 1)$, manifold with various colliders, *J. Phys. D: Appl. Phys.* **50**(11), 114003 (2017).

[34] M. McGee. Earth's CO_2 Home Page (2021). https://www.co2.earth/ [accessed on 14 April 2021].

[35] Core Writing Team, R. K. Pachauri, and L. A. Meyer. IPCC, 2014: Climate Change 2014: Synthesis Report. Contribution of Working Groups I, II and III to the Fifth Assessment Report of the Intergovernmental Panel on Climate Change. Geneva, Switzerland (2014).

[36] G. J. van Rooij, H. N. Akse, W. A. Bongers, and M. C. M. Van De Sanden, Plasma for electrification of chemical industry: A case study on CO_2 reduction, *Plasma Phys. Control. Fusion.* **60**(1), 014019 (2017).

[37] A. Bogaerts and E. C. Neyts, Plasma technology: An emerging technology for energy storage, *ACS Energy Lett.* **3**(4), 1013–1027 (2018).

[38] P. Mehta, P. Barboun, D. B. Go, J. C. Hicks, and W. F. Schneider, Catalysis enabled by plasma activation of strong chemical bonds: A review, *ACS Energy Lett.* **4**(5), 1115–1133 (2019).

[39] R. S. Abiev, D. A. Sladkovskiy, K. V. Semikin, D. Y. Murzin, and E. V. Rebrov, Non-thermal plasma for process and energy intensification in dry reforming of methane, *Catalysts.* **10**(11), 1358 (2020).

[40] A. Bogaerts and G. Centi, Plasma technology for CO_2 conversion: A personal perspective on prospects and gaps, *Front. Energy Res.* **8**, 111 (2020).

[41] A. Bogaerts, X. Tu, J. C. Whitehead, G. Centi, L. Lefferts, O. Guaitella, F. Azzolina-Jury, H.-H. Kim, A. B. Murphy, W. F. Schneider, T. Nozaki, J. Hicks, A. Rousseau, F. Thevenet, A. Khacef, and M. Carreon, The 2020 plasma catalysis roadmap, *J. Phys. D: Appl. Phys.* **53**(44), 443001 (2020).

[42] A. W. van de Steeg, T. Butterworth, D. C. M. van den Bekerom, A. F. Silva, M. C. M. van de Sanden, and G. J. van Rooij, Plasma activation of N_2, CH_4 and CO_2: An assessment of the vibrational non-equilibrium time window, *Plasma Sources Sci. Technol.* **29**(11), 115001 (2020).

[43] R. Snoeckx and A. Bogaerts, Plasma technology – a novel solution for CO_2 conversion?, *Chem. Soc. Rev.* **46**(19), 5805–5863 (2017).

[44] M. Aresta, I. Karimi, and S. Kaw, *An Economy Based on Carbon Dioxide and Water*. Springer, Cham, Switzerland (2019).

[45] U. Kogelschatz, Dielectric-barrier discharges: Their history, discharge physics, and industrial applications, *Plasma Chem. Plasma Process.* **23**(1), 1–46 (2003).

[46] H.-H. Kim, Nonthermal plasma processing for air-pollution control: A historical review, current issues, and future prospects, *Plasma Processes Polym.* **1**(2), 91–110 (2004).

[47] R. Morent, N. De Geyter, J. Verschuren, K. De Clerck, P. Kiekens, and C. Leys, Non-thermal plasma treatment of textiles, *Surf. Coat. Technol.* **202**(14), 3427–3449 (2008).

[48] H. C. M. Knoops, T. Faraz, K. Arts, and W. M. M. E. Kessels, Status and prospects of plasma-assisted atomic layer deposition, *J. Vac. Sci. Technol. A.* **37**(3), 030902 (2019).

[49] M. A. Sobolewski, Electrical characterization of radio-frequency discharges in the gaseous electronics conference reference cell, *J. Vac. Sci. Technol. A.* **10**(6), 3550–3562 (1992).

[50] K. Takashima, Y. Zuzeek, W. R. Lempert, and I. V. Adamovich, Characterization of a surface dielectric barrier discharge plasma sustained by repetitive nanosecond pulses, *Plasma Sources Sci. Technol.* **20**(5), 055009 (2011).

[51] M. Scapinello, L. M. Martini, P. Tosi, A. Maranzana, and G. Tonachini, Molecular growth of PAH-like systems induced by oxygen species: Experimental and theoretical study of the reaction of naphthalene with $HO(^2\Pi_{3/2})$, $O(^3P)$, and $O_2(^3\Sigma_g^-)$, *RSC Adv.* **5**(48), 38581–38590 (2015).

[52] L. M. Martini, A. Maranzana, G. Tonachini, G. Bortolotti, M. Scapinello, M. Scotoni, G. Guella, G. Dilecce, and P. Tosi, Reactivity of fatty acid methyl esters under atmospheric pressure plasma jet exposure: An experimental and theoretical study, *Plasma Processes Polym.* **14**(10), 1600254 (2017).

[53] C. Daunton, L. E. Smith, J. D. Whittle, R. D. Short, D. A. Steele, and A. Michelmore, Plasma parameter aspects in the fabrication of stable amine functionalized plasma polymer films, *Plasma Processes Polym.* **12**(8), 817–826 (2015).

[54] T. Huiskamp, Nanosecond pulsed streamer discharges Part I: Generation, source-plasma interaction and energy-efficiency optimization, *Plasma Sources Sci. Technol.* **29**(2), 023002 (2020).

[55] A. Khomenko, V. Podolsky, and X. Wang, Different approaches of measuring high-voltage nanosecond pulses and power delivery in plasma systems, *Electr. Eng.* **103**(1), 57–66 (2021).

[56] B. M. Goldberg, T. L. Chng, A. Dogariu, and R. B. Miles, Electric field measurements in a near atmospheric pressure nanosecond pulse discharge with picosecond electric field induced second harmonic generation, *Appl. Phys. Lett.* **112**(6), 064102 (2018).

[57] C. Ding, A. Y. Khomenko, S. A. Shcherbanev, and S. M. Starikovskaia, Filamentary nanosecond surface dielectric barrier discharge. experimental comparison of the streamer-to-filament transition for positive and negative polarities, *Plasma Sources Sci. Technol.* **28**(8), 085005 (2019).

[58] K. Grosse, M. Falke, and A. von Keudell, Ignition and propagation of nanosecond pulsed plasmas in distilled water – Negative vs positive polarity applied to a pin electrode, *J. Appl. Phys.* **129**(21), 213302 (2021).

[59] H. S. Fogler, *Essentials of Chemical Reaction Engineering*. Pearson Education, Inc., Upper Saddle River, NJ (2011).

[60] R. Snoeckx, A. Rabinovich, D. Dobrynin, A. Bogaerts, and A. Fridman, Plasma-based liquefaction of methane: The road from hydrogen production to direct methane liquefaction, *Plasma Processes Polym.* **14**(6), 1600115 (2017).

[61] A. George, B. Shen, M. Craven, Y. Wang, D. Kang, C. Wu, and X. Tu, A review of non-thermal plasma technology: A novel solution for CO_2 conversion and utilization, *Renew. Sustain. Energy Rev.* **135**, 109702 (2021).

[62] L. D. Pietanza, O. Guaitella, V. Aquilanti, I. Armenise, A. Bogaerts, M. Capitelli, G. Colonna, V. Guerra, R. Engeln, E. Kustova, and A. Lombardi, Advances in non-equilibrium CO_2 plasma kinetics: A theoretical and experimental review, *Eur. Phys. J. D.* **75**(9), 237 (2021).

[63] M. Scapinello, L. M. Martini, and P. Tosi, CO_2 hydrogenation by CH_4 in a dielectric barrier discharge: Catalytic effects of nickel and copper, *Plasma Processes Polym.* **11**(7), 624–628 (2014).

[64] L. M. Martini, G. Dilecce, G. Guella, A. Maranzana, G. Tonachini, and P. Tosi, Oxidation of CH_4 by CO_2 in a dielectric barrier discharge, *Chem. Phys. Lett.* **593**, 55–60 (2014).

[65] X. Tu and J. C. Whitehead, Plasma-catalytic dry reforming of methane in an atmospheric dielectric barrier discharge: Understanding the synergistic effect at low temperature, *Appl. Catal. B: Environ.* **125**, 439–448 (2012).

[66] X. Tu and J. C. Whitehead, Plasma dry reforming of methane in an atmospheric pressure ac gliding arc discharge: Co-generation of syngas and carbon nanomaterials, *Int. J. Hydrogen Energy.* **39**(18), 9658–9669 (2014).

[67] X. Tao, M. Bai, X. Li, H. Long, S. Shang, Y. Yin, and X. Dai, CH_4–CO_2 reforming by plasma — challenges and opportunities, *Prog. Energy Combust. Sci.* **37**(2), 113–124 (2011).

[68] J. Martin del Campo, S. Coulombe, and J. Kopyscinski, Influence of operating parameters on plasma-assisted dry reforming of methane in a rotating gliding arc reactor, *Plasma Chem. Plasma Process.* **40**(4), 857–881 (2020).

[69] The Pacific Northwest National Laboratory and U.S. Department of Energy's Office of Energy. H2 Tools (2021). https://h2tools.org/hyarc/calculator-tools/lower-and-higher-heating-values-fuels [accessed on 29 October 2021].

[70] N. Pinhão, A. Moura, J. B. Branco, and J. Neves, Influence of gas expansion on process parameters in non-thermal plasma plug-flow reactors: A study applied to dry reforming of methane, *Int. J. Hydrogen Energy.* **41**(22), 9245–9255 (2016).

[71] R. Snoeckx, S. Heijkers, K. Van Wesenbeeck, S. Lenaerts, and A. Bogaerts, CO_2 conversion in a dielectric barrier discharge plasma: N_2 in the mix as a helping hand or problematic impurity?, *Energy Environ. Sci.* **9**(3), 999–1011 (2016).

[72] M. Ramakers, G. Trenchev, S. Heijkers, W. Wang, and A. Bogaerts, Gliding arc plasmatron: Providing an alternative method for carbon dioxide conversion, *ChemSusChem.* **10**(12), 2642–2652 (2017).

[73] Y. Uytdenhouwen, S. Van Alphen, I. Michielsen, V. Meynen, P. Cool, and A. Bogaerts, A packed-bed DBD micro plasma reactor for CO_2 dissociation: Does size matter?, *Chem. Eng. J.* **348**, 557–568 (2018).

[74] S. Kelly and J. A. Sullivan, CO_2 decomposition in CO_2 and CO_2/H_2 spark-like plasma discharges at atmospheric pressure, *ChemSusChem.* **12**(16), 3785–3791 (2019).

[75] G. Niu, Y. Qin, W. Li, and Y. Duan, Investigation of CO_2 splitting process under atmospheric pressure using multi-electrode cylindrical DBD plasma reactor, *Plasma Chem. Plasma Process.* **39**(4), 809–824 (2019).

[76] B. Raja, R. Sarathi, and R. Vinu, Development of a swirl-induced rotating glow discharge reactor for CO_2 conversion: Fluid dynamics and discharge dynamics studies, *Energy Technol.* **8**(12), 2000535 (2020).

[77] P. Kaliyappan, A. Paulus, J. D'Haen, P. Samyn, Y. Uytdenhouwen, N. Hafezkhiabani, A. Bogaerts, V. Meynen, K. Elen, A. Hardy, and M. K. Van Bael, Probing the impact of material properties of core-shell SiO_2@TiO_2 spheres on the plasma-catalytic CO_2 dissociation using a packed bed DBD plasma reactor, *J. CO2 Util.* **46**, 101468 (2021).

[78] B. L. M. Klarenaar, R. Engeln, D. C. M. Van Den Bekerom, M. C. M. Van De Sanden, A. S. Morillo-Candas, and O. Guaitella, Time evolution of vibrational temperatures in a CO_2 glow discharge measured with infrared absorption spectroscopy, *Plasma Sources Sci. Technol.* **26**(11), 115008 (2017).

[79] B. L. M. Klarenaar, M. Grofulović, A. S. Morillo-Candas, D. C. M. Van Den Bekerom, M. A. Damen, M. C. M. Van De Sanden, O. Guaitella, and R. Engeln, A rotational Raman study under non-thermal conditions in a pulsed CO_2 glow discharge, *Plasma Sources Sci. Technol.* **27**(4), 045009 (2018).

[80] B. L. M. Klarenaar, A. S. Morillo-Candas, M. Grofulović, M. C. M. Van De Sanden, R. Engeln, and O. Guaitella, Excitation and relaxation of the asymmetric stretch mode of CO_2 in a pulsed glow discharge, *Plasma Sources Sci. Technol.* **28**(3), 035011 (2019).

[81] M. Grofulović, B. L. M. Klarenaar, O. Guaitella, V. Guerra, and R. Engeln, A rotational Raman study under non-thermal conditions in pulsed CO_2-N_2 and CO_2-O_2 glow discharges, *Plasma Sources Sci. Technol.* **28**(4), 045014 (2019).

[82] M. A. Damen, D. A. C. M. Hage, A. W. van de Steeg, L. M. Martini, and R. Engeln, Absolute CO number densities measured using TALIF in a non-thermal plasma environment, *Plasma Sources Sci. Technol.* **28**(11), 115006 (2019).

[83] M. A. Damen, L. M. Martini, and R. Engeln, Vibrational quenching by water in a CO_2 glow discharge measured using quantum cascade laser absorption spectroscopy, *Plasma Sources Sci. Technol.* **29**(9), 095017 (2020).

[84] A. H. Khoja, M. Tahir, and N. A. S. Amin, Dry reforming of methane using different dielectric materials and DBD plasma reactor configurations, *Energy Convers. Manage.* **144**, 262–274 (2017).

[85] R. Lee, R. Labrecque, and J. M. Lavoie, Correction to inline analysis of the dry reforming process through fourier transform infrared spectroscopy and use of nitrogen as an internal standard for online gas chromatography analysis, *Energy Fuels.* **29**(2), 1266–1267 (2015).

[86] M. Scapinello, L. M. Martini, G. Dilecce, and P. Tosi, Conversion of CH_4/CO_2 by a nanosecond repetitively pulsed discharge, *J. Phys. D: Appl. Phys.* **49**(7), 075602 (2016).

[87] M. Scapinello, E. Delikonstantis, and G. D. Stefanidis, Direct methane-to-ethylene conversion in a nanosecond pulsed discharge, *Fuel.* **222**, 705–710 (2018).

[88] E. Delikonstantis, M. Scapinello, and G. D. Stefanidis, Low energy cost conversion of methane to ethylene in a hybrid plasma-catalytic reactor system, *Fuel Process. Technol.* **176**, 33–42 (2018).

[89] M. Scapinello, E. Delikonstantis, and G. D. Stefanidis, A study on the reaction mechanism of non-oxidative methane coupling in a nanosecond pulsed discharge reactor using isotope analysis, *Chem. Eng. J.* **360**, 64–74 (2019).

[90] E. Delikonstantis, M. Scapinello, O. Van Geenhoven, and G. D. Stefanidis, Nanosecond pulsed discharge-driven non-oxidative methane coupling in a plate-to-plate electrode configuration plasma reactor, *Chem. Eng. J.* **380**, 122477 (2020).

[91] C. Montesano, S. Quercetti, L. M. Martini, G. Dilecce, and P. Tosi, The effect of different pulse patterns on the plasma reduction of CO_2 for a nanosecond discharge, *J. CO_2 Util.* **39**, 101157 (2020).

[92] C. Montesano, M. Faedda, L. M. Martini, G. Dilecce, and P. Tosi, CH_4 reforming with CO_2 in a nanosecond pulsed discharge. The importance of the pulse sequence, *J. CO_2 Util.* **49**, 101556 (2021).

[93] I. Gulko, E. R. Jans, C. Richards, S. Raskar, X. Yang, D. C. M. van den Bekerom, and I. V. Adamovich, Selective generation of excited species in ns pulse/RF hybrid plasmas for plasma chemistry applications, *Plasma Sources Sci. Technol.* **29**(10), 104002 (2020).

[94] N. Britun, T. Silva, G. Chen, T. Godfroid, J. van der Mullen, and R. Snyders, Plasma-assisted CO_2 conversion: Optimizing performance via microwave power modulation, *J. Phys. D: Appl. Phys.* **51**(14), 144002 (2018).

[95] T. D. Butterworth, B. Amyay, D. V. D. Bekerom, A. V. D. Steeg, T. Minea, N. Gatti, Q. Ong, C. Richard, C. van Kruijsdijk, J. T. Smits, A. P. van Bavel, V. Boudon, and G. J. van Rooij, Quantifying methane vibrational and rotational temperature with Raman scattering, *J. Quant. Spectrosc. Radiat. Transf.* **236**, 106562 (2019).

[96] D. C. M. Van Den Bekerom, A. Van De Steeg, M. C. M. Van De Sanden, and G. J. Van Rooij, Mode resolved heating dynamics in pulsed microwave

CO$_2$ plasma from laser Raman scattering, *J. Phys. D: Appl. Phys.* **53**(5), 054002 (2020).

[97] J. Benedikt, H. Kersten, and A. Piel, Foundations of measurement of electrons, ions and species fluxes toward surfaces in low-temperature plasmas, *Plasma Sources Sci. Technol.* **30**(3), 033001 (2021).

[98] T. Faraz, K. Arts, S. Karwal, H. C. Knoops, and W. M. Kessels, Energetic ions during plasma-enhanced atomic layer deposition and their role in tailoring material properties, *Plasma Sources Sci. Technol.* **28**(2), 024002 (2019).

[99] R. Engeln, B. Klarenaar, and O. Guaitella, Foundations of optical diagnostics in low-temperature plasmas, *Plasma Sources Sci. Technol.* **29**(6), 063001 (2020).

[100] N. den Harder, D. C. van den Bekerom, R. S. Al, M. F. Graswinckel, J. M. Palomares, F. J. Peeters, S. Ponduri, T. Minea, W. A. Bongers, M. C. van de Sanden, and G. J. van Rooij, Homogeneous CO$_2$ conversion by microwave plasma: Wave propagation and diagnostics, *Plasma Processes Polym.* **14**(6, SI), 1600120 (2017).

[101] M. Šimek, P. F. Ambrico, T. Hoder, V. Prukner, G. Dilecce, S. De Benedictis, and V. Babický, Nanosecond imaging and emission spectroscopy of argon streamer micro-discharge developing in coplanar surface DBD, *Plasma Sources Sci. Technol.* **27**(5), 055019 (2018).

[102] T. Furusato, Y. Inada, M. Sasaki, Y. Matsuda, and T. Yamashita, Shock-wave propagation in supercritical CO$_2$ induced by nanosecond-pulsed arc plasma, *J. Phys. D: Appl. Phys.* **53**(40), 40LT01 (2020).

[103] D. Wang and T. Namihira, Nanosecond pulsed streamer discharges: II. Physics, discharge characterization and plasma processing, *Plasma Sources Sci. Technol.* **29**(2), 023001 (2020).

[104] G. S. Settles, *Schlieren and Shadowgraph Techniques: Visualizing Phenomena in Transparent Media.* Springer, Berlin, Heidelberg, New York (2001).

[105] E. Traldi, M. Boselli, E. Simoncelli, A. Stancampiano, M. Gherardi, V. Colombo, and G. S. Settles, Schlieren imaging: A powerful tool for atmospheric plasma diagnostic, *EPJ Tech. Instrum.* **5**(1), 4 (2018).

[106] M. Castela, S. Stepanyan, B. Fiorina, A. Coussement, O. Gicquel, N. Darabiha, and C. O. Laux, A 3-D DNS and experimental study of the effect of the recirculating flow pattern inside a reactive kernel produced by nanosecond plasma discharges in a methane-air mixture, *Proc. Combust. Inst.* **36**(3), 4095–4103 (2017).

[107] B. Singh, P. Vlachos, and S. P. Bane, Shock generated vorticity in spark discharges, *J. Phys. D: Appl. Phys.* **54**(31), 315202 (2021).

[108] S. Heijkers, L. M. Martini, G. Dilecce, P. Tosi, and A. Bogaerts, Nanosecond pulsed discharge for CO$_2$ conversion: Kinetic modeling to elucidate the chemistry and improve the performance, *J. Phys. Chem. C.* **123**(19), 12104–12116 (2019).

[109] T. Silva, N. Britun, T. Godfroid, and R. Snyders, Optical characterization of a microwave pulsed discharge used for dissociation of CO$_2$, *Plasma Sources Sci. Technol.* **23**(2), 025009 (2014).

[110] A. S. Morillo-Candas, C. Drag, J. P. Booth, T. C. Dias, V. Guerra, and O. Guaitella, Oxygen atom kinetics in CO_2 plasmas ignited in a DC glow discharge, *Plasma Sources Sci. Technol.* **28**(7), 075010 (2019).

[111] U. Fantz, Basics of plasma spectroscopy, *Plasma Sources Sci. Technol.* **15**(4), S137–S147 (2006).

[112] G. Dilecce, L. M. Martini, P. Tosi, M. Scotoni, and S. De Benedictis, Laser induced fluorescence in atmospheric pressure discharges, *Plasma Sources Sci. Technol.* **24**(3), 034007 (2015).

[113] G. Dilecce, Optical spectroscopy diagnostics of discharges at atmospheric pressure, *Plasma Sources Sci. Technol.* **23**(1), 015011 (2014).

[114] M. Ceppelli, A. Salden, L. M. Martini, G. Dilecce, and P. Tosi, Time-resolved optical emission spectroscopy in CO_2 nanosecond pulsed discharges, *Plasma Sources Sci. Technol.* **30**(11), 115010 (2021).

[115] G. Dilecce, P. F. Ambrico, G. Scarduelli, P. Tosi, and S. De Benedictis, $CN(B^2\Sigma^+)$ formation and emission in a N_2-CH_4 atmospheric pressure dielectric barrier discharge, *Plasma Sources Sci. Technol.* **18**(1), 015010 (2009).

[116] H. R. Griem, *Principles of Plasma Spectroscopy.* Cambridge University Press, Cambridge, UK (2005).

[117] H. J. Kunze, *Introduction to Plasma Spectroscopy.* Springer, Berlin Heidelberg (2009).

[118] A. Kramida, Y. Ralchenko, J. Reader, and NIST ASD Team. NIST Atomic Spectra Database (ver. 5.8), [Online]. National Institute of Standards and Technology, Gaithersburg, MD. (2020). https://physics.nist.gov/PhysRef Data/ASD/LIBS/libs-form.html. [accessed on 14 April 2021].

[119] G. Dilecce, P. F. Ambrico, and S. De Benedictis, New $N_2(C^3\Pi_u,v)$ collision quenching and vibrational relaxation rate constants: 2. PG emission diagnostics of high-pressure discharges, *Plasma Sources Sci. Technol.* **16**(1), S45–S51 (2007).

[120] M. Šimek, Optical diagnostics of streamer discharges in atmospheric gases, *J. Phys. D: Appl. Phys.* **47**(46), 463001 (2014).

[121] G. Raposo, A. W. van de Steeg, E. R. Mercer, C. F. A. M. van Deursen, H. J. L. Hendrickx, W. A. Bongers, G. J. van Rooij, M. C. M. van de Sanden, and F. J. J. Peeters, Flame bands: CO + O chemiluminescence as a measure of gas temperature, *J. Phys. D: Appl. Phys.* **54**(37), 374005 (2021).

[122] T. Urbanietz, M. Böke, V. Schulz-von Der Gathen, and A. Von Keudell, Non-equilibrium excitation of CO_2 in an atmospheric pressure helium plasma jet, *J. Phys. D: Appl. Phys.* **51**(34), 345202 (2018).

[123] T. Silva, M. Grofulović, B. L. M. Klarenaar, A. S. Morillo-Candas, O. Guaitella, R. Engeln, C. D. Pintassilgo, and V. Guerra, Kinetic study of low-temperature CO_2 plasmas under non-equilibrium conditions. I. Relaxation of vibrational energy, *Plasma Sources Sci. Technol.* **27**(1), 015019 (2018).

[124] M. Grofulović, T. Silva, B. L. M. Klarenaar, A. S. Morillo-Candas, O. Guaitella, R. Engeln, C. D. Pintassilgo, and V. Guerra, Kinetic study

of CO_2 plasmas under non-equilibrium conditions. II. Input of vibrational energy, *Plasma Sources Sci. Technol.* **27**(11), 115009 (2018).

[125] M. A. Damen, L. M. Martini, and R. Engeln, Temperature evolution in a pulsed CO_2-N_2 glow discharge measured using quantum cascade laser absorption spectroscopy, *Plasma Sources Sci. Technol.* **29**(6), 065016 (2020).

[126] W. Demtröder, *Laser Spectroscopy 2.* Springer, Heidelberg New York Dordrecht London (2015).

[127] R. N. Zare, My life with LIF: A personal account of developing laser-induced fluorescence, *Annual Rev. Anal. Chem.* **5**(1), 1–14 (2012).

[128] G. Dilecce, L. M. Martini, M. Ceppelli, M. Scotoni, and P. Tosi, Progress on laser induced fluorescence in a collisional environment: The case of OH molecules in ns pulsed discharges, *Plasma Sources Sci. Technol.* **28**(2), 025012 (2019).

[129] M. Ceppelli, L. M. Martini, G. Dilecce, M. Scotoni, and P. Tosi, Non-thermal rate constants of quenching and vibrational relaxation in the OH $\left(A^2\Sigma^+, v' = 0, 1\right)$ manifold, *Plasma Sources Sci. Technol.* **29**(6), 065019 (2020).

[130] L. M. Martini, N. Gatti, G. Dilecce, M. Scotoni, and P. Tosi, Laser induced fluorescence in nanosecond repetitively pulsed discharges for CO_2 conversion, *Plasma Phys. Control. Fusion.* **60**(1), 014016 (2017).

[131] L. M. Martini, S. Lovascio, G. Dilecce, and P. Tosi, Time-resolved CO_2 dissociation in a nanosecond pulsed discharge, *Plasma Chem. Plasma Process.* **38**(4), 707–718 (2018).

[132] G. Dilecce, P. F. Ambrico, and S. De Benedictis, OODR-LIF direct measurement of $N_2(C^3\Pi_u, v = 0\text{-}4)$ electronic quenching and vibrational relaxation rate coefficients by N_2 collision, *Chem. Phys. Lett.* **431**(4), 241–246 (2006).

Chapter 2

Plasma Processes for the Surface Engineering of Polymers[*]

Fabio Palumbo[†,§], Fiorenza Fanelli[†,¶], Antonella Milella[‡,†,‖],
Pietro Favia[‡,†,**] and Francesco Fracassi[‡,†,††]

[†] *CNR-Istituto di Nanotecnologia, Unità di Bari,
Consiglio Nazionale delle Ricerche, c/o Dipartimento di Chimica,
Università degli Studi di Bari Aldo Moro,
via Orabona 4, 70125, Bari (ITALY)*
[‡] *Dipartimento di Chimica, Università degli Studi di Bari Aldo Moro,
via Orabona 4, 70125, Bari (ITALY)*
[§] *fabio.palumbo@cnr.it*
[¶] *fiorenza.fanelli@cnr.it*
[‖] *antonella.milella@uniba.it*
[**] *pietro.favia@uniba.it*
[††] *francesco.fracassi@uniba.it*

Plasma processing has reached a paramount importance in many manufacturing applications. Applications can be found in very different fields, ranging from corrosion protection to packaging, and from microelectronics to biomaterials. Our main goal is dealing with the optimization and development of plasma processes based on a unique material chemistry approach. In all the applications we considered, the focal point has been a detailed chemical description of the process and the material under study, with a particular emphasis on its final properties.

This chapter summarizes some of the plasma processes developed in our laboratory in the last twenty years for the film deposition, surface treatment and etching of polymers. For each kind of process, both low-pressure and atmospheric pressure plasma solutions are reviewed and discussed.

---

[*]This chapter is dedicated to the memory of our mentor, Prof. Riccardo d'Agostino, for having shared with us his deep knowledge, for having encouraged us to believe in Science, for having pursued good ideas with perseverance and passion.

Contents

1. Introduction

Plasma is ionized gas matter with equal density of positive and negative particles [1] and identifies "the fourth state of matter", the most abundant visible matter in the Universe. The core of the stars is made of thermonuclear plasma, as well as solar wind, ionosphere and Aurora Borealis. However, the former is an example of thermal (hot) plasma, while the latter are low-temperature, non-equilibrium (cold) ones.

Thermal plasmas are currently used in applications like metallurgy, welding, metal cutting and in the production of ceramic coatings. Nowadays, cold plasma is the main technology for the production of plasma TVs, integrated circuits in microelectronics and solar cells, but newer and clean applications are foreseen for innovative materials, communications, energy, catalysis, environment, food, healthcare and agriculture.

Cold plasma discharges are usually ignited at low (usually 1–100 Pa) or at atmospheric pressure by application of an electric field to a flowing or static gas/vapor mixture in reactors with a suitable configuration. Chemical bond breaking, ionization and excitation may occur, accompanied by the formation of active species such as free electrons, atoms, radicals and ions in ground and excited levels, and visible-UV photons. By controlling the relative abundance of these species, different processes can be designed for many existing or forecasted technological applications.

Most applications require that the gas temperature remains the closest possible to room temperature. Due to the low efficiency and the low number of elastic collisions at low pressure, the energy transfer from the free electrons to heavier species is negligible, and, as a matter of fact, it is relatively easy to produce cold gas discharges. On increasing pressure the electron–species collision frequency increases and the energy transfer becomes more efficient, with consequent gas heating and arcing. To keep the gas cold in the presence of atmospheric pressure discharges, several approaches can be used, namely: Sharp electrodes as in corona discharges; by pulsing the electric field with micro-nanosecond wide pulses; plasma confinement into sub-millimeter size gaps; and by reducing the current upon covering the electrodes with dielectric layers, as in dielectric barrier discharges (DBDs)[2].

Cold plasmas are usually ignited with AC electric fields in the kHz (audio frequency, AF), MHz (radio frequency, RF) and GHz (microwaves, MWs) ranges. The field can be applied continuously or in pulsed mode to electrodes, coils in contact with the gas, or through MW applicators.

Three kinds of surface modifications can be addressed by plasma processing, i.e., ablation, plasma deposition and plasma treatment. Ablation is the removal of substrate material by formation of volatile compounds with active species in the plasma. Plasma deposition or plasma-enhanced chemical vapor deposition (PECVD), allows for decoration of substrates with coatings of many possible chemical compositions, cross-linkings and properties. In plasma treatments, materials are exposed to reactive (e.g., O_2, N_2, H_2, NH_3, H_2O vapor, N_2/H_2, etc.) or inert (Ar, He, etc.) non-polymerizable feeds. The surface modifications obtained in this way are extremely shallow, and include cleaning from surface contaminants, oxidation, cross-linking and grafting of chemical groups at the surface of polymers.

A number of architectures of plasma sources are available, and the choice depends on the application and on the pressure regime of the selected process. The simplest and most common configuration for material surface processing with low-pressure (LP) plasma is the parallel-plate reactor sketched in Fig. 1. It can be designed in different configurations depending on the substrate (large/small objects, particles, batch, etc.), the throughput (laboratory and industrial scales) and the selected application. For web (polymers, textiles) processing, for example, the reactor can be implemented with winding/unwinding roll-to-roll equipment.

Besides corona discharges, DBDs are among the most widely used experimental set-ups to generate cold atmospheric pressure plasmas [4].

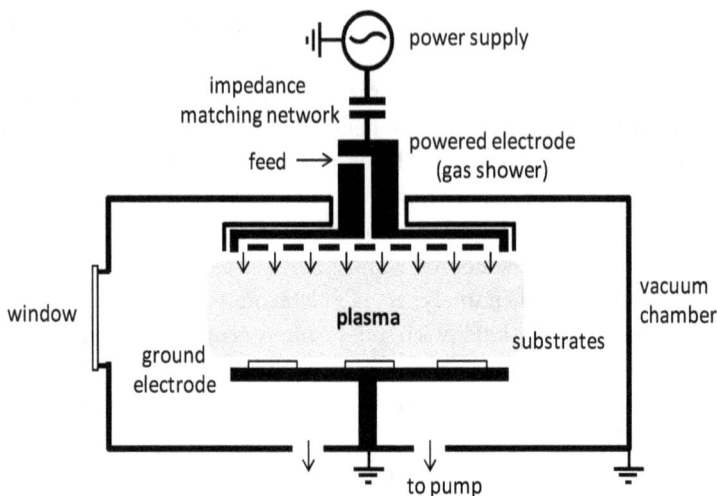

Fig. 1. General sketch of an LP parallel plate plasma reactor. Reprinted with permission from [3]. © 2018 Elsevier.

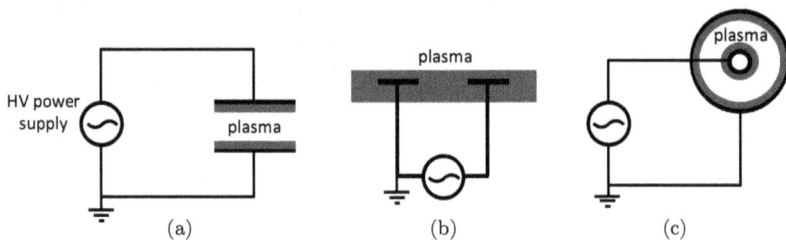

Fig. 2. Three possible configurations of atmospheric pressure DBD (APDBD): (a) parallel-plate DBD; (b) co-planar surface DBD; (c) co-axial DBD. The dielectric layer that covers the electrodes is represented in gray. Reprinted with permission from [3]. © 2018 Elsevier.

In DBDs, a layer of dielectric material (alumina, quartz, etc.) is used to cover one or both electrodes, in order to reduce the current forming a plasma at room temperature. The inter-electrode gap is usually few millimeters wide for the higher breakdown voltage of gases at atmospheric pressure. This feature may result in a drawback with respect to low-pressure processes, were much larger electrode gaps are common, but this limitation is eliminated in plasma jet configurations.

At atmospheric pressure, discharges are fed with higher gas flow rates (L/min) with respect to low-pressure plasma, often with a buffer noble gas

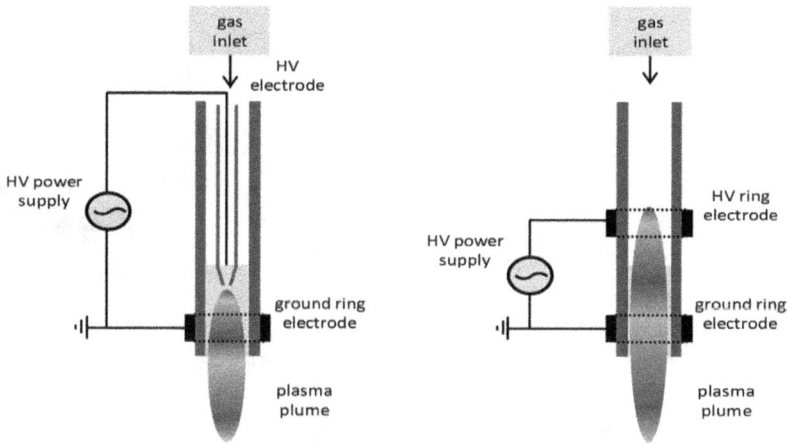

Fig. 3. Design of two APPJ plasma sources. Glass (dielectric) tubes are in gray. Reprinted with permission from [3]. © 2018 Elsevier.

(He, Ar) to get low breakdown voltage [5]. The sketches of three common DBD configurations are described in Fig. 2.

Atmospheric pressure plasma can be ignited in jet configuration, where plasma plumes or afterglow effluents invest the sample placed at some distance (millimeters–centimeters) from the source itself [6,7]. Among other possible configurations, this approach allows for the treatment of large 3D objects even with DBD. Two of the many possible atmospheric pressure plasma jet (APPJ) designs [8] are shown in Fig. 3.

This chapter is intended to provide the reader with fundamental and applicative aspects of plasma technology by reviewing several examples of processes developed in our laboratory over the years on the surface engineering of polymers, with the aid of low pressure and atmospheric pressure plasmas (Fig. 4). The developments achieved in our laboratory are discussed in relation to the advancements pursued in related fields by similar scientific groups.

2. Plasma-Enhanced Chemical Vapor Deposition of Polymers

2.1. *Low-pressure PECVD of fluorocarbon films*

Plasma-deposited fluorocarbon films (also reported as Teflon-like) have attracted great attention over the years for several applications due to their properties, such as hydrophobicity, biocompatibility and low friction.

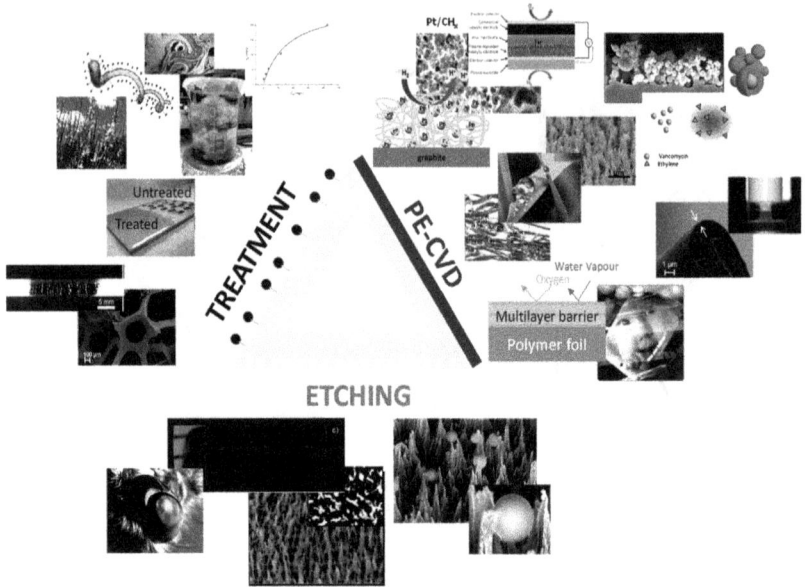

Fig. 4. A collection of plasma processes for the surface modification of materials reviewed in this chapter.

In paper and textile industries, fluorocarbon films are used to impart hydrophobicity and/or oleophobicity and stain resistance since tuning the fluorine/carbon (F/C) ratio in the film results in tuned hydrophobic character and surface energy [9–14].

In the field of biomaterials, fluorocarbon coatings have been plasma deposited on the inside walls of Dacron vascular grafts and on the surface of other conventional polymers [15–19], and positive results have been published due to their non-thrombogenicity, resistance to platelet adhesion and activation, blood compatibility and adsorption–retention properties toward blood proteins such as fibrinogen and albumin. Similar coatings gave interesting results when deposited on intra-ocular lenses (IOLs), since they induced a reduced damage by loss of endothelial cells in the cornea of rabbits [20]. A comprehensive discussion on the low-pressure plasma deposition of fluorocarbon coatings can be found in [21].

In continuous discharge (CD) processes, the discharge is kept continuously switched on and the substrate is directly exposed to the glow, thus it undergoes the direct interaction with neutral (chemical reactions), ionic (positive ion bombardment) species, generated in the plasma, and radiation emitted. Both chemical reactions and ion bombardment play an important role in CD deposition kinetics, and lead to coatings with tunable composition in terms of F/C ratio and cross-linking, where the composition and structure of the starting monomer are almost completely lost. Generally, a maximum F/C = 1.6 ratio can be obtained in this regime by finely tuning the deposition parameters. Plasma diagnostics [22–24] revealed that a fluorocarbon plasma is populated by CF_x ($1 \leq x \leq 3$) radicals, F and C atoms and ions produced by the fragmentation of the monomer, as well as by heavier species originated by recombination reactions among different fragments and monomer molecules. All these active species are present in different excitation states, and their distribution highly depends on the experimental parameters of the discharge. The distribution of species in the plasma and in its vicinity drives the interactions with the substrate along with its nature, temperature, ion bombardment and position in the reactor, and controls structure and composition of the coating or of the etched/fluorinated layer. Since F atoms and CF_x radicals have been demonstrated to be the main etching and deposition precursor species in CD fluorocarbon plasmas, the fluorine to radicals F/CF_x density ratio in the plasma plays a key role in describing the deposition of fluoropolymers [22, 25, 26].

Homogeneous and heterogeneous reactions in CD-PECVD processes of Teflon-like coatings were rationalized by d'Agostino [22, 25, 27–29], and the ion activated growth model (AGM) of deposition was proposed. AGM, which has been utilized also to rationalize other PECVD processes, like those from organosilicon monomers, combines the contribution of low-energy ion bombardment to activate the substrate surface, and of CF_x radicals formed in the glow discharge as *building blocks* of the coating.

The energy, driven by the ions bombarding the growing film (or the substrate, in the early deposition stages), which depends on the process parameters (e.g., pressure, power, bias, geometry), also plays a key role on film chemistry and structure. At low energy, the ions create surface defective sites (e.g., dangling bonds) that act as preferential chemisorption sites for the precursor CF_x radicals from the plasma. If the energy of the ions overcomes a certain threshold, however, the desorption of the precursors is induced. When the ion energy becomes very high, the sputtering of the coating may also be activated.

Substrate temperature affects the adsorption–desorption equilibrium of the CF_x radicals on the substrate (coating), and the following reaction of polymerization with the surface active sites originated by the ion-bombardment. As an overall effect, the increased substrate temperature, in the 25–100°C range, was found to result in a reduced deposition rate and F/C ratio of the deposited coating [17, 27].

To obtain coatings with very high Teflon character (F/C$\sim$2), it is possible to work in the afterglow (AG) mode. In this case, the substrates are positioned some centimeters downstream from the glow, where the density and the energy of electrons and ions is negligible or absent. The absence of charged particles in the AG region excludes the AGM mechanism and high retention of the monomer structure in the coating is achievable due to a deposition mechanism similar to the conventional radical polymerization. In AG processes, species with a long life-time (e.g., CF_2 radicals) have the highest probability to interact with the substrate and, eventually, to deposit a coating. The optimum plasma–substrate distance has to be found, for each single case, in order to deposit coatings with the requested composition, cross-linking and monomer structure retention degree at an acceptable deposition rate. A short distance from the glow may result in the coatings being very similar to those deposited in the glow, or in substrate etching, depending on the density of species in that particular position of the reactor. The AG deposition of Teflon-like films with highly oriented-CF_2-$(CF_2)_n$-CF_3 was demonstrated in the literature from C_2F_4 [16,30,31], C_2F_6 [17,31], C_2F_6/H_2 (80/20, [17,31]) and C_3F_6O feeds [32].

Since in these processes the flow-dynamics of the reactor can play a major role, AG-PECVD processes are investigated mostly in tubular reactors, with many different plasma sources. The distance from the glow in each condition is the key parameter to select between cross-linked films very similar to those obtained by continuous mode, and AG Teflon-like coatings.

The main drawback of the AG-PECVD is the low deposition rate, in the orders of a few Å/min, which can lead to a partial coverage of the substrate with the coating. Moreover, likely due to the low surface energy of the terminal –CF_3 groups, the process is self-limited, i.e., after a certain deposition time, in fact, the coating does not grow anymore, albeit the discharge is still on.

Due to their peculiar structure, AG chain-oriented Teflon-like coatings display an hydrophobic character very close to that of Polytetrafluoroethylene (PTFE) (water contact angle, WCA $\sim 115°$). Intranuovo *et al.* [33], reported the deposition of micro/nano-structured fluorocarbon coatings by AG-PECVD with hexafluoropropylene oxide (C_3F_6O, HFPO). The extent of micro/nano-structuring, as well as the F/C ratio of the coatings, could be precisely tuned with both substrate position and deposition time. "Stone roses" morphological features were obtained, by controlling the plasma parameters, which imparted superhydrophobic (WCA $= 165°$) and slippery behavior to the film surface.

Micro/nanostructured Teflon-like films can also be deposited from modulated glow (MD) discharges. In this case, the power input is delivered periodically to the plasma for t_{on} (μ/ms), and switched off, or reduced to a certain fraction, during off time t_{off} [34–40]. The substrates are positioned in the region where the glow is activated, thus they experience ion bombardment and interact with unstable species from the plasma during t_{on}, with long-lived radicals/atoms and, in some cases, with unreacted monomer molecules during t_{off}, since ions and electrons extinguish, being characterized by shorter lifetimes. MD-PECVD processes can be investigated as a function of the period, $t_{on} + t_{off}$, and of the duty cycle, $DC[\%] = 100\,t_{on}/(t_{on} + t_{off})$, of the discharge. Most of the experiments are investigated keeping the power at zero value during t_{off}.

With all the other parameters constant, the power delivered to the discharge and the duty cycle determines the degree of fragmentation of the monomer in the gas phase and, consequently, the degree of retention of the monomer structure in the coating. In this context the deposition rate is frequently expressed in terms of coating thickness per "unit input energy" or per pulse. Many experiments performed by different authors at constant t_{on} and increasing t_{off} values showed that the coating can grow also during t_{off}. This is due to heterogeneous reactions of the radicals formed during t_{on} with the substrate and, eventually, of the monomer molecules with the activated substrate, depending on the reactivity of the monomer.

Various researchers [31,34–43] found that lowering the duty cycle results in a progressively less branched polymeric film structure with higher F/C ratio that is more similar to PTFE. This was explained by a reduced extent of fragmentation of the monomer in the gas phase and a less energetic ion bombardment at the film surface during the on time. Ion bombardment may, in fact, lead to structural rearrangement, cross-linking, etching or defluorination of the film, resulting in a more damaged polymer surface. Many authors deposited PTFE-like films from hexafluoropropylene oxide (HFPO, C_3F_6O) modulated glow discharges with an F/C ratio as high as F/C=(1.8–1.9) [36, 44, 45]. Moreover, Fisher and co-workers [43] demonstrated that films deposited in modulated discharges fed with HFPO at lower duty cycle (i.e., 5%) contain chains perpendicular to the substrate surface.

Surface morphology of films deposited from modulated plasmas of hexafluoropropylene oxide (HFPO), (1,1,2,2)-tetrafluoroethane ($C_2H_2F_4$), and difluoromethane (CH_2F_2) revealed nodular growth (cauliflower-like appearance), with the size and distribution of the nodules being dependent on the precursor, the degree of surface modification to which the growing film is exposed, and the substrate surface temperature.

A systematic study of the morphology of fluorocarbon films deposited both from continuous and modulated discharges fed with tetrafluoroethylene was carried out in our group [39, 46–49]. Films obtained in continuous plasmas always showed very smooth surfaces with root mean square (RMS) roughness of the order of 1 nm. The sample deposited from modulated plasma at very low DC (5%) and long off times (320 ms) showed a complex morphology characterized by ribbon-like structures typically several micrometers long and hundreds of nanometers in width. These ribbons are randomly distributed over the entire surface and twisted in an intricate way. At higher magnification, the image shows that the surface area between these structures is populated by islands of nanometric size, also randomly distributed. Imaging SIMS analyses demonstrated that ribbon-like structures are more Teflon-like with respect to the rest of the sample, and from XRD analyses, such a film was also found to be partially crystalline. Highly structured films exhibit a super-water repellent surface with contact angles up to 170° and slippery behavior [47]. Increasing the duty cycle leads to loss of ribbon-like structures and appearance of globular structures of increased density at higher duty cycles.

A mechanism for the formation of different morphologies was proposed, based on different radical mobility in the different deposition regimes [39]. At low duty cycles and long periods, the deposition rate is low enough for the radicals to rearrange in ribbons, which appear to constitute a thermodynamically more stable form. However, when the duty cycle is increased, the process becomes kinetically driven, and the anisotropy is progressively lost. The number of active sites on the sample surface increases, together with the radical concentration. As a consequence, the probability that each active site is reached by radicals coming from all possible directions increases, too, and this results in the formation of globules over the entire surface. Moreover, since the time for the radical migration decreases (shorter off-time) and ion bombardment increases, the average heights of the nuclei as well as their height differences are lowered.

2.2. *Atmospheric pressure PECVD of fluorocarbon films*

We thoroughly investigated the atmospheric pressure PECVD of fluorocarbon films in DBDs. The main aims of our studies were to elucidate fundamental aspects of fluorocarbon plasma chemistry at atmospheric

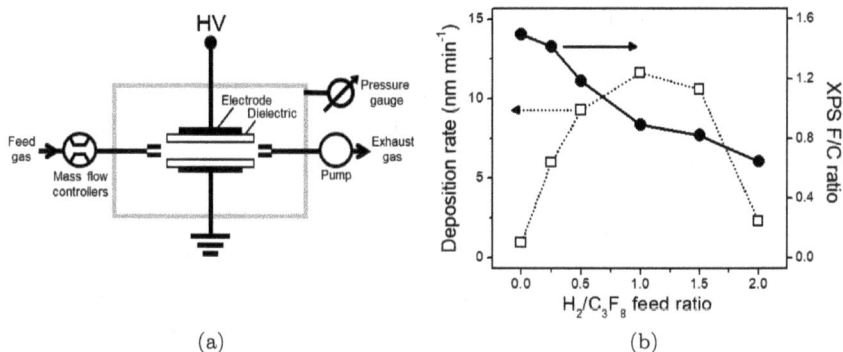

(a) (b)

Fig. 5. (a) Simplified scheme of the atmospheric pressure DBD reactor used for thin film deposition. (b) Effect of the H_2-C_3F_8 feed ratio on deposition rate and X-ray photoelectron spectroscopy (XPS) F/C ratio of fluorocarbon films deposited in DBDs fed with He–C_3F_8–H_2 mixtures. The experimental set-up values were: 20 kHz frequency, $2.8 kV_{p-p}$ peak–peak voltage, 4 slm He flow rate, $[C_3F_8] = 0.01\%$ and H_2/C_3F_8 feed ratio = (0–2) [50,51]. Adapted with permission from [50]. © 2007 WILEY–VCH Verlag GmbH & Co. KGaA, Weinheim.

pressure (AP), to evaluate the potential of DBDs for the preparation of fluorocarbon coatings and to establish a comparison with the well-established low-pressure plasma technology.

In particular, we focused on DBDs fed with noble gases (i.e., helium or argon) in mixture with low molecular-weight fluorocarbon monomers (i.e., tetrafluoromethane, hexafluoropropene, octafluoropropane) and additives (e.g., H_2 and O_2) [50–52]. Deposition processes were carried out using DBD reactors with parallel-plate electrode configuration and lateral gas injection (Fig. 5(a)). The plasma was generated by applying a sinusoidal high voltage (2.5–$6.0 kV_{p-p}$) in the frequency range 15–30 kHz. The DBD was fed with a high flow rate of noble gas (4–6 slm), while the concentrations of fluorocarbon monomers and additives were kept below 0.2%.

Our first study was devoted to the investigation of DBDs fed with He–C_3F_6 and He–C_3F_8–H_2 mixtures [50,51]. Overall, it was demonstrated that the deposition process can be controlled by varying both the feed mixture composition and the excitation frequency of the DBD (i.e., the input power). He–C_3F_6-fed DBDs allow depositing thin films with XPS F/C ratio of about 1.5. While the chemical composition of the coatings deposited from C_3F_6 does not change appreciably with process conditions, the deposition rate can be increased by increasing the frequency. On the other hand, in case of He–C_3F_8–H_2-fed DBDs, it is possible to change the XPS F/C ratio and the cross-linking degree of the fluorocarbon layers by varying the H_2/C_3F_8 feed ratio (Fig. 5(b)). Hydrogen addition also promotes the increase of the deposition rate, which is maximized at $[H_2]/[C_3F_8]$ feed ratio of 1 (Fig. 5(b)). These results were found to be in agreement with those reported for low-pressure plasmas [25].

In a successive study, we focused on the deposition and etching of fluorocarbon films in DBDs fed with $Ar/CF_4/H_2$ and $Ar/CF_4/O_2$ mixtures, respectively [53, 54]. The results indicated that, as widely reported in the literature on LP fluorocarbon plasmas [22, 55], the contemporaneous presence of F atoms (fluoropolymer etchant species) and CF_x fragments (fluoropolymer precursor species) leads to a competition between etching and deposition also in atmospheric pressure DBDs. However, experiments revealed that an important difference between LP and AP plasma operation lies in the lack of energetic ion bombardment at atmospheric pressure [4,56].

We also carried out an explorative study on the influence of contaminants such as air and water vapor on the PECVD of fluorocarbon films at atmospheric pressure [54, 57, 58]. The aim of this study was to evaluate whether the presence of these contaminants could have a detrimental effect on the overall deposition process, causing, for instance, a change of the discharge regime, a variation of the fluoropolymer composition, oxygen and nitrogen uptake, as well as a decrease of the deposition rate. We undertook this study also keeping in mind that the possibility of obtaining fluorocarbon films with the desired properties in highly contaminated environments could lead to the reduction of costs of AP plasma processes and reactors. Experiments were carried out in an airtight AP reactor, using a Ar-0.2% C_3F_6-fed DBD and adding controlled amounts of air and water vapor to the feed mixture. We observed that contaminants' addition is responsible for a decrease of the deposition rate and an increase of surface roughness of the deposited layers, however, it does not affect significantly their chemical composition and wettability. Interestingly, we found that even at water and air concentrations of 0.05% and 0.1%, respectively (corresponding to air/C_3F_6 and H_2O/C_3F_6 feed ratios of 0.5 and 0.25, respectively), the XPS F/C ratio of the coating remains as high as 1.7, the O and N surface atomic concentrations are less than 1%, and the advancing and receding WCAs are about 125° and 105°, respectively.

More recently, we investigated the PECVD of fluorocarbon coatings onto open-cell polymer foams to determine the potential of DBDs for thin film deposition on the outer and inner surfaces of complex three-dimensional (3D) porous materials [59, 60]. In particular, an He–C_3F_6-fed DBD was used to deposit a fluorocarbon coating on a commercial polyurethanane (PU) foam characterized by porosity of about 97% and pore density of 30 ppi (pores per linear inch). During the deposition process, the sample was sandwiched between the dielectric-covered electrodes of a parallel-plate DBD reactor, so that the discharge could ignite both outside the foam and throughout its 3D interconnected porous structure (Fig. 6(a)). This enabled the uniform coverage of the entire sample with a fluorocarbon layer. XPS analyses revealed very moderate changes in the surface chemical composition moving from the exterior to the interior of the plasma-treated samples (representative high-resolution C 1s spectrum in Fig. 6(b)). SEM observations showed that plasma ignition does not alter the porous structure of the sample (Fig. 6(c)) and allowed estimating the thickness of the coating deposited on the foam struts (Fig. 6(d)). As expected, the plasma-treated samples showed hydrophobic behavior and were characterized by a static WCA of $\sim$135°, significantly greater than the one of the pristine foam ($\sim$90°).

2.3. *Low-pressure PECVD of organosilicon films*

Organosilane and organosiloxane gases in mixtures with oxygen are widely utilized for PECVD of thin films with a tunable organic/inorganic character which, in turn, allows tailoring of their physical and chemical properties (e.g., density, refractive index, dielectric constant, surface energy, internal stress, hardness, etc.). Silicon dioxide-like films are employed in integrated circuits as dielectrics and in micro-electro-mechanical system devices as sacrificial layers. They also find applications in optics as scratch-resistant coatings and in packaging as gas/vapor/aroma barrier layers on polymeric substrates [61–70]. Carbon-containing silicon dioxide-like films are being extensively investigated as low-k inter-metal and inter-layer dielectrics in microelectronics [71, 72].

Excellent fundamental studies on organosilicon plasmas were carried out by Segui *et al.* [73–75] on the infrared diagnostics of gas phase products, Flamm *et al.* [76, 77] on the deposition mechanism of conformal silicon dioxide films, and d'Agostino *et al.* [78, 79] on the effect of substrate bias. A comprehensive review work by Wròbel and Wertheimer can be found in [61].

The deposition mechanism of organosilicon films from hexamethyld-isiloxane (HMDSO)-oxygen plasmas under high monomer fragmentation regime (i.e., high plasma power) in RF plasmas was investigated by

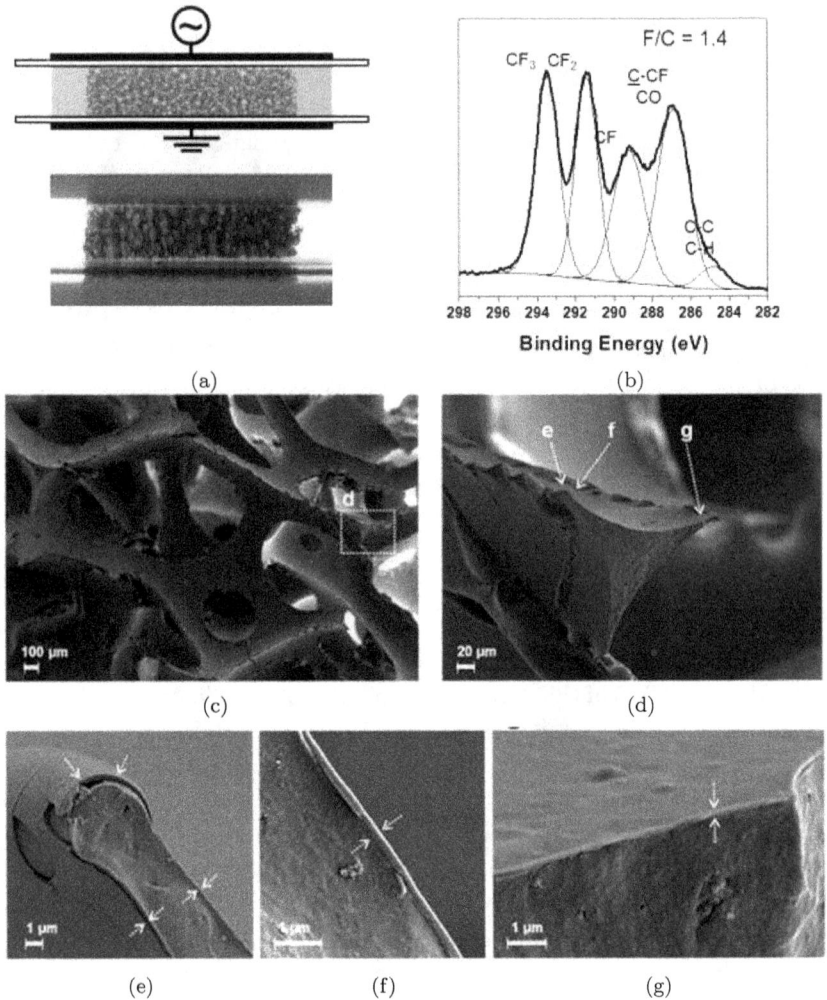

Fig. 6. (a) Diagram showing the side view of the parallel-plate DBD cell and photograph of the He–C$_3$F$_6$ DBD during thin film deposition on a PU foam sample. Reproduced with permission from [58]. © 2011 WILEY-VCH Verlag GmbH & Co. KGaA, Weinheim. (b) Representative high-resolution C 1s XPS signal of the fluorocarbon layer deposited on the inner surfaces of the foam. (c–g) SEM images corresponding to the cross-section of a plasma-treated foam sample: (c) low-magnification image, (d) cross-sectioned ligament, (e–g) different regions of the cross-sectioned ligament where the deposited coating is clearly visible (indicated by white arrows). Reproduced with permission from [58]. © 2011 WILEY-VCH Verlag GmbH & Co. KGaA, Weinheim.

Lamendola *et al.* [80, 81]. They coupled the plasma composition, as determined by optical emission spectroscopy, to film chemistry and to water vapor and oxygen barrier performances. Briefly summarizing, under high fragmentation regime with an excess of oxygen in the plasma, silicon dioxide-like films can be deposited, while low oxygen/monomer ratio results in C-rich films named as silicone-like. The use of pure HMDSO plasma leads to low fragmentation regime and a C-rich siloxane film with a partial monomer retention is obtained.

Creatore *et al.* [63, 82] also demonstrated that in C-free SiO_2-like coatings further optimization passes through condensation of silanol groups by increasing the plasma power during deposition, thus improving film barrier performances.

As reported above, silicon dioxide-like films find application in many technological fields which require dense films with low stress. For this purpose, ion bombardment-assisted deposition with low-to-medium energy ions has been proven to be of particular interest because it leads to film matrix densification at low temperature [83–85]. The latter is crucial when thermally sensitive substrates, such as polymers, are under investigation. Bombardment of material surface by low-energy ions can lead to breakage of chemical bonds, surface diffusion and heating, which contributes to the film densification. However, excessive ion energies can lead to structural damage and high internal stress.

In the literature, a comprehensive study on the densification mechanism in silicon dioxide thin films, based on molecular-dynamics simulations and ion-assisted deposition experiments, was carried out by Martinu *et al.* [86]. Their results show that in films grown in the presence of ion bombardment, structural changes occur at an intermediate length scale, i.e., the structural unit of the network, the $Si(O_{1/2})_4$ tetrahedron, stays unchanged while the ring statistics is affected. The presence of small rings (three- and four-membered) is associated with stress and defects in the network, while large rings (e.g., nine-membered) correspond to film porosity. Ion bombardment induces the structural densification of the SiO_2-like films because small rings rearrange to produce larger and less strained ones and too large rings turn into five-, six-, and seven-membered ones, reducing the free volume and causing the density to increase. The densification process can be identified as a relaxation mechanism resulting in silicon dioxide-like films with low stress and low defect concentration.

Starting from this study, we investigated the deposition of silicon dioxide-like barrier films from different siloxane and silane precursors, highly diluted in oxygen and argon. The variation of the precursor was investigated as a route to obtain silicon dioxide-like films with different structures, densities and hence barrier performances. Although the films were characterized by the same elemental composition, differences in film density and porosity were evidenced from optical properties' measurements and infrared absorption spectroscopy. These differences were correlated with the film microstructure and in turn with the barrier performances. The results confirmed that films with high density and low porosity performed better as single inorganic barrier layers for food-packaging, allowing a decrease of the water vapor transmission rate (WVTR) of one order of magnitude with respect to the bare polyethylene terephthalate (PET) substrate [87].

Results discussed so far showed a correlation between the barrier performance of a film and its chemical composition. Density and defects like pinholes and grain boundaries are important as well in determining the minimum value of the gas permeation rate through the barrier layer. Da Silva Sobrinho *et al.* [68] showed that the oxygen transmission rate and the defect density are highly correlated: When the barrier film thickness reaches the critical value, a sharp decrease not only in the oxygen transmission rate (OTR) but also in defect density is observed. Defects in the barrier film are caused, for example, by dust particles on the polymer substrate prior to deposition, the surface roughness (which is influenced by plasma–polymer interaction [88]) or imperfections of the polymer such as anti-block particles, geometric shadowing during the deposition and scratches or cracks in the film due to sample handling. Why the defect density decreases drastically in correspondence to the critical thickness is not yet fully understood. It is possible that at this thickness the deposited layer is thick enough to cover the substrate roughness and irregularities. Due to the presence of defects in the films, the lower limit of the WVTR of a single-barrier layer is estimated to be $5 \, 10^{-2} \, gm^{-2} \, day^{-1}$ [89]. This is enough to protect polymers in the food packaging industry, where the barrier layer is a thin inorganic film, usually of aluminium, aluminium oxide, silicon oxide or silicon nitride.

However, single-barrier layers are not sufficient for applications such as flexible electronics. The minimum allowable WVTR in this case is estimated, in fact, to be $(10^{-5}\text{--}10^{-6}) \, gm^{-2} \, day^{-1}$ for flexible OLEDs [89, 90]. The only way to meet this stringent demand is the application of a stack of alternating organic and inorganic layers. Although the organic layers do not possess barrier properties, their presence appears to be significant. In the literature, different theories explaining the beneficial effects of the organic interlayer are found. Affinito [91] suggests that the organic material fills up defects in the inorganic layer, thus repairing them and decreasing the gas transmission rate. Burrows *et al.* [89] suggest that planarization is the main action of the inorganic layer: Irregularities at the surface are planarized by a non-conformal, ultra-smooth film effectively decoupling the non-uniformities of the barrier layers. The model suggested by Graff *et al.* [92] also states that the organic interlayer decouples the defects, the effect being an increased diffusion path length and thus an increased diffusion time.

Instead of distinct organic/inorganic transitions, a film of gradually changing composition may be used. In this way, the tendency of introducing stress in the film, which generally leads to easy delamination, is avoided. Kim *et al.* [93] deposited such a stack in an RF parallel plate reactor. The WVTR of a polycarbonate substrate coated with the graded-composition barrier was less than $10^{-5}\,\mathrm{gm}^{-2}\,\mathrm{day}^{-1}$, which is a remarkable result. The inorganic layers are dense silicon oxynitride films, while the organic zones are organosilicate films.

In our laboratory, multi-stacks with high-barrier performances were deposited onto Melinex ST504 and Teonex Q65 from vinyltrimethylsilane (VTMS)/O_2 plasmas. In detail, it was found that a 50-nm thick single layer was enough to ensure barrier performance comparable with the best values in the literature. Furthermore, in the case of Melinex, the assembling of a tri-layer (SiO_x/SiC_xH_y/SiO_x) could further lower the WVTR by nearly one order of magnitude [94].

Besides food packaging application, several studies have also been carried out by our group on the plasma deposition of corrosion protective coatings for metals. Tetraethoxysilane (TEOS), hexamethyldisiloxane (HMDSO) hexamethyldisiloxane (HMDSO) and hexamethydisilazane (HMDSN) have been investigated as monomers for the deposition of protective SiO_x coatings. High oxygen/monomer ratio and RF power are necessary to deposit inorganic silanol-free dense layers with enhanced barrier properties.

Furthermore, the deposition of SiO_x coatings from modulated plasmas fed with TEOS/O_2/Ar mixture has been also investigated in relation to the barrier performances. It was found that the charge transfer resistance for the pulsed deposited coating is at least one order of magnitude higher than for the film deposited in continuous plasma, thus providing better protection capabilities. The deposition of inorganic layers through pulsed discharges was believed to result in more uniform and low-stress films [95].

Our studies further demonstrated that a dense coating without silanol groups is not enough to achieve sufficient metal protection, but a substrate pre-treatment in plasma is necessary to prepare the surface for the subsequent coating process. Such a process must be optimized choosing the right chemistry. In general, oxygen plasma pre-treatment is not useful for those metals that form oxides with poor mechanical and protective properties (Mg and silver alloys); in this case, hydrogen pre-treatment is suggested. Oxygen plasma is suitable if a passivating surface is formed (low carbon steel and Al). In all cases, however, the action of the plasma pre-treatment is also that of cleaning the surface from organic contaminants, enhancing the coating performance through adhesion improvement. Since stainless steel has a well-passivated surface, both hydrogen and oxygen plasmas can successfully enhance the protective performance of SiO_x coatings, through surface cleaning [95]. In a further study we explored the alternated deposition of silicon oxide and organosilicon layers and demonstrated that the deposition of 600-nm thick organosilicon multilayers ensured effective protection of low carbon steel substrates. In the multilayer the presence of the organic hydrophobic medium between the two inorganic layers increases the electrolyte path, allowing to nearly halve the total thickness of the protective coating though maintaining high corrosion resistance [96].

If, from the side of barrier applications, high-density films are required to ensure excellent performances, porous low-density organosilicate films have been widely investigated as low dielectric constant (k) materials for advanced interconnects in integrated circuits.

Enhancement in integrated circuit performance over the past four decades stems mainly from the periodic scaling down of device features, however, increasing latency of global wiring (RC delay) limits performances as feature sizes are reduced below 1 micron. Efforts to minimize delay through new materials' introductions are well underway. With respect to wire materials, the migration from Al to Cu metallization reduced line resistance by $\sim30\%$ and it is expected to remain the interconnect of choice for the near future. Further reduction of RC delay can also be achieved with low-permittivity dielectrics [71].

Intense research efforts have focused on developing inorganic and organic low-k materials that meet the stringent requirements for successful integration, but few exhibit the proper combination of low-permittivity plus thermal, chemical and mechanical stability. In this framework, PECVD solutions are preferable over other synthetic approaches because they have shorter integration cycles and do not require new equipments [97].

Starting from SiO_2, the easiest way to decrease the k value is to introduce terminal groups that cannot network, such as methyl groups, to form a so-called organosilicate glass (OSG). Methyl groups with their steric hindrance create additional free space in the polymer structure and the fact that Si–C is a less polarizable bond with respect to Si–O is an additional factor that contributes to decrease in the dielectric constant from about 4.3 (typical of quartz-like SiO_2) down to 2.8 [98–101]. The introduction of methyl groups creates pores in the material of diameters typically in the range 1.0–1.5 nm. To achieve $k < 2.5$ in the next generation of dielectrics, pores must be introduced into the network, as air, with $k \sim 1$, reduces the total dielectric constant in accordance with Lichtenecker's law of mixed dielectric materials.

Several approaches to the synthesis of ultralow-k films by PECVD have been reported in the literature. The first involves the deposition of composite organosilicon/porogen thin films from a mixture of an organosilicon monomer usually employed for the synthesis of low-k materials and a porogen as a sacrificial material. Upon annealing, the porogen is removed leaving behind nanopores, and cross-linking occurs within the matrix. The best result obtained with this method was $k = 2.3$ with a film thickness loss upon annealing of 36% [102]. The latter is a very important parameter to be considered since it is related to the poor mechanical properties of the silicate matrix as it can indicate a collapse of the matrix itself.

In a second approach, a dual phase film was deposited from a mixture of an organosilicon compound, which generally leads to low-k values, and an organic precursor. Upon annealing, a film with a dielectric constant of 2.05 was obtained and a thickness loss of 28%. By increasing the annealing time, the k value could be further decreased to 1.95. During the thermal treatment, hydrocarbon fragments were released resulting in the formation of pores with a diameter of 5 nm, as detected by X-ray porosimetry (XRP) [103–106].

The third approach consists in the use of PECVD from monomers containing a labile side group. In this case, thus, the organic precursor is already embedded in the monomer structure. Rhee and coworkers [107,108] studied the synthesis of ultralow-k films from both direct and remote PECVD and the lowest k values achieved were 2.0 and 2.1, respectively. No information about thickness loss upon annealing or hardness was provided. By adding CO_2 as oxidant to the gas mixture, the dielectric constant was lowered down to 1.9 [109]. Burkey and Gleason [110] obtained a film with a dielectric constant of 2.6 and a thickness loss of 30%, by working with silanes containing labile side groups. Though, the best result they could achieve was by far the one with cyclic organosiloxane as a precursor, which resulted in a film with $k = 2.4$ and a thickness loss of about 4%.

We also investigated the deposition of ultralow-k films from divinylte-tramethyldisiloxane (DVTMDSO)-oxygen plasmas which allowed to reach a minimum k value of 2.3 and a thickness loss of 11%. The effect of all process parameters on the chemical composition and on the electrical properties of deposited films can be found in [111]. Films with good thermal stability (thickness loss of 6%) and still very low permittivity ($k = 2.3$) were obtained from allyltrimethylsilane (ATMS)-oxygen plasmas in very low monomer fragmentation regimes [112]. Thermogravimetric analysis coupled with mass spectrometry revealed that, during post-deposition thermal treatment, silicon containing fragments were lost from the film matrix, leading to partial collapse, along with hydrocarbon ones. Loss of thermally labile organic groups are instead believed to be responsible for formation of pores [111]. In a further study, we explored several organosilicons with different structures (siloxane, silane) and substituents (methyl, vinyl, allyl, isopropyl) as precursors for the low-k films. Silane-based precursors allowed to obtain films with permittivity values as low as 2.3, with a limited thickness loss of 6% upon thermal annealing at 400°C. Films deposited from the siloxane monomer, though having the same dielectric constant, were characterized by an increased thickness shrinkage of 11%. Thermal stability was correlated not only to the cross-linking degree induced by oxidation but also to the presence of methylene bridges in the polymer backbone, which mainly accounted for the better thermal stability of silane-based films. Film thermal stability also correlated with deposition temperature. Substrate heating (150°C) during deposition ensured the best balance between very low permittivity values and good thermal stability [113].

2.4. *Atmospheric pressure PECVD of organosilicon films*

Recently, the interest of both the academic and industrial community has considerably expanded to the use of atmospheric pressure plasmas for the deposition of organosilicon coatings. DBD technology has proven to be the most popular approach for the deposition of this class of thin films at atmospheric pressure [5, 54]. As in low-pressure plasmas, HMDSO has been the most widely used monomer, due to its non-toxic character, and relatively high vapor pressure at room temperature [5, 104, 114]. However, over the years, the use of other organosilicons has been also proposed (e.g., tetraethoxysilane, hexamethyldisilazane, etc. [115]).

Published studies have been mainly focused on the PECVD process performances in terms of deposition rates, chemical composition and structure of the films, with the aim of tailoring the final properties of the layer (e.g., mechanical, electrical, optical and gases/moisture diffusion properties) for different applications. Our efforts have been directed instead toward the investigation of the plasma chemistry of organosilicon-containing DBDs to gain insights into the deposition mechanism. In particular, in our work, we have attempted to correlate results from the chemical and morphological characterization of the films with those obtained from the analysis of the DBD exhaust by gaschromatography-mass spectrometry (GC-MS). We have exploited GC-MS as a powerful indirect diagnostic technique of the plasma phase. GC-MS can in fact contribute to a better understanding of the plasma chemistry in PECVD processes, since it enables the evaluation of the monomer reactivity as well as the qualitative and quantitative determination of stable byproducts formed by plasma activation [116–118].

The employment of this diagnostic technique strictly requires the optimization of a suitable exhaust sampling procedure, which generally involves the use of a discrete sampling system, such as a liquid nitrogen (LN2) cold trap located between the plasma chamber and the pump. The sampling of the exhaust gas is performed for a fixed time during the deposition process. Then, the trap is isolated from the system, the condensate is dissolved in a suitable solvent and analyzed by GC-MS. During GC-MS analysis, the different compounds are first separated by the GC column, then ionized by electron impact and analyzed using a mass filter (e.g., a quadrupole) to obtain the mass spectra, which can be used as fingerprints for compounds' identification by comparison with available MS libraries. The identification can be confirmed by comparison with the gas chromatographic retention time and mass spectrum of standard compounds. The construction of calibration curves with the internal standard method allows the quantitative analysis of identified species.

We first investigated the deposition of organosilicon films in DBDs fed with $Ar/HMDSO/O_2$ mixtures by comparing the FTIR spectra of the deposits with the GC-MS analyses of the exhaust gas [119]. We found that, under the explored experimental conditions, O_2 addition does not enhance the activation of the monomer, while it highly influences the chemical composition and structure of the deposited coating as well as the quali-quantitative distribution of byproducts present in the exhaust gas.

As expected, without O_2 addition a coating with high retention of the monomer structure is obtained and the exhaust gas contains several byproducts such as silanes, silanols, linear and cyclic siloxanes. The dimethylsiloxane unit ($-Si(CH_3)_2O-$) seems to be the most important building block of oligomerization byproducts. Oxygen addition to the feed is responsible for an intense reduction of the organic character of the film as well as for a steep decrease, below the quantification limit, of the concentration of all byproducts except silanols. We also found that, due to the lack of energetic ion bombardment, the concentration of silanol groups in the coatings deposited at atmospheric pressure cannot be effectively reduced as in LP plasmas [117].

Later on we carried out a comprehensive study of thin film deposition in DBDs fed with argon, oxygen and different methyldisiloxanes (MDSOs), i.e., hexamethyldisiloxane, pentamethyldisiloxane (PMDSO) and 1,1,3,3-tetramethyldisiloxane (TMSO) [54, 120]. FTIR and XPS analyses revealed that the carbon content of the films depends on both the O_2/MDSO feed ratio and the number of methyl groups in the monomer molecule. For instance, as reported in Fig. 7(a), the FTIR absorption integral ratio $Si(CH_3)_x$/SiOSi decreases with the O_2/MDSO ratio for all monomers. In addition, at fixed O_2/MDSO ratio, the integral ratio decreases in the order HMDSO>PMDSO>TMDSO. The data also suggested that, in the case of coatings deposited from PMDSO and TMDSO, carbon removal is further enhanced by the presence of Si–H bonds.

The GC-MS investigation of the exhaust gas showed that the number and concentration of heavy oligomers (containing from 3 to 5 silicon atoms) decrease with decreasing the number of methyl groups in the monomer molecule (i.e., in the order HMDSO>PMDSO>TMDSO). Specifically, methylpentasiloxanes and methyltetrasiloxanes are observed only with HMDSO, various methyltrisiloxanes are detected both with HMDSO and PMDSO, while 1,1,3,3,5,5-hexamethyltrisiloxane is the sole compound containing more than two Si atoms detected in the exhaust gas of TMDSO-fed DBDs. O_2 addition in the feed mixture results in the steep decrease of the concentration of silanes and siloxanes below the quantification limit, due to oxidation reactions. Figure 7(b) displays, for instance, GC-MS results for octamethyltrisiloxane. This trend well correlates with the variation of the chemical composition of the deposited film as a function of the O_2/MDSO feed ratio (Fig. 7(a)). The almost complete absence of oligomerization by-products when TMDSO is used as a monomer could be explained considering the globular morphology of the coatings revealed by SEM

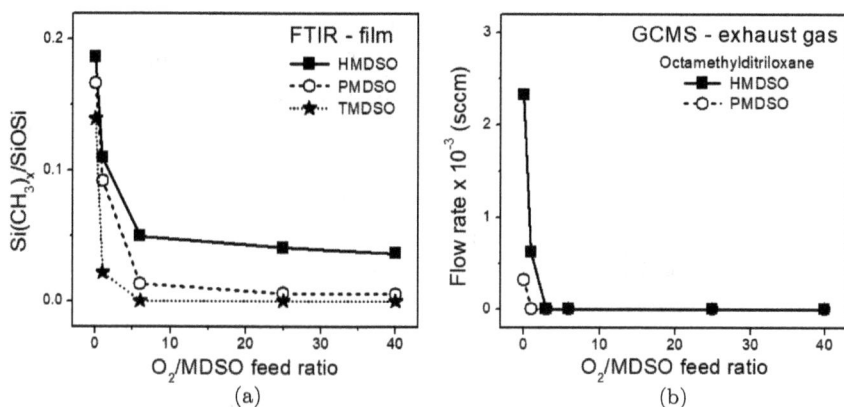

Fig. 7. AP PECVD of organosilicon films in DBDs fed with mixtures of Ar, O_2 and different methyldisiloxanes (HMDSO, PMDSO and TMDSO). (a) FTIR absorption integrals ratio $Si(CH_3)_x/SiOSi$ of the deposited coatings. (b) Flow rate of octamethyltrisiloxane in the exhaust gas as a function of the $O_2/MDSO$ feed ratio. Adapted from [120]. © 2012 WILEY-VCH Verlag GmbH & Co. KGaA, Weinheim.

observations. It seems reasonable that, due to the high reactivity of organosilicon fragments formed by TMDSO activation, oligomerization reactions are mainly pushed toward the production of high molecular weight compounds, which are then incorporated in the coating in the form of globules.

2.5. *Low-pressure PECVD of nanocomposite films*

The possibility of combining the physico-chemical properties of metals and polymers in a single material and of controlling their properties in a continuous way has over the years stimulated a growing interest in many fields of chemistry, physics and materials science. Numerous materials have been produced that are unique in terms of their mechanical, electrophysical, magnetic, optical and catalytic properties.

Basic vacuum deposition techniques for nanocomposite films include: plasma polymerization of metal organic compounds, simultaneous plasma polymerization and sputtering of a metal, RF sputtering from the composite target or targets of a polymer and a metal, metal cluster beam deposition and simultaneous plasma polymerization, and simultaneous evaporation of a metal and a polymer [121]. The pioneering approach developed by Kay *et al.* [122–125] consisting of a simultaneous PECVD of the polymeric matrix and sputtering from a metal target offers several advantages:

(i) Easy control of the metal content in the film, ranging from few percent up to 100%; (ii) homogeneous dispersion of metal clusters in the organic matrix; and (iii) control of particle size [124]. Many papers have been published on this approach using fluorocarbons [122,124,125], hydrocarbons [124,125], organosilicons [126,127] and other organic compounds as matrix precursors [128], and Au [124, 127], Ag [125, 126], W [126], and other elements [122,124,125] to generate metallic clusters.

For the deposition of such metal–polymer nanocomposite films, the metal target is attached to the RF-powered electrode. To promote sputtering, the target is much smaller than the grounded one. Thus, the system configuration is highly asymmetric and the self-bias potential on the target is highly negative. Positive ions formed in the plasma are then accelerated to the metal surface and sputter metal atoms and clusters which are included in the growing film. At very low metal volume fraction (filling factor, f), nearly spherical metal clusters are isolated in the polymer matrix. Increasing the metal filling factor by increasing the power to the discharge, metal clusters become larger and more irregular in shape. A further increase of the filling factor results in coalescence of the metal clusters (percolation threshold) and, eventually, in the development of interconnected worm-like metal structures [129]. Post-deposition thermal treatment speeds up coalescence of small particles and formation of even larger aggregates. When the RF power is kept constant and filling factor is varied by changing the monomer content in the plasma, the size of metal inclusions does not vary [129]. The amount of incorporated metal and its microstructural arrangement determine the electrical conductivity of metal-containing polymeric films. When the filling factor is changed, the conductivity varies in a broad range, from a dielectric regime (metal particles are dispersed in polymeric matrix) to a metallic regime (isolated metal grains start to touch and percolation occurs) [126,128,129].

The structure of composite films is inherently metastable and aging effects may be observed. The plasma polymers may contain a great concentration of free radicals affecting changes in the chemical structure over time because of post-deposition oxidation in the open air. However, the post-deposition migration of small metal particles is supposed to play the dominant role in structural rearrangements, due to their relatively easy motion in the less rigid plasma polymer matrix. Increasing substrate temperature or ion bombardment during the nanocomposite film growth was found to decrease aging effects in Au/fluorocarbon films, as particle mobility is inhibited in a cross-linked polymer matrix [130].

In our laboratory, we have investigated polymer/metal nanocomposite films for several applications. For biomedical devices, Ag-containing polyethyleneoxide (PEO)-like films are interesting candidates as bacterial resistant coatings, because they merge the non-fouling properties of PEO [131–133] with the anti-bacterial properties of silver [134, 135]. Preserving the PEO structure, $(CH_2CH_2O)_n$, of the monomer in the coatings is of primary importance for the non-fouling character of PECVD PEO-like films and this can be accomplished by working in monomer low-fragmentation regime (i.e., low-plasma power). However, more drastic conditions are needed to sputter the metal. Then, a trade-off exists between preservation of the PEO character in the film matrix and the amount of Ag embedded. An estimation of the PEO-like character can be derived from the curve fitting of the XPS C1s signal of the Ag/PEO-like coatings. The PEO character and the Ag percentage in the deposited films could be tailored by properly tuning process parameters such as the plasma power and the monomer flow rate in the gas feed. In particular, the increase in the RF power applied to the Ag electrode decreased the PEO character while increasing the amount of Ag in the film. The opposite effect was observed on increasing the monomer flow rate in the discharge.

When considering the potential application of such Ag/PEO-like coatings in biomedical devices, it is crucial to study the water stability and the Ag^+ ions release mechanism [136,137]. Such studies revealed that deposited films were stable in water and release of Ag^+ ions in water occurs according to a reservoir-type mechanism similar to that of other drug delivery systems where the therapeutic agent is uniformly dispersed/dissolved in an inert polymer matrix. Ag/PEO-like coatings were found very effective in reducing the proliferation of bacteria, e.g., Psaeudomonas Aeuruginosa, at any loading of silver [138]. In a further study, PEO-like barrier thin films were deposited on top of the Ag-containing coatings to control the release of silver in such a way as to preserve its bactericidal activity but limiting its cytotoxicity [139]. It was found that the amount of released Ag^+ ions could be tightly controlled through the PEO-like diffusion barriers, resulting in being effective against Staphylococcus Epidermis, while still keeping cell viability.

Another application we have explored is the use of Pt-containing thin films as electrocatalysts for proton exchange membrane (PEM) fuel cells. Proton exchange membrane fuel cells are simple electrochemical devices able to provide electricity with virtually zero emissions. In hydrogen-based fuel cells, the oxidation of H_2 occurs at the anode, while O_2 is reduced at the cathode. Both these reactions heavily rely on the use of Pt as catalyst. The overall reaction is the recombination of molecular hydrogen and oxygen, which produces water and energy. The core of a PEM fuel cell is the membrane electrode assembly (MEA), which consists of a proton conductive membrane sandwiched between two catalytic electrodes. Nanoparticles of platinum and its alloys, supported on conductive porous carbonaceous substrates, are mainly used as catalysts, while Nafion, a perfluorosulfonic polymer, is used as a proton conductive material.

The commercial availability of such devices is, however, still hampered mainly by the high cost of these highly active materials. Therefore, extensive research efforts are currently being devoted, from the catalyst side, to a reduction and a more efficient utilization of the Pt amount of the fuel cell electrodes [140]. In this regard, the use of nanosized platinum clusters improves the catalytic activity per platinum mass due to the higher surface area to-volume ratio, and therefore reduces the amount of the expensive metal needed. We proposed the plasma deposition of Pt–hydrocarbon nanocomposite films as electrocatalysts in micro fuel cells for portable applications [141–143].

Results showed that the Pt content in the film can be continuously varied as a function of the plasma input power and the monomer flow rate. Platinum is incorporated into the hydrocarbon matrix in its metallic form and the film shows negligible aging (i.e., oxidation) during air exposure. Through transmission electron microscopy, it was found that metal is uniformly dispersed in the polymeric matrix and forms nanocrystalline clusters with size varying as a function of Pt content. Morphological evaluations revealed that deposited films are characterized by columnar structures with column size, orientation and spacing depending on the plasma conditions used during deposition. Device testing proved the suitability of such coatings as anodic as well as cathodic catalysts, with a maximum power density as high as $300 \, \mathrm{mW \, cm^{-2}}$ [141]. In a further study, we also investigated the deposition of Pt/fluorocarbon nanocomposite coatings, though the catalytic activity was lower [144].

2.6. *Atmospheric pressure cold plasma deposition of nanocomposite thin films*

Combining atmospheric pressure plasma sources with atomization systems allows to extend their applications in different fields, from high-tech agriculture to medicine and the administration of pesticides or drugs [145].

An aerosol is a mixture where liquid or solid particles are suspended in a gaseous medium for a time long enough to be observed. The size of the suspended particles is typically below tens of microns, with a density below 1% [146]. Artificial aerosols formed by liquid droplets are prepared through three main techniques: ultrasonic generation, pneumatic jet and electrostatic atomization.

A main concern in PECVD is the selection and delivery of suitable chemical precursors, when dealing with solid and, in general, low vapour pressure chemicals. Coupling an aerosol atomizer with a plasma source working at atmospheric pressure offers a way to overcome such an issue. In fact, with this approach, any stable precursors can be used in coating technology, since the precursor, liquid or solid, can be injected pure or dissolved in a solvent via atomization. Such a coating process can be named aerosol-assisted plasma deposition (AAPD). Other advantages of AAPD are the high deposition rate due to the high mass transport rate of the precursor and the easy multi-precursor injection with good stoichiometric control.

When coupling plasma with an aerosol, two approaches to inject a feed can be barely identified, with the aim of depositing a composite coating. In the first approach, the aerosol feed consists of a liquid monomer, bringing the matrix building blocks, and an active compound (e.g., a drug) is dispersed inside (a solvent can be added to stabilize the dispersion). In this case, all the components for the coating deposition are introduced through the aerosol; there is no need for an auxiliary line for the addition of a film precursor. Another approach involves the atomization of a solution (or suspension) typically not useful for film formation (e.g., an aqueous solution of a drug), hence, the main film precursor is injected through an auxiliary line (gas line, or bubbling system or aerosol generator). Organic bio-composite coatings, consist of polymer matrix embedding bioactive components such as proteins, nucleic acids, lipids, drugs, and biopolymers. Such coatings can find applications in different fields, such as biosensing, as antibacterial or non-fouling films, inhibiting the formation of biofilm by coupling suitable peptides or enzymes, in drug release and, in general, in the biomedical field [147–154].

To handle such complex and often sensitive molecules, preserving their bioactivity in a harsh environment as plasma is an issue. Aerosol-assisted plasma processing offers this possibility, in fact, in the aerosol droplets the solvent has a shielding role, forming a shell around the active compounds and acting as a barrier to the energetic species present in the plasma.

O'Hare *et al.* [155] demonstrated the deposition of antibiotics containing composite coatings. They dissolved cetalkonium chloride, benzalkonium chloride or cetylpyridinium chloride in acrylic acid, PEG or their mixture, and atomized this solution in a pilot line to coat fabric substrates. The coatings resisted washing at pH in the range 2–12, and provided significant protection against fungal contamination of fabrics. The nebulization of biomolecules (proteins, in particular) in DBD sources was reported earlier by Heyse *et al.* [156, 157], demonstrating the plasma deposition of acetylene and pyrrole coatings embedding allophycocyanin, a fluorescent protein, and other enzymes, such as glucose oxidase, lipase and alkaline phosphatase.

We proposed the plasma deposition of lysozyme-containing coatings based on a hydrocarbon matrix [158]. Matrix-assisted laser desorption/ ionization analyses recognized the presence of native lysozyme, though part of methionine and tryptophan aminoacidic residues presented some oxidation. Also high performance liquid chromatography analysis of water-releasing solution confirmed the presence of lysozyme ($14\,\mu g/cm^2$), and agar diffusion test demonstrated antibacterial activity close to that of the native enzyme.

Similarly, Hsiao *et al.* [159, 160] replicated these experiments with lysozyme and albumine depositing films at different distances from the inlet. Besides confirming the activity of the embedded proteins, they found that spherical structures are formed depending on the sample position in the reactor. Such a result, the formation in some circumstances of spherical structures when a solution is atomized in atmospheric pressure plasma, deserves further discussion, and more details can be found elsewhere [161]. In fact, besides the mentioned papers on lysozymes and BSA, spherical nanostructures have been found also when depositing with APPD vancomycin and gentamicin nanocomposite coatings [162–165]. These studies further confirmed that the drugs retain their antibacterial activity once embedded in a plasma polymerized film and were gradually released in the aqueous medium. However, as it can be observed in Fig. 8, the SEM investigation revealed a distinctive morphology of the coatings. Spherical features appeared on the surface, with size dependent on the experimental conditions. Moving from continuous to pulse mode, the morphology of the coatings passed from larger and more homogeneous in size (Fig. 8(a)) to taller aggregates of slightly smaller spheres (Fig. 8(b)). Furthermore, after 1 h of water immersion, the spherical features become empty or collapse (Fig. 8(c)).

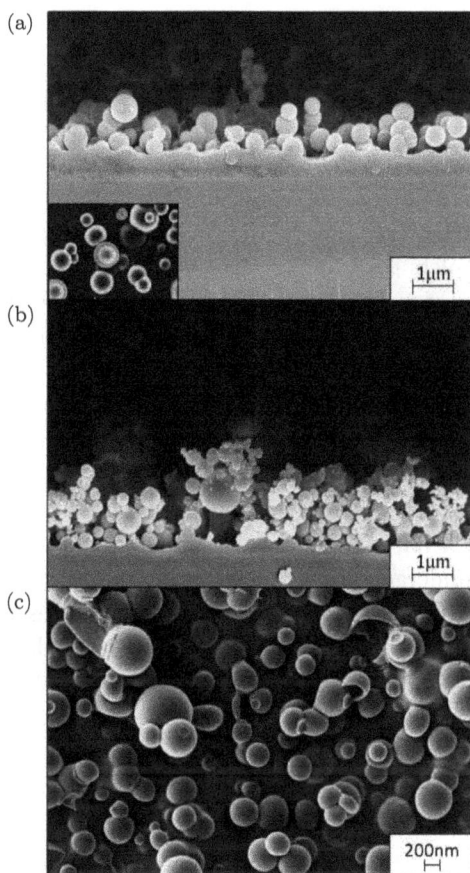

Fig. 8. Cross-sectional SEM images of plasma-deposited coatings (20 sccm ethylene, 5 slm He, 15 mg/mL of vancomycin solution) in: (a) Continuous mode (CM, inset top view), (b) Pulse mode condition. (c) Top view: SEM image of a CM plasma-deposited coating after 60 min immersion in water. Reprinted with permission from [164]. © 2018 John Wiley & Sons Inc.

Thanks to this result, and confirmation by investigation with confocal fluorescence microscopy [163], and a following systematic study carried out on aerosol solutions fed with different chemical compounds [165], it was demonstrated that the nanosphere had a capsule structure. More in detail, the core is made of the solute of the nebulized droplets and the shell originates from the precursor added in the feed, in this case, ethylene. It has been hypothesized that the atomized droplets enter the plasma zone, becoming negatively charged and fragments of the matrix precursor surround them. Hence, water evaporation occurs, and solid aggregates start to form. Radical film precursors adsorb on the solid aggregates, giving rise to the coating wrapping the solute, thus creating the nanocapsules, which then lean on the substrate surface. The formation of the nanocapsules is easier for heavier solid solutes, because of their lower diffusivity and for ionic species, which is likely due to the presence of a negative charge on the surface of the droplets [161].

Over the last decade, numerous studies have reported on the preparation of organic-inorganic nanocomposite coatings by using aerosol-assisted plasma processes in which a dispersion containing preformed inorganic nanoparticles (NPs) and the liquid precursor of the polymeric component is injected in aerosol form in an atmospheric pressure cold plasma. The research in this field has been motivated by two main reasons: (i) The versatility of this strategy, which can enable the combination of a wide range of NPs and, in principle, of any organic precursor that when introduced in the plasma can undergo fragmentation and consequent plasma polymerization [166]; (ii) the fact that aerosol-assisted processes appear to be very promising for large-scale thin film production [167]. To date, the adoption of this deposition strategy has afforded many different hybrid nanocomposite films combining hydrocarbon or organosilicon plasma-polymers with, for instance, metal (e.g., Pt) [168] or metal oxide (TiO_2, ZnO, etc. [169,170] NPs, polyhedral oligomeric silsesquioxane (POSS) [171] or inorganic salt nanoinclusions (e.g., $AgNO_3$ [172]).

We recently investigated the growth and structure of hydrocarbon polymer/ZnO NPs nanocomposite films deposited by using a parallel-plate DBD (Fig. 9(a)) fed with helium and the aerosol of a dispersion of ZnO NPs in hydrocarbon solvents (i.e., n-octane and 1,7-octadiene) [60,169,173].

The plasma was generated by applying a sinusoidal AC high voltage, (22.0 ± 0.2) kHz with (2.6 ± 0.2) kV$_{rms}$, in pulsed mode (20 ms period, 13 ms plasma on-time, 65% duty cycle). Commercial ZnO nanoparticles (average particle size of 36 nm, hexagonal crystal structure) were utilized. The surface of the NPs was functionalized with oleate (capping) using a simple wet-chemistry procedure, to obtain stable dispersions in hydrocarbon solvents. The concentration of NPs in the dispersions was varied in the range 0.5–5.0 wt.%, while the concentration of 1,7-octadiene in the binary n-octane/1,7-octadiene mixture was increased up to 10 vol%.

As it can be observed in the representative ATR-FTIR (attenuated total reflectance-Fourier transform infrared spectroscopy) spectrum in Fig. 9(b),

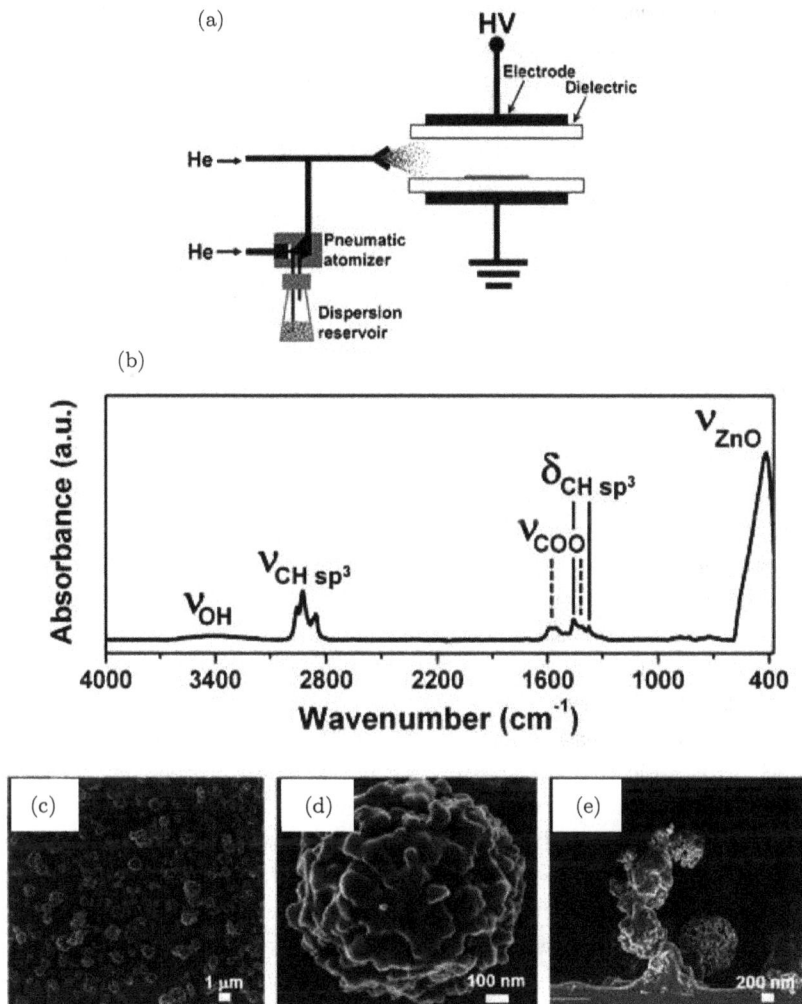

Fig. 9. (a) Schematic representation of the DBD reactor used for the aerosol-assisted plasma deposition of nanocomposite coatings. (b) ATR-FTIR spectrum of a typical hydrocarbon polymer/ZnO nanoparticles nanocomposite coating deposited from a DBD fed with He and the aerosol of a dispersion of oleate-capped ZnO NPs in an n-octane. Representative SEM images of an NC coating: (c) Low-magnification image; (d) High-magnification image evidencing the organic layer on the NPs agglomerate; (e) Cross-sectional image. Reproduced with permission from [166]. © 2014 Springer Science Business Media New York.

the deposited coatings present the typical absorptions of both the oleate-capped ZnO NPs and the polyethylene-like organic component originated from the plasma polymerization of n-octane and 1,7-octadiene. On the other hand, as assessed by XPS analyses, the surface composition of the coatings is dominated by the hydrocarbon polymer. In addition, SEM observations showed that the nanocomposite film contains quasi-spherical agglomerates of ZnO nanoparticles (diameter in the range 200–2200 nm) covered and held together by the polymer (Fig. 9(c)). The NP agglomerates lead to a hierarchical multiscale surface roughness. In particular, while the NP agglomerates provide the micrometer-scale roughness, the NPs on the top of the agglomerates provide the nanoroughness.

We carried out a comprehensive study on the influence of the dispersion composition, to demonstrate the possibility of tuning the structure and wetting properties of the hybrid layers. We found that the increase of the NPs' concentration in the starting dispersion leads to a continuous increase of the ZnO loading in the coatings [169, 173]. On the other hand, 1,7-octadiene addition favors the immobilization of NP agglomerates on the substrate surface at low concentrations in the hydrocarbon mixture (≤ 2 vol%), while if enhances the growth of the organic component at higher concentrations [173].

We also found that the increase of the ZnO loading in the coating results in a significant increase of the RMS roughness and ultimately leads to superhydrophobic surfaces (advancing and receding water contact angles higher than 150°) with very low WCA hysteresis (less than 10°). This wettability behavior is due to the coexistence of the low surface energy conferred by the hydrocarbon polymer and the surface texture induced by NPs incorporation. Interestingly, in addition to superhydrophobicity, these plasma-deposited coatings exhibit also the peculiar photocatalytic properties of ZnO, as revealed from our recent study of the photocatalytic degradation of a dye molecule (i.e., methylene blue) in water [174].

Figure 10 summarizes results obtained by increasing the concentration of oleate-capped ZnO NPs in the starting dispersion from 0.5 to 5 wt.%, at constant composition of the n-octane/1,7-octadiene solvent mixture ([1,7-octadiene] = 0.5 vol%). First of all, we observed that the increase of the NPs concentration in the dispersion leads to a remarkable increase of the ZnO loading in the coatings from (26 ± 4) wt.% to (78 ± 5) wt.% (Fig. 10(a)).

Cross-sectional SEM images in Fig. 10(c) show that the increase of the NPs concentration in the dispersion causes a transition from a

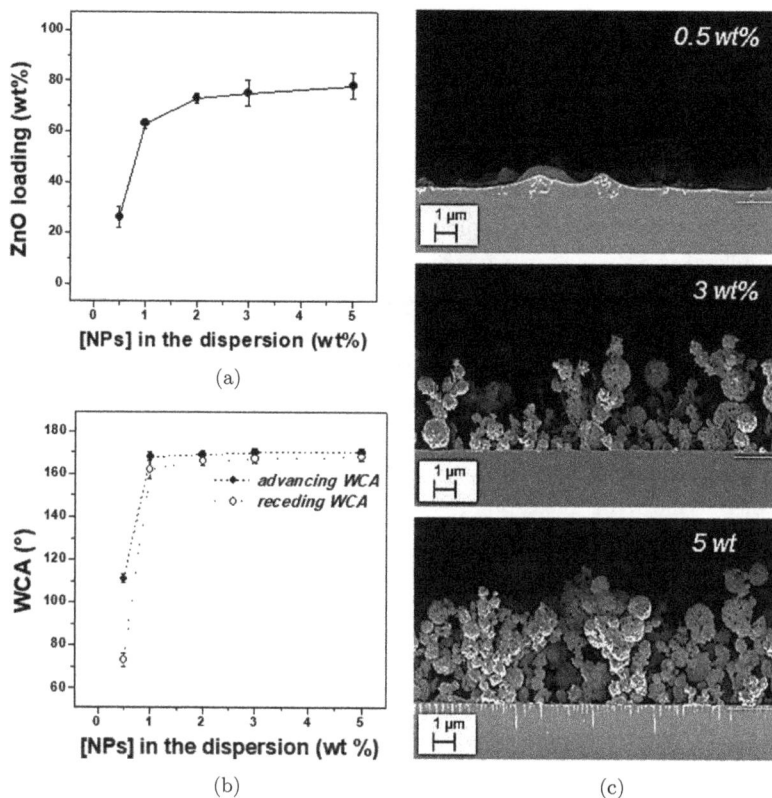

Fig. 10. (a) ZnO loading, and (b) advancing and receding WCAs of the NC coatings as a function of the concentration of oleate-capped ZnO NPs in the n-octane/1,7-octadiene solvent mixture ([1,7-octadiene] = 0.5 vol%, deposition time = 10 min) [173]. (c) Cross-sectional SEM images of NC coatings deposited from dispersions characterized by different concentrations of oleate-capped ZnO NPs (0.5, 3 and 5 wt.%).

polymer-dominated to a nanoparticles-dominated morphology. While the coating deposited from a 0.5 wt.% NPs dispersion presents a relatively smooth morphology, a hierarchical morphology appears for higher NP concentrations. This change in coatings' morphology is also accompanied by a considerable increase of the root-mean-square (RMS) roughness from ~200 nm to ~1000 nm. The NPs content in the starting dispersion significantly influences also the wettability of the coatings. In particular, with increasing the NPs concentration, advancing and receding WCAs increase from 111° to 170°, and from 73° to 168°, respectively (Fig. 10(b)).

In particular, superhydrophobic surfaces with very low hysteresis are obtained for NP concentrations in the dispersion, greater or equal to 1 wt.%.

3. Plasma Treatment of Polymers

3.1. *Low pressure plasma treatment of polymers*

In many applications in technological fields such as automotive, aerospace and packaging polymers, the use of polymers is hindered by their frequently found low surface energy [175]. Plasma treatments can be effectively carried out to modify polymer surfaces leading to improved adhesion, wettability, printability, dye uptake, etc. Unlike PECVD, that leads to material addition in significant amount, plasma treatments consist in the modification of the composition and of the structure of a few molecular layers at the polymer surface. In this kind of process, the plasma is fed with inert (e.g., He or Ar) or non-polymerizing gases (e.g., N_2, NH_3, O_2, H_2O, etc.).

During plasma treatments, four major events can be observed on surfaces, each always present, but one may be favored, depending on the substrate, the plasma chemistry, the reactor configuration and the operating parameters. These effects are:

- **Surface cleaning:** The removal of organic contamination from the surface. Often commercial polymers have additives, or contaminants, such as anti-oxidants, mould release agents, anti-block agents. Typically they are oily or wax-like and form a layer (1–10) nm thick. Oxygen plasmas, among others, readily get rid of this contamination.
- **Etching:** Etching is not different from surface cleaning, except for the action onto the polymer itself more than on the contaminants. If the plasma treatment is carried out in harsh conditions, plasma etching contribution (removal of substrate material) can be higher than desired and be detrimental for the properties of the material.
- **Cross-linking:** It occurs in polymer surfaces even in inert gases feeds, which are effective in creating free radicals without addition of any new chemical group [176]. Ion bombardment, vacuum ultraviolet (VUV) photons and radicals can break C-C or C-X bonds, and the formed dangling bond can react with other surface radicals or with other chains in chain. The surface reactions then leads to cross-linking of the polymer chains, but also to unsaturation and branching [175]. In low pressure plasma VUV radiation can be important in plasma treatments, and low wavelength UV can bring enough energy to break organic bonds and to

initiate rapid free-radical chemistry. It is reported that VUV radiation can account for as much as 80% of the surface plasma reactions [175,177].

- **Chemical functionalization:** Finally, plasma treatments change the chemical composition of the polymer surface as grafting of a variety of chemical moieties is promoted. Mostly, polar groups are grafted, such as -OH, -COOH and -NH$_2$. This is the more investigated effect since it is known to enhance metal-polymer adhesion [178], dyeability [179], printability, cell affinity in biomaterials [180] and the grafted functional groups can serve to anchor bioactive molecules such as proteins, peptides and drugs [181–183].

Air, oxygen or water vapor plasmas can be considered to form oxygen containing polar moieties at the surface of polymers such as alcohol, carbonyl, ether and carboxylic acid [175,184–197]. However, attention must be addressed to possible over-treatment of the polymer surface since long processing time can result in excessive bond breakage and oxidation. The consequence is the formation of so-called low-molecular-weight oxidized materials (LMWOMs) weakly bonded to the polymer backbone, badly affecting polymer performances.

The ability of plasma processing to oxidize the different polymers varies with their chemical nature: Low density polyethylene (PE) undergoes limited oxidation with respect to polyetheretherketone, or polyimide. This can be ascribed to a higher etch rate for PE occurring simultaneously with the grafting. In fact, the strong backbone bonds and aromatic components decrease removal rates [189].

Plasma treatments in oxygen or similar oxidizing feeds have been proposed for the sterilization of surgical tools and biomedical devices [180]. Polyethylene has been functionalized with –COOH groups for following wet immobilization reactions of heparin and highly sulphated hyaluronic acid (HyalS$_x$, $x = 3.5$), two anti-thrombotic molecules. A *bis*-amino compound was used in order to anchor the bioactive molecule to the surface via covalent bond, but preserving the active conformation and keeping the natural biological activity unaltered. According to thrombin time and platelet-activation tests, heparin- and HyalS$_{3.5}$-immobilized PE samples presented an anti-thrombotic surface [190].

Grafting of nitrogen-containing groups, such as imine, imide, cyano or amine, is done by means of N$_2$ or NH$_3$ plasmas. Amino groups are the most abundant species, when ammonia is the processing gas [191–195]. The grafting of amine groups is particularly attractive because they are good nucleophiles suitable either for the covalent or ionic binding of macromolecules or for adhesive bonding. The selectivity of NH_3/H_2 plasma treatments in grafting $-NH_2$ groups rather than other N-groups has been investigated. XPS assisted with $-NH_2$ derivatization by 4-trifluoromethylbenzaldehyde was carried out with this aim. It was found that high plasma power, high hydrogen content, short process time and, chiefly, the position of the substrate in the afterglow region of the reactor pushed the process toward a high NH_2/N ratio [195, 196].

To have a more hydrophobic polymer surface, e.g. a textile, plasma can be fed with fluorocarbon gas such as CF_4, C_2F_6 or SF_6. However, often a competition between grafting, deposition and etching subsists [197–199]. In a very recent application, we investigated CF_4 plasma treatments to improve the affinity of Spanish Broom (SB) cellulose fiber toward hydrocarbons for the remediation of polluted water. Increasing the plasma power resulted in less fluorinated grafted group and a progressively more damaged fiber surface. Batch experiments were performed with the aim of studying kinetic and thermodynamic aspects of the adsorption process, as a function of the initial total hydrocarbon load and of the adsorbent amount. The kinetics data showed that the fiber removal efficiency ranged between 80% and 90% after one minute of contact time, depending on the initial hydrocarbon/fiber weight ratio, 20–240 mg/g. A maximum adsorption capacity larger than 270 mg/g was estimated, demonstrating that the functionalized fiber is capable of performing a significant hydrocarbon removal action compared to other cellulosic materials reported in the literature. Furthermore, one single washing step in cyclohexane regenerates the adsorption capacity to 90%, which is comparable to the one of the fresh grafted fiber. Best results were obtained with the SB fiber treated at low plasma power (10 W), which resulted in a superhydrophobic surface, with a maximum amount of highly fluorinated moieties such as CF_3 and CF_2 and minimum etching degrading effects [200].

Since plasma treatments are designed mostly to drive the wettability of polymers, contact angle measurement of various liquid before and after the treatment is a common method to test the effectiveness of the plasma process. Since the modified polymer surfaces are not chemically and morphologically homogeneous surfaces, both advancing and receding contact angles are measured [184, 186, 194, 201].

An important issue in plasma treatment of polymers is the stability of modified surfaces. In air, highly hydrophilic surfaces are unstable hence they undergo "hydrophobic recovery", sometimes indicated as aging, the trend to reach the equilibrium by decreasing the solid/air interfacial energy [201]. This phenomenon follows different mechanisms, namely:

- Reorientation of polymer chains to bring the low energy groups facing atmosphere and the high energy ones submerged in the bulk of the polymer.
- Absorption of hydrocarbon contaminations, naturally present in the atmosphere.

The former mechanism is particularly relevant in polymers since the molecular chains/bonds are particularly mobile and, in fact, demonstrated to be favored by temperature increase [202].

Such instability in hydrophilic surfaces, especially when dealing with polymer modifications, can have serious technological concerns, as in industrial fields involving dyes adhesion, metal polymer adhesion or printing onto polymer surfaces. For instance, in the case of ammonia plasma-treated PTFE, as-treated samples are quite hydrophilic with WCA less than 20°, however, in about 10 days the surface rearranges reaching a WCA as high as 55°. To obtain a permanently wettable polymer surface, a polymer pre-treatment in H_2 plasma can be used. The latter, inhibits the surface restructuring by cross-linking the PTFE surface, thanks to the combined scavenging action of H atoms toward F atoms from the polymer surface and the UV radiation produced in H_2 plasma. It is important reporting that, to observe the benefits given by the H_2 pre-treatment, a mild ammonia treatment has to be carried out to avoid destruction of the thin hydrogenated/cross-linked interface generated on PTFE by H_2 plasma [203].

Another approach to produce stable highly hydrophilic surfaces is the deposition of an inorganic hydrophilic coating, such as an OH-rich SiO_x film, onto an O_2 plasma-treated surface, with a treatment inducing also natotexturing of the surface as described in the "plasma etching" section [204]. Carbon nanotubes (CNTs) are an interesting molecular form of carbon usually described as hollow tubular channels of one or more graphene layers, designated, respectively, as single wall (SWCNT) or multiwall (MWCNT) [205–207].

For many applications, CNTs must be finely dispersed in solvents or within a matrix (in nanocomposite) and strongly bond to it. CNTs, instead, due to their large surface areas, intertubular van der Waals interactions and high chemical inertness, lead to inhomogeneous composites in common matrices [208, 209]. For the same reason, carbon nanotubes have low or no solubility in most solvents and form unstable suspension in polar solvents, hindering their use in several fields of applications [210]. Hence, it is important to study the chemical functionalization of nanotubes. The most common functionalization consists in the harsh wet chemical oxidation with inorganic acids that often results in structural damage and properties degradation [211]. On the other hand, plasma treatment is considered more appropriate for producing functionalized CNTs with minimal structural damage, thanks to the mild conditions, in terms of treatment time and temperature [212].

Garzia Trulli *et al.* [213] demonstrated that O_2 low-pressure plasma processes can tune the surface functionality of commercially available low cost MWCNTs, by grafting of surface oxygenated chemical groups. The main issue in plasma treatment of particulate is the homogeneity of the surface modification. Hence, plasma reactors need to be implemented with suitable shaking methods. In this work, the process has been carried out by properly stirring the CNTs in vials, thus providing a homogeneous and efficient functionalization process. Their dispersion and stability in aqueous media was significantly improved, as shown in Fig. 11, due to the grafting of acid sites (up to 5.3%, as revealed by DLS analyses and acid–base titrations). The presence of polar ionizable groups limits the agglomeration between adjacent CNTs by repulsive forces. The dispersion of plasma-treated CNT powders in water (0.35 mg/mL) remained stable for at least one month, which is a highly significant result for the application in nanocomposite materials.

3.2. *Atmospheric pressure plasma treatments*

The utilization of AP non-equilibrium plasmas for the treatment of a large variety of polymers has received enormous attention and a wide range of processes has been proposed for many different applications since the 1980s [7, 214]. Optimized treatments have exploited many different direct and remote AP plasma sources, which mainly present corona discharge and DBD electrode configurations [7, 214]. AP plasma treatments have been primarily used to increase the surface energy of the polymeric surfaces and, therefore, to improve their wettability as well as their adhesion to inks, glues, adhesives and metal coatings [215, 216].

The studies on the subject have widely showed that the AP non-equilibrium plasmas are able to effectively modify the surface chemical

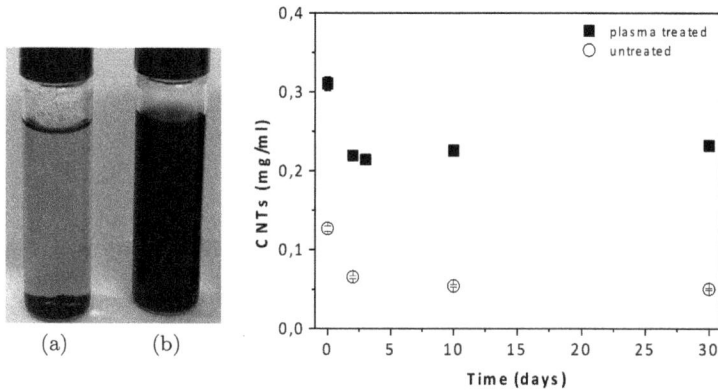

Fig. 11. (Left panel) (a) Untreated CNTs, and (b) CNTs dispersed in water and treated with plasma, one month after ultra-sonication. (Right panel) NIR absorbance (850 nm) trend of untreated CNTs and plasma treated CNTs dispersed in water (0.35 mg/mL) as a function of settling time. Adapted with permission from [213]. © 2017 Elsevier.

composition of the polymers (i.e., through the grafting of polar oxygen- and nitrogen-containing chemical functionalities and the change of the cross-linking degree) as well as to induce a variation in surface roughness (e.g., an increase of the roughness) [7]. Although wettability and adhesion enhancement represent a well-established application of AP plasmas, the research in this field is still active nowadays, as demonstrated by the presence of many recent fundamental and applicative publications on the treatment of both synthetic [7, 217–219] and natural polymers [220, 221]. AP plasma processes are commonly carried out using air or nitrogen as feed gas [7,214]. This is without doubt the less expensive option, which renders AP plasmas very attractive compared to the low-pressure counterpart, in particular for in-line open-air surface processing.

The use of noble gases (e.g., He and Ar) with possible admixture of reactive additives (e.g., oxygen, water, ammonia, etc.) has been also widely reported [7,217,218]. Unfortunately, feed mixture dilution with noble gases remarkably increases the process cost, however it offers the advantages of a better control of the plasma chemistry. Recently, we investigated the plasma treatment of open-cell polyurethane foams by DBDs fed with He-O$_2$ mixtures [219]. XPS analyses revealed the uniform chemical functionalization of both the exterior and interior surfaces of the foams with oxygen-containing groups. In addition, SEM observations and WCA measurements showed that the plasma-treated foams exhibit increased surface roughness and wettability, respectively. It was also observed that aging of the plasma-treated samples is more pronounced in water than in air, due to the inevitable dissolution of oxidized polymer fragments formed during plasma exposure. Since oxygen-containing functional groups (e.g., hydroxyls, carboxyls, etc.) formed at polymer surfaces can interact with heavy metal ions [222, 223], the He–O_2 DBD treatment of PU foams could represent a viable alternative for the preparation of adsorbent materials able to remove these pollutants from water. For this purpose, the ability of plasma-treated foams to adsorb cadmium ions (Cd^{2+}) from dilute aqueous solutions (500 ppb) was explored. In particular, it was found that the capability of the plasma-treated foams to remove Cd^{2+} from water can be improved by a simple post-processing step involving the immersion of the porous samples in NaOH-NaCl aqueous solution with pH 11 and NaCl concentration of 1 mol/L, for 15 min under continuous stirring. This adsorption capacity enhancement seemed to be related to the increase of the XPS surface concentration of carboxylic groups after plasma exposure.

We also studied the fluorination of polypropylene (PP) and polyethyleneterephtalate (PET) by DBD fed with helium and tetrafluoromethane (CF_4) mixtures [224]. We found that the plasma treatment leads to an extensive surface fluorination due to the grafting of F atoms and CF_x radicals. The study of the effect of different process parameters revealed that the fluorination degree and hydrophobicity of the polymers increases with the CF_4 concentration in the feed gas, the treatment duration and the DBD excitation frequency. Overall, in this study, XPS F/C ratios and static WCA values as high as about 1.2° and 110°, respectively, were obtained for both PP and PET.

4. Plasma Etching of Polymers

Plasma etching of polymers is a topic of crucial importance in microelectronic fabrication technology [225]. The main advantage of plasma in etching processes lies in the ability to define high aspect ratio patterns in different kinds of substrates: Sub-micron features can be transferred into resist polymer films several microns thick, with anisotropic etching processes, to form high-resolution lithographic masks with multilayer resist schemes [189]. Another typical application is the plasma etching of polyimide films to define openings for electrical contact between layers of metallization [225].

Plasma etching of polymers is mostly carried out by oxygen plasmas [189, 225]. Some papers report a linear correlation between concentration of atomic oxygen in the plasma and etch rates of various polymers. Oxygen atoms can add to unsaturated groups to form radical sites or can scavenge hydrogen from saturated hydrocarbons forming a polymer radical site. Once a dangling bond is formed, it can react with atomic or molecular oxygen, weakening the adjacent C–C bond. When the oxygen atom has reacted with the polymer backbone formation of volatile species such as CO and CO_2, it undergoes subsequent etching. Significant increase in etching rate is obtained with assistance of ion bombardment and addition of a small amount of fluorine-containing gases, such as CF_4, C_2F_6, SF_6.

An interesting application of plasma etching developed in the last 10 years is the nanotexturing of polymers, mainly with the aim of driving extreme wettability behavior, but not limited to it [226–232].

Often, living organisms respond to environment stimuli thanks to nano-, micro-texturing-dependent properties [233]. Some plant leaves, such as those of lotus, rice and Salvinia, have been deeply investigated because of their water repellency and self-cleaning behavior due to the presence of nano- and micro-protuberances, coated with a hydrophobic wax [234–237]. Similarly, hydrophobic surface texture is present in some animals, as sharks or water striders, for drag reduction [238]. On the other hand, the nanotexturing can lead to "sticky" surfaces, like in rose petals, and in some cases can lead to super-adhesive behavior, as in the case of Gecko legs [239, 240]. Furthermore, insects effectively use texturing to control not only wettability but also some optical properties, such as light reflection [241] and structural coloration [242].

Mimicking nature, producing chemically defined nano- and micro-textured surfaces can improve the performance of smart materials and devices, which can find application in fog collection or resistance, anti-icing, efficient energy conversion and biosensors. Plasma etching can be advantageously used as a direct top-down route to nanotexturing. Plasma nano-texturing is the mask-less process forming a texture in the submicrometric scale on the surface of a material by plasma etching [243, 244].

In nanotexturing, a polymer substrate is placed onto the bottom electrode of a plasma reactor in a Reactive Ion Etching configuration, though other coupling configurations can be found in the literature [245]. Then the plasma is fed with an etchant gas such as oxygen, fluorocarbon or hydrogen [244, 246]. Plasma nanotexturing was carried out to produce

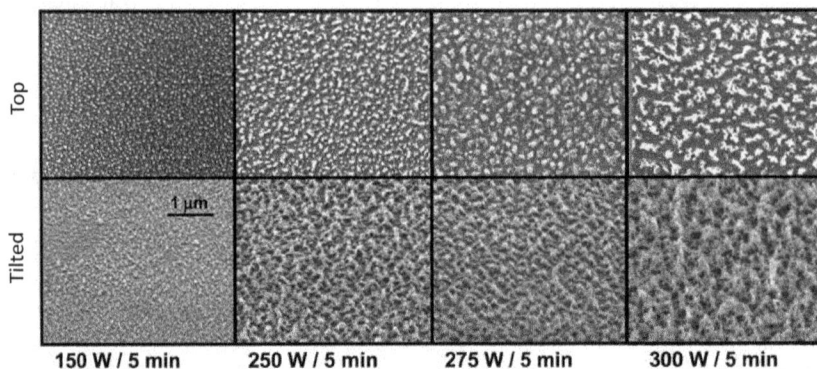

Fig. 12. SEM images of the polystyrene (PS) samples treated with a $CF_4/(17\%)O_2$-fed discharge as a function of the power input. Adapted with permission from [247]. © 2008 American Chemical Society.

super-hydrophobic nanotextured polystyrene (PS) pieces in one step, optimizing the CF_4/O_2 gas flow rate ratio [247,248]. When the oxygen content was 17%, submicrometric pillars were obtained in a few minutes (Fig. 12), leading to a water-repellent PS surface. Plasma nanotextured samples maintained their transparency to visible light, however they become opaque for longer treatment time because of the formation of larger micrometric features, responsible for light scattering. Also the effect of chemistry onto the wettability of textured PS surfaces was investigated. It was found that by depositing films by PECVD using a ethylene/c-C_4F_8 mixture a tuning of the hydrophobic behavior was achieved. In particular, as expected, the best super-hydrophobic character was obtained with the most fluorinated coating, even though the chemical composition effect was less important for taller nanofeatures.

Plasma nanotexturing has been investigated also for polycarbonate (PC) commonly used in optical lenses [244,246]. As illustrated in Fig. 13, O_2 plasma engraved pillars in the PC surface, with height increasing with time, and also power input (not shown). The pillars were (20–80) nm in the lateral dimension, and as high as about 1 μm. Soon after the process began the polymer became very hydrophilic, because of the grafting of polar moieties and the presence of nanotexturing.

However, such behavior was not permanent. In three weeks, the WCA increased, losing the hydrophilic character according to the hydrophobic recovery phenomenon. However, the plasma deposition of a suitable coating

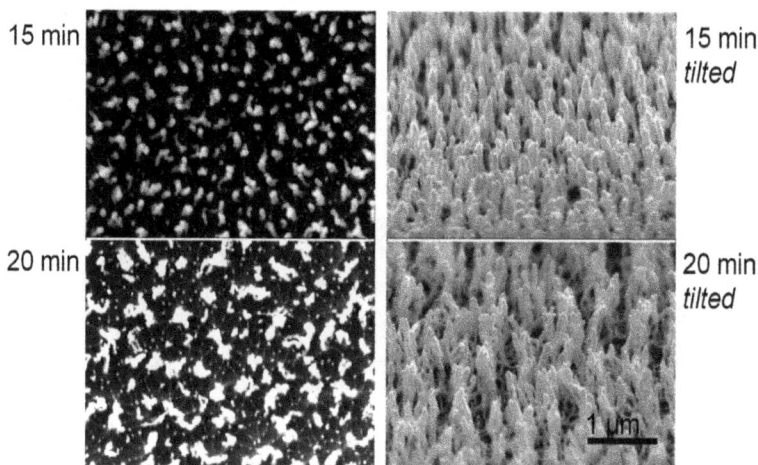

Fig. 13. SEM pictures of polycarbonate plasma etched in oxygen plasma for different process times. Adapted with permission from [204]. © 2011 Wiley-VCH GmbH, Weinheim.

served as an instrument for tuning the durable wettability of the nanotextured surfaces. In particular, the Teflon-like or silicone-like coating contributed to produce water-repellent surfaces, with water contact angles as high as 172° and negligible hysteresis.

More interestingly, addressing the plasma deposition toward an hydrophilic SiO_2-like coating allowed for the preparation of stably (more than 6 months) hydrophilic surfaces, able to drive water condensation. Such surfaces were capable of displaying durable anti-fog surfaces as illustrated in Fig. 14. In fact, the hydrophilic property lead to the formation of a liquid film and not to water droplet condensation, that is the source of opaque appearance. With the idea to produce durable super-hydrophobic polymer surfaces, a bulky hydrophobic polymer as PTFE was considered as substrate for single plasma nanotexturing [249–252].

In Fig. 15, it can be observed that changing RF power and process time had an effect on the size and on the shape of the submicrometric features as well. Specifically, increasing time (or power), the nanostructures went from a conic shape with hairy filaments (FS) on the apex, to cones bearing spherical beads on top (S-o-C). All these surfaces expressed a sharp super-hydrophobic behavior. Furthermore, water droplets bouncing experiments were carried out, to study the dynamic behavior of water contact. It was shown that the contact time of the falling drop with the surface was very

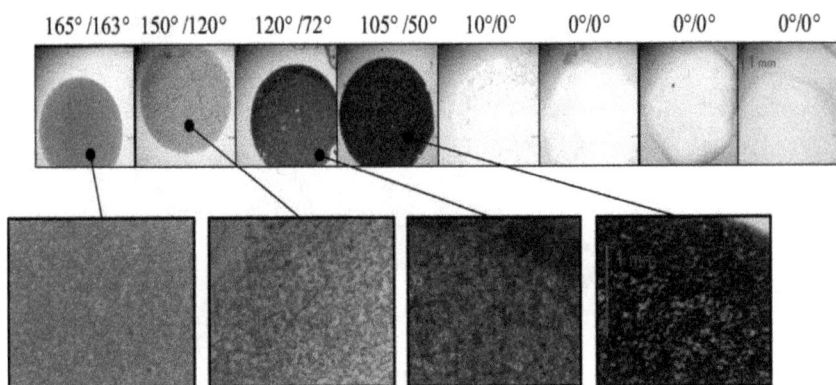

Fig. 14. Images of water condensed onto PC samples nanotextured with plasma etching and PECVD, coated with different organosilicon-based films. Higher magnification images of the dropwise condensing (fogged) surfaces are shown in what follows. Values of the advancing/receding water contact angle are reported at the top. Reproduced with permission from [246]. © 2014 American Chemical Society.

Fig. 15. SEM images of Teflon surfaces treated by a self-masked, oxygen-fed plasma process. The effect of increasing the process time at constant power (200 W) is visible on the diagonal from the upper-left to the lower-right image. The effect of increasing the power at a constant process time (15 min) is shown on the diagonal from the lower-left to the upper-right image. Reproduced with permission from [250]. © 2016 Elsevier.

short for both the FS (200 W/10 min) and the S-o-C (300 W/15 min), being around 10.7 ms. However, increasing the impact speed, by increasing the falling height, the contact time increased for the S-o-C samples, since the impact pressure increased, whereas for the FS ones, it remained low.

This unique result was explained with the formation of stable air cavities underneath the filaments, a property that can be found in nature for Salvinia leaves [236], which have a shape very similar to the submicrometric features of the FS samples. The presence of the spherical nano-objects has been related to a simultaneous thermal effect and better explained in [252].

Some comments can be added on the mechanism. What is believed is that during the etching process the surface is contaminated by unvolatile species. More in detail, when plasma is ignited, etching starts by the formation of volatile species, and the polymer surface is eroded. Simultaneously, ions from the plasma are accelerated towards electrodes and walls causing the sputtering of inorganic species that implying onto the sample surface. Such species cannot be ablated in the experimental conditions used and they assemble a kind of nano-mask with respect to the arriving etchant species. Hence, the etching process of the polymer can proceed only through the openings between the clusters, in a random but reproducible way [252,253].

References

[1] I. Langmuir, Oscillations in Ionized Gases, *Proc. Natl. Acad. Sci. USA.* **14**(8), 627–637 (1928).

[2] P. J. Bruggeman. Atmospheric Pressure Plasmas. In P. K. Chu and X. Lu (eds.), *Low Temperature Plasma Technology: Methods and Applications*, pp. 13–38. CRC Press (2013).

[3] I. Trizio, M. Garzia Trulli, C. Lo Porto, D. Pignatelli, G. Camporeale, F. Palumbo, E. Sardella, R. Gristina, and P. Favia. Plasma processes for life sciences. In *Reference Module in Chemistry, Molecular Sciences and Chemical Engineering*, pp. 1–24. Elsevier (2018). doi: https://doi.org/10.1016/B978-0-12-409547-2.12271-1.

[4] U. Kogelschatz, Dielectric-barrier discharges: Their history, discharge physics, and industrial applications, *Plasma Chem. Plasma Process.* **23**, 1–46 (2003).

[5] F. Massines, C. Sarra-Bournet, F. Fanelli, N. Naudé, and N. Gherardi, Atmospheric pressure low temperature direct plasma technology: Status and challenges for thin film deposition, *Plasma Process. Polym.* **9**(11), 1041–1073 (2012).

[6] X. P. Lu and M. Laroussi, Temporal and spatial emission behaviour of homogeneous dielectric barrier discharge driven by unipolar submicrosecond square pulses, *J. Phys. D: App. Phys.* **39**(6), 1127 (2006).

[7] F. Fanelli and F. Fracassi, Atmospheric pressure non-equilibrium plasma jet technology: General features, specificities and applications in surface processing of materials, *Surf. Coat. Technol.* **322**, 174–201 (2017).

[8] P. Favia, M. Creatore, F. Palumbo, V. Colaprico, and R. d'Agostino, Process control for plasma processing of polymers, *Surf. Coat. Technol.* **142**, 1–6 (2001).

[9] V. Panchalingam, X. Chen, H. H. Huo, C. R. Savage, R. B. Timmons, and R. C. Eberhart, Pulsed plasma discharge polymer coatings, *ASAIO J.* **39**(3), M305–M309 (1993).

[10] S. R. Coulson, I. S. Woodward, J. P. S. Badyal, S. A. Brewer, and C. Willis, Plasmachemical functionalization of solid surfaces with low surface energy perfluorocarbon chains, *Langmuir.* **16**(15), 6287–6293 (2000).

[11] W. Chen, A. Y. Fadeev, M. C. Hsieh, D. Öner, J. Youngblood, and T. J. McCarthy, Ultrahydrophobic and ultralyophobic surfaces: Some comments and examples, *Langmuir.* **15**(10), 3395–3399 (1999).

[12] S. R. Coulson, I. S. Woodward, J. P. S. Badyal, S. A. Brewer, and C. Willis, Ultralow surface energy plasma polymer films, *Chem. Mater.* **12**(7), 2031–2038 (2000).

[13] R. d'Agostino, I. Corzani, P. Favia, R. Lamendola, and G. Palumbo, Modulated plasma glow discharge treatments for making superhydrophobic substrates, https://patents.google.com/patent/WO2000014297A1/en (2000).

[14] J. P. S. Badyal, S. R. Coulson, R. C. Willis, and S. A. Brewer, Surface coatings, https://patents.google.com/patent/WO1998058117A1/en (1998).

[15] D. Kiaei, A. S. Hoffman, and S. R. Hanson, Ex vivo and in vitro platelet adhesion on RFGD deposited polymers, *J. Biomed. Mat. Res.* **26**(3), 357–372 (1992).

[16] D. Kiaei, A. S. Hoffman, and T. A. Horbett, Tight binding of albumin to glow discharge treated polymers, *J. Biomater. Sci. Polym. Ed.* **4**(1), 35–44 (1993).

[17] P. Favia, V. H. Perez-Luna, T. Boland, D. G. Castner, and B. D. Ratner, Surface chemical composition and fibrinogen adsorption-retention of fluoropolymer films deposited from an RF glow discharge, *Plasmas Polym.* **1**(4), 299–326 (1996).

[18] G. Clarotti, F. Schue, J. Sledz, A. A. B. Aoumar, K. E. Geckeler, A. Orsetti, and G. Paleirac, Modification of the biocompatible and haemocompatible properties of polymer substrates by plasma-deposited fluorocarbon coatings, *Biomaterials.* **13**(12), 832–840 (1992).

[19] J. L. Bohnert, B. C. Fowler, T. A. Horbett, and A. S. Hoffman, Plasma gas discharge deposited fluorocarbon polymers exhibit reduced elutability of adsorbed albumin and fibrinogen, *J. Biomater. Sci. Polym. Ed.* **1**(4), 279–297 (1989).

[20] B. D. Ratner, A. Chilkoti, and G. P. Lopez. Plasma deposition and treatment for biomaterial applications. In R. d'Agostino (ed.), *Plasma Deposition, Treatment and Etching of Polymers*, pp. 463–516. Academic Press, San Diego, USA (1990).

[21] A. Milella. Plasma Processing of Polymers. In *Encyclopedia of Polymer Science and Technology*. (Doi:10.1002/0471440264.pst560). John Wiley and Sons Inc (2008).

[22] R. d'Agostino, F. Cramarossa, F. Fracassi, and F. Illuzzi. Plasma deposition and treatment for biomaterial applications. In R. d'Agostino (ed.), *Plasma Deposition, Treatment and Etching of Polymers*, pp. 95–162. Academic Press, San Diego, USA (1990).

[23] P. Favia. Process control in plasma-deposition and plasma-treatments of polymeric biomaterials. In R. d'Agostino, P. Favia, and F. Fracassi (eds.), *Plasma Processing of Polymers, NATO ASI Series E: Appl. Science, Vol. 346*, pp. 487–504. Kluwer Acad. Publ., Dordrecht, The Netherlands (1997).

[24] J. W. Coburn and M. Chen, Optical emission spectroscopy of reactive plasmas: A method for correlating emission intensities to reactive particle density, *J. App. Phys.* **51**(6), 3134–3136 (1980).

[25] R. d'Agostino. PECVD of polymer films: Mechanisms, chemistry and diagnostics. In R. d'Agostino, P. Favia, and F. Fracassi (eds.), *Plasma Processing of Polymers, NATO ASI Series E: Appl. Science, Vol. 346*, pp. 3–46. Kluwer Acad. Publ., Dordrecht, The Netherlands (1997).

[26] J. W. Coburn and H. F. Winters, Plasma etching: A discussion of mechanisms, *J. Vac. Sci. Technol.* **16**(2), 391–403 (1979).

[27] R. d'Agostino, F. Cramarossa, and F. Illuzzi, Mechanisms of deposition and etching of thin films of plasma-polymerized fluorinated monomers in radio frequency discharges fed with C_2F_6–H_2 and C_2F_6–O_2 mixtures, *J. Phys. D: App. Phys.* **61**(8), 2754–2762 (1987).

[28] R. d'Agostino, F. Cramarossa, and S. De Benedictis, Diagnostics and decomposition mechanism in radio-frequency discharges of fluorocarbons utilized for plasma etching or polymerization, *Plasma Chem. Plasma Process.* **2**(3), 213–231 (1982).

[29] R. d'Agostino, P. Favia, F. Fracassi, and F. Illuzzi, The effect of power on the plasma-assisted deposition of fluorinated monomers, *J. Polym. Sci. A Polym. Chem.* **28**(12), 3387–3402 (1990).

[30] D. G. Castner, K. B. Lewis Jr, D. A. Fischer, B. D. Ratner, and J. L. Gland, Determination of surface structure and orientation of polymerized tetrafluoroethylene films by near-edge X-ray absorption fine structure, X-ray photoelectron spectroscopy, and static secondary ion mass spectrometry, *Langmuir.* **9**(2), 537–542 (1993).

[31] D. G. Castner, P. Favia, and B. D. Ratner. Deposition of Fluorocarbon Films by Remote RF Glow Discharges. In B. D. Ratner and D. G. Castner (eds.), *Surface Modification of Polymeric Biomaterials*, pp. 45–52. Springer, Boston, MA (1997).

[32] C. I. Butoi, N. M. Mackie, J. L. Barnd, E. R. Fisher, L. J. Gamble, and D. G. Castner, Control of surface film composition and orientation with

downstream pecvd of hexafluoropropylene oxide, *Chem. Mater.* **11**(4), 862–864 (1999).

[33] F. Intranuovo, E. Sardella, P. Rossini, R. d'Agostino, and P. Favia, PECVD of fluorocarbon coatings from hexafluoropropylene oxide: Glow vs. Afterglow, *Chem. Vap. Deposition.* **15**(4–6), 95–100 (2009).

[34] G. Cicala, M. Losurdo, P. Capezzuto, and G. Bruno, Time-resolved optical emission spectroscopy of modulated plasmas for amorphous silicon deposition, *Plasma Sources Sci. Technol.* **1**(3), 156 (1992).

[35] V. Panchalingam, B. Poon, H. H. Huo, C. R. Savage, R. B. Timmons, and R. C. Eberhart, Molecular surface tailoring of biomaterials via pulsed RF plasma discharges, *Plasma Sources Sci. Technol.* **5**(1–2), 131–145 (1994).

[36] S. J. Limb, K. K. Lau, D. J. Edell, E. F. Gleason, and K. K. Gleason, Molecular design of fluorocarbon film architecture by pulsed plasma enhanced and pyrolytic chemical vapor deposition, *Plasmas Polym.* **4**(1), 21–32 (1999).

[37] V. Panchalingam, X. Chen, C. R. Savage, R. B. Timmons, and R. C. Eberhart, Molecular tailoring of surfaces via pulsed RF plasma depositions, *J. Appl. Polym. Sci.: Appl. Polym. Symp.* **54**, 123–141 (1994).

[38] A. Carletto and J. P. S. Badyal, Mechanistic reaction pathway for hexafluoropropylene oxide pulsed plasma deposition of PTFE-like films, *J. Phys. Comm.* **1**(5), 055024 (2017).

[39] A. Milella, F. Palumbo, P. Favia, G. Cicala, and R. d'Agostino, Deposition mechanism of nanostructured thin films from tetrafluoroethylene glow discharges, *Pure Appl. Chem.* **77**(2), 399–414 (2005).

[40] A. Milella, F. Palumbo, and R. d'Agostino. Fundamentals on plasma deposition of fluorocarbon films. In R. d'Agostino, P. Favia, Y. Kawai, H. Ikegami, N. Sato, and F. Arefi-Khonsari (eds.), *Advanced Plasma Technology*, pp. 175–195. Wiley-VCH, Weinheim, Germany (2007).

[41] C. B. Labelle, S. M. Karecki, R. Reif, and K. K. Gleason, Fourier transform infrared spectroscopy of effluents from pulsed plasmas of 1, 1, 2, 2-tetrafluoroethane, hexafluoropropylene oxide, and difluoromethane, *J. Vac. Sci. Tech. A.* **17**(6), 3419–3428 (1999).

[42] B. A. Cruden, K. K. Gleason, and H. H. Sawin, Time resolved ultraviolet absorption spectroscopy of pulsed fluorocarbon plasmas, *J. Appl. Phys.* **89**(2), 915–922 (2001).

[43] C. I. Butoi, N. M. Mackie, L. J. Gamble, D. G. Castner, J. Barnd, A. M. Miller, and E. R. Fisher, Deposition of highly ordered CF_2-rich films using continuous wave and pulsed hexafluoropropylene oxide plasmas, *Chem. Mater.* **12**(7), 2014–2024 (2000).

[44] S. J. Limb, D. J. Edell, E. F. Gleason, and K. K. Gleason, Pulsed plasma-enhanced chemical vapor deposition from hexafluoropropylene oxide: Film composition study, *J. Appl. Polym. Sci.* **67**(8), 1489–1502 (1998).

[45] C. R. Savage, R. B. Timmons, and J. W. Lin, Molecular control of surface film compositions via pulsed radio-frequency plasma deposition of perfluoropropylene oxide, *Chem. Mater.* **3**(4), 575–577 (1991).

[46] A. Milella, F. Palumbo, P. Favia, G. Cicala, and R. d'Agostino, Deposition mechanism of nanostructured thin films from tetrafluoroethylene glow discharges, *Pure Appl. Chem.* **77**(2), 399–414 (2005).

[47] G. Cicala, A. Milella, F. Palumbo, P. Favia, and R. d'Agostino, Morphological and structural study of plasma deposited fluorocarbon films at different thicknesses, *Diamond Relat. Mater.* **12**(10–11), 2020–2025 (2003).

[48] G. Cicala, A. Milella, F. Palumbo, P. Rossini, P. Favia, and R. d'Agostino, Nanostructure and composition control of fluorocarbon films from modulated tetrafluoroethylene plasmas, *Macromolecules.* **35**(24), 8920–8922 (2002).

[49] P. Favia, G. Cicala, A. Milella, F. Palumbo, P. Rossini, and R. d'Agostino, Deposition of super-hydrophobic fluorocarbon coatings in modulated RF glow discharges, *Surf. Coat. Technol.* **169**, 609–612 (2003).

[50] F. Fanelli, F. Fracassi, and R. d'Agostino, Atmospheric pressure PECVD of fluorocarbon coatings from glow dielectric barrier discharges, *Plasma Process. Polym.* **4**(S1), S430–S434 (2007).

[51] F. Fanelli, R. d'Agostino, and F. Fracassi, Atmospheric pressure PECVD of fluorocarbon thin films by means of glow dielectric barrier discharges, *Plasma Process. Polym.* **4**(9), 797–805 (2007).

[52] F. Fanelli, Optical emission spectroscopy of argon-fluorocarbon-oxygen fed atmospheric pressure dielectric barrier discharges, *Plasma Process. Polym.* **6**(9), 547–554 (2009).

[53] F. Fanelli, F. Fracassi, and R. d'Agostino, Deposition and etching of fluorocarbon thin films in atmospheric pressure DBDs fed with Ar-CF_4-H_2 and Ar-CF_4-O_2 mixtures, *Surf. Coat. Technol.* **204**(11), 1779–1784 (2010).

[54] F. Fanelli, Thin film deposition and surface modification with atmospheric pressure dielectric barrier discharges, *Surf. Coat. Technol.* **205**(5), 1536–1543 (2010).

[55] F. Fracassi, E. Occhiello, and J. W. Coburn, Effect of ion bombardment on the plasma-assisted etching and deposition of plasma perfluoropolymer thin films, *J. Appl. Phys.* **62**, 3980–3981 (1987).

[56] P. J. Bruggeman, F. Iza, and R. Brandenburg, Foundations of atmospheric pressure non-equilibrium plasmas, *Plasma Sources Sci. Technol.* **26**(12), 123002 (2017).

[57] F. Fanelli, G. Di Renzo, F. Fracassi, and R. d'Agostino, Recent advances in the atmospheric pressure PECVD of fluorocarbon films: Influence of air and water vapour impurities, *Plasma Process. Polym.* **6**(S1), S503–S507 (2009).

[58] F. Fanelli, R. d'Agostino, and F. Fracassi, Effect of gas impurities on the operation of dielectric barrier discharges fed with He, Ar, and Ar-C_3F_6, *Plasma Process. Polym.* **8**(6), 557–567 (2011).

[59] F. Fanelli and F. Fracassi, Thin film deposition on open-cell foams by atmospheric pressure dielectric barrier discharges, *Plasma Process. Polym.* **13**(4), 470–479 (2016).

[60] F. Fanelli, P. Bosso, A. M. Mastrangelo, and F. Fracassi, Thin film deposition at atmospheric pressure using dielectric barrier discharges: Advances on three-dimensional porous substrates and functional coatings, *Jpn. J. Appl. Phys.* **55**(7S2), 07LA01 (2016).

[61] A. M. Wròbel and M. R. Wertheimer. Plasma-polymerized organosilicones and organometallics. In R. d'Agostino (ed.), *Plasma Deposition, Treatment and Etching of Polymers*, pp. 163–268. Academic Press, Boston, USA (1990).

[62] A. A. Bright, Helium plasma enhanced chemical vapor deposited oxides and nitrides: Process mechanisms and applications in advanced device structures, *J. Vac. Sci. Technol. A.* **9**(3), 1088–1093 (1991).

[63] M. Creatore, F. Palumbo, R. d'Agostino, and P. Fayet, RF plasma deposition of SiO_2-like films: Plasma phase diagnostics and gas barrier film properties optimisation, *Surf. Coat. Technol.* **142–144**, 163–168 (2001).

[64] L. Agres, Y. Segui, R. Delsol, and P. Raynaud, Oxygen barrier efficiency of hexamethyldisiloxane/oxygen plasma-deposited coating, *J. Appl. Polym. Sci.* **61**(11), 2015–2022 (1996).

[65] C. Vallée, A. Goullet, and A. Granier, Direct observation of water incorporation in PECVD SiO_2 films by UV-visible ellipsometry, *Thin Solid Films.* **311**(1–2), 212–217 (1997).

[66] A. S. da Silva Sobrinho, J. Chalse, G. Dennler, and M. R. Wertheimer, Characterization of defects in PECVD-SiO_2 coatings on PET by confocal microscopy, *Plasmas Polym.* **3**(4), 231–247 (1998).

[67] A. S. da Silva Sobrinho, G. Czeremuszkin, M. Latrèche, G. Dennler, and M. R. Wertheimer, A study of defects in ultra-thin transparent coatings on polymers, *Surf. Coat. Technol.* **116–119**, 1204–1210 (1999).

[68] A. S. da Silva Sobrinho, G. Czeremuszkin, M. Latrèche, and M. R. Wertheimer, Defect-permeation correlation for ultrathin transparent barrier coatings on polymers, *J. Vac. Sci. Technol. A.* **18**(1), 149–157 (2000).

[69] G. Czeremuszkin, M. Latrèche, M. R. Wertheimer, and A. S. da Silva Sobrinho, Ultrathin silicon-compound barrier coatings for polymeric packaging materials: An industrial perspective, *Plasmas Polym.* **6**(1–2), 107–120 (2001).

[70] L. Martinu and D. Poitras, Plasma deposition of optical films and coatings: A review, *J. Vac. Sci. Technol. A.* **18**(6), 2619–2645 (2000).

[71] K. Maex, M. R. Baklanov, D. Shamiryan, F. Lacopi, S. H. Brongersma, and Z. S. Yanovitskaya, Low dielectric constant materials for microelectronics, *J. Appl. Phys.* **93**(11), 8793–8841 (2003).

[72] B. D. Hatton, K. Landskron, W. J. Hunks, M. R. Bennett, D. Shukaris, D. D. Perovic, and G. A. Ozin, Materials chemistry for low-k materials, *Materials Today.* **9**(3), 22–31 (2006).

[73] S. Sahli, Y. Segui, S. H. Moussa, and M. A. Djouadi, Growth, composition and structure of plasma-deposited siloxane and silazane, *Thin Solid Films.* **217**(1–2), 17–25 (1992).

[74] S. Sahli, Y. Segui, S. Ramdani, and Z. Takkouk, RF plasma deposition from hexamethyldisiloxane-oxygen mixtures, *Thin Solid Films.* **250**(1–2), 206–212 (1994).

[75] Y. Segui and P. Raynaud. Plasma diagnostic by infrared absorption spectroscopy. In R. d'Agostino, P. Favia, and F. Fracassi (eds.), *Plasma Processing of Polymers, NATO ASI Series E: Appl. Science, Vol. 346*, pp. 81–100. Kluwer Acad. Publ., Dordrecht, The Netherlands (1997).

[76] N. Selamoglu, J. A. Mucha, D. E. Ibbotson, and D. L. Flamm, Silicon oxide deposition from tetraethoxysilane in a radio frequency downstream reactor: Mechanisms and step coverage, *J. Vac. Sci. Technol. B.* **7**(6), 1345–1351 (1989).

[77] C. P. Chang, D. L. Flamm, D. E. Ibbotson, and J. A. Mucha, Frequency effects and properties of plasma deposited fluorinated silicon nitride, *J. Vac. Sci. Technol. B.* **6**(2), 524–532 (1988).

[78] F. Fracassi, R. d'Agostino, P. Favia, and M. Van Sambeck, Thin film deposition in glow discharges fed with hexamethyldisilazane-oxygen mixtures, *Plasma Sources Sci. Technol.* **2**(2), 106–111 (1993).

[79] F. Fracassi, R. d'Agostino, and P. Favia, Plasma-enhanced chemical vapor deposition of organosilicon thin films from tetraethoxysilane-oxygen feeds, *J. Electrochem. Soc.* **139**(9), 2636–2644 (1992).

[80] R. Lamendola and R. d'Agostino, Process control of organosilicon plasmas for barrier film preparations, *Pure Appl. Chem.* **70**(6), 1203–1208 (1998).

[81] R. Lamendola and R. d'Agostino. Mechanism in plasma enhanced chemical vapour deposition from organosilicon feeds. In R. d'Agostino, P. Favia, and F. Fracassi (eds.), *Plasma Processing of Polymers, NATO ASI Series E: Appl. Science, Vol. 346*, pp. 321–334. Kluwer Acad. Publ., Dordrecht, The Netherlands (1997).

[82] M. Creatore, F. Palumbo, and R. d'Agostino, Diagnostics and insights on PECVD for gas-barrier coatings, *Pure Appl. Chem.* **74**(3), 407–411 (2002).

[83] L. Martinu, J. E. Klemberg-Sapieha, O. M. Küttel, A. Raveh, and M. R. Wertheimer, Critical ion energy and ion flux in the growth of films by plasma-enhanced chemical-vapor deposition, *J. Vac. Sci. Technol. A.* **12**(4), 1360–1364 (1994).

[84] R. Vernhes, A. Amassian, J. E. Klemberg-Sapieha, and L. Martinu, Plasma treatment of porous SiNx:H films for the fabrication of porous-dense multilayer optical filters with tailored interfaces, *J. Appl. Phys.* **99**(11), 114315 (2006).

[85] A. Milella, M. Creatore, M. A. Blauw, and M. C. M. van de Sanden, Remote plasma deposited silicon dioxide-like film densification by means of RF substrate biasing: Film chemistry and morphology, *Plasma Process. Polym.* **4**(6), 621–628 (2007).

[86] A. Lefèvre, L. Lewis, L. Martinu, and M. R. Wertheimer, Structural properties of silicon dioxide thin films densified by medium-energy particles, *Phys. Rev.. B.* **64**(11), 115429 (2001).

[87] A. M. Coclite, A. Milella, R. d'Agostino, and F. Palumbo, On the relationship between the structure and the barrier performance of plasma deposited silicon dioxide-like films, *Surf. Coat. Technol.* **204**(24), 4012–4017 (2010).

[88] A. Bergeron, J. E. Klemberg-Sapieha, and L. Martinu, Structure of the interfacial region between polycarbonate and plasma-deposited $SiN_{1.3}$ and SiO_2 optical coatings studied by ellipsometry, *J. Vac. Sci. Technol. A.* **16**(6), 3227–3234 (1998).

[89] P. E. Burrows, G. L. Graff, M. E. Gross, P. M. Martin, M. K. Shi, M. Hall, E. Mast, C. Bonham, W. Bennett, and M. B. Sullivan, Ultra barrier flexible substrates for flat panel displays, *Displays.* **22**(2), 65–69 (2001).

[90] J. S. Lewis and M. S. Weaver, Thin-film permeation-barrier technology for flexible organic light-emitting devices, *IEEE J. Selected Topics in Quantum Electronics.* **10**(1), 45–57 (2004).

[91] J. Affinito. A new class of ultra-barrier materials. In *Proceedings of 47th Annual Technical Conference*, pp. 563–593, Dallas, Texas, USA (24–29 April, 2004).

[92] G. L. Graff, R. E. Williford, and P. E. Burrows, Mechanisms of vapor permeation through multilayer barrier films: Lag time versus equilibrium permeation, *J. Appl. Phys.* **96**(4), 1840–1849 (2004).

[93] T. W. Kim, M. Yan, A. G. Erlat, P. A. McConnelee, M. Pellow, J. Deluca, T. P. Feist, A. R. Duggal, and M. Schaepkens, Transparent hybrid inorganic/organic barrier coatings for plastic organic light-emitting diode substrates, *J. Vac. Sci. Technol. A.* **23**(4), 971–977 (2005).

[94] A. Francescangeli, F. Palumbo, and R. d'Agostino, Deposition of barrier coatings from vinyltrimethylsilane-fed glow discharges, *Plasma Process. Polym.* **5**(7), 708–717 (2008).

[95] F. Palumbo, R. d'Agostino, F. Fracassi, S. Laera, A. Milella, E. Angelini, and S. Grassini, On low pressure plasma processing for metal protection, *Plasma Process. Polym.* **6**(S1), S684–S689 (2009).

[96] A. M. Coclite, A. Milella, F. Palumbo, C. Le Pen, and R. d'Agostino, Plasma deposited organosilicon multistacks for high-performance low-carbon steel protection, *Plasma Process. Polym.* **7**(9–10), 802–812 (2010).

[97] A. Grill, PECVD low and ultralow dielectric constant materials: From invention and research to products, *J. Vac. Sci. Technol. B.* **34**(2), 020801 (2016).

[98] M. Creatore, W. M. M. Kessels, Y. Barrell, J. Benedikt, and M. C. M. van de Sanden, Expanding thermal plasma for low-k dielectrics: Engineering the film chemistry by means of specific dissociation paths in the plasma, *Mat. Sci. Semiconductor Processing.* **7**(4–6), 283–288 (2004).

[99] J. Y. Kim, M. S. Hwang, Y. H. Kim, H. J. Kim, and Y. Lee, Origin of low dielectric constant of carbon-incorporated silicon oxide film deposited by plasma enhanced chemical vapor deposition, *J. Appl. Phys.* **90**(5), 2469–2473 (2001).

[100] J. Lubguban Jr, T. Rajagopalan, N. Mehta, B. Lahlouh, S. L. Simon, and S. Gangopadhyay, Low-k organosilicate films prepared by tetravinyltetramethylcyclotetrasiloxane, *J. Appl. Phys.* **92**(2), 1033–1038 (2002).

[101] A. D. Ross and K. K. Gleason, Effects of condensation reactions on the structural, mechanical, and electrical properties of plasma-deposited organosilicon thin films from octamethylcyclotetrasiloxane, *J. Appl. Phys.* **97**(11), 113707 (2005).

[102] D. D. Burkey and K. K. Gleason, Temperature-resolved fourier transform infrared study of condensation reactions and porogen decomposition in hybrid organosilicon-porogen films, *J. Vac. Sci. Technol. A.* **22**(1), 61–70 (2004).

[103] A. Grill and V. Patel, Ultralow-k dielectrics prepared by plasma-enhanced chemical vapor deposition, *Appl. Phys. Lett.* **79**(6), 803–805 (2001).

[104] A. Grill, Plasma enhanced chemical vapor deposited SiCOH dielectrics: From low-k to extreme low-k interconnect materials, *J. Appl. Phys.* **93**(3), 1785–1790 (2003).

[105] A. Grill and D. A. Neumayer, Structure of low dielectric constant to extreme low dielectric constant SiCOH films: Fourier transform infrared spectroscopy characterization, *J. Appl. Phys.* **94**(10), 6697–6707 (2003).

[106] A. Grill, V. Patel, K. P. Rodbell, E. Huang, M. R. Baklanov, K. P. Mogilnikov, M. Toney, and H. C. Kim, Porosity in plasma enhanced chemical vapor deposited SiCOH dielectrics: A comparative study, *J. Appl. Phys.* **94**(5), 3427–3435 (2003).

[107] S. K. Kwak, K. H. Jeong, and S. W. Rhee, Nanocomposite low-k SiCOH films by direct PECVD using vinyltrimethylsilane, *J. Electrochem. Soc.* **151**(2), F11 (2004).

[108] J. M. Park and S. W. Rhee, Remote plasma-enhanced chemical vapor deposition of nanoporous low-dielectric constant SiCOH films using vinyltrimethylsilane, *J. Electrochem. Soc.* **149**(8), F92 (2002).

[109] K. H. Jeong, S. G. Park, and S. W. Rhee, Nanocomposite low-k SiCOH films by plasma-enhanced chemical vapor deposition using vinyltrimethyl-silane and CO_2, *J. Vac. Sci. Technol. B.* **22**(6), 2799–2803 (2004).

[110] D. D. Burkey and K. K. Gleason, Organosilicon thin films deposited from cyclic and acyclic precursors using water as an oxidant, *J. Electrochem. Soc.* **151**(5), F105 (2004).

[111] A. Milella, J. L. Delattre, F. Palumbo, F. Fracassi, and R. d'Agostino, From low-k to ultralow-k thin-film deposition by organosilicon glow discharges, *J. Electrochem. Soc.* **153**(6), F106 (2006).

[112] A. Milella, F. Palumbo, J. L. Delattre, F. Fracassi, and R. d'Agostino, Deposition and characterization of dielectric thin films from allyltrimethyl-silane glow discharges, *Plasma Process. Polym.* **4**(4), 425–432 (2007).

[113] A. M. Coclite, A. Milella, F. Palumbo, F. Fracassi, and R. d'Agostino, A chemical study of plasma-deposited organosilicon thin films as low-k dielectrics, *Plasma Process. Polym.* **6**(8), 512–520 (2009).

[114] P. A. Premkumar, S. A. Starostin, M. Creatore, H. de Vries, R. M. J. Paffen, P. M. Koenraad, and M. C. M. van de Sanden, Smooth and self-similar SiO_2-like films on polymers synthesized in roll-to-roll atmospheric pressure-PECVD for gas diffusion barrier applications, *Plasma Process. Polym.* **7**(8), 635–639 (2010).

[115] D. Trunec, Z. Navrátil, P. Stahel, L. Zajíková, V. Buríková, and J. Cech, Deposition of thin organosilicon polymer films in atmospheric pressure glow discharge, *J. Phys. D: Appl. Phys.* **37**(15), 2112–2120 (2004).

[116] F. Fracassi, R. d'Agostino, F. Fanelli, A. Fornellu, and F. Palumbo, GC-MS investigation of hexamethyldisiloxane-oxygen fed plasmas, *Plasma Polym.* **8**(4), 259–269 (2003).

[117] F. Fanelli, R. d'Agostino, and F. Fracassi, GC-MS investigation of hexamethyldisiloxane-oxygen fed cold plasmas: Low pressure versus atmospheric pressure operation, *Plasma Process. Polym.* **8**(10), 932–941 (2011).

[118] F. Fanelli, F. Fracassi, S. Lovascio, and R. d'Agostino, GC-MS investigation of organosilicon and fluorocarbon fed plasmas, *Contrib. Plasma Phys.* **51**(2–3), 137–142 (2011).

[119] F. Fanelli, S. Lovascio, R. d'Agostino, F. Arefi-Khonsari, and F. Fracassi, Ar/HMDSO/O$_2$ fed atmospheric pressure DBDs: Thin film deposition and GC-MS investigation of by-products, *Plasma Process. Polym.* **7**(7), 535–543 (2010).

[120] F. Fanelli, S. Lovascio, R. d'Agostino, and F. Fracassi, Insights into the atmospheric pressure plasma-enhanced chemical vapor deposition of thin films from methyldisiloxane precursors, *Plasma Process. Polym.* **9**(11–12), 1132–1143 (2012).

[121] H. Biederman, Nanocomposites and nanostructures based on plasma polymers, *Surf. Coat. Technol.* **205**, S10–S14 (2011).

[122] E. Kay and A. A. Dilks, Metal-containing plasma polymerized fluorocarbon films: Their synthesis, structure, and polymerization mechanism, *J. Vac. Sci. Technol.* **16**, 428–430 (1979).

[123] E. Kay, F. Parmigiani, and W. Parrish, Microstructure of sputtered metal films grown in high-and low-pressure discharges, *J. Vac. Sci. Technol. A.* **6**(6), 3074–3081 (1988).

[124] C. Laurent and E. Kay, Properties of metal clusters in polymerized hydrocarbon versus fluorocarbon matrices, *J. Appl. Phys.* **65**(4), 1717–1723 (1989).

[125] A. Heilmann, J. Werner, M. Kelly, B. Holloway, and E. Kay, XPS depth profiles and optical properties of plasma polymer multilayers with embedded metal particles, *Appl. Surf. Sci.* **115**(4), 365–376 (1997).

[126] F. Fracassi, R. d'Agostino, F. Palumbo, F. Bellucci, and T. Monetta, Plasma-assisted deposition of tungsten-containing siloxane thin films, *Thin Solid Films.* **264**(1), 40–45 (1995).

[127] H. Biederman and L. Martinu. Plasma polymer-metal composite films. In R. d'Agostino (ed.), *Plasma Deposition, Treatment and Etching of Polymers*, pp. 269–320. Academic Press, San Diego, USA (1990).

[128] P. Favia, M. Vulpio, R. Martino, R. d'Agostino, R. P. Mota, and M. Catalano, Plasma-deposition of Ag-containing polyethyleneoxide-like coatings, *Plasma Process. Polym.* **5**(1), 1–14 (2000).

[129] H. Biederman, D. Slavínská, P. Bílková, and V. Stundžia. Composite films metal/plasma polymer, recent development, current and possible

applications. In R. d'Agostino, P. Favia, and F. Fracassi (eds.), *Plasma Processing of Polymers, NATO ASI Series E: Appl. Science*, Vol. 346, pp. 365–378. Kluwer Acad. Publ., Dordrecht, The Netherlands (1997).

[130] R. d'Agostino, L. Martinu, and V. Pische, Effect of bias and temperature on the bulk and surface properties of gold-containing plasma-polymerized fluorocarbons, *Plasma Chem. Plasma Process.* **11**(1), 1–13 (1991).

[131] K. L. Prime and G. M. Whitesides, Adsorption of proteins onto surfaces containing end-attached oligo (ethylene oxide): A model system using self-assembled monolayers, *J. Am. Chem. Soc.* **115**(23), 10714–10721 (1993).

[132] B. D. Johnston, E. E. Ratner and J. D. Bryers. RF plasma deposited PEO-like films: Surface characterization and inhibition of pseudomonas aeruginosa accumulation. In R. d'Agostino, P. Favia, and F. Fracassi (eds.), *Plasma Processing of Polymers, NATO ASI Series E: Appl. Science, Vol. 346*, pp. 465–476. Kluwer Acad. Publ., Dordrecht, The Netherlands (1997).

[133] G. P. Lopez, B. D. Ratner, C. D. Tidwell, C. L. Haycox, R. J. Rapoza, and T. A. Horbett, Glow discharge plasma deposition of tetraethylene glycol dimethyl ether for fouling-resistant biomaterial surfaces, *J. Biomed. Mat. Res.* **26**(4), 415–439 (1992).

[134] A. Oloffs, C. Grosse-Siestrup, S. Bisson, M. Rinck, R. Rudolph, and U. Gross, Biocompatibility of silver-coated polyurethane catheters and silvercoated dacron® material, *Biomaterials.* **15**(10), 753–758 (1994).

[135] J. M. Schierholz, L. J. Lucas, A. Rump, and G. Pulverer, Efficacy of silver-coated medical devices, *J. Hosp. Infection.* **40**(4), 257–262 (1998).

[136] E. Sardella, R. Gristina, R. d'Agostino, and P. Favia. Micro- and nanostructuring in plasma processes for biomaterials: Micro- and nano-features as powerful tools to address selective biological responses. In R. d'Agostino, P. Favia, Y. Kawai, H. Ikegami, N. Sato, and F. Arefi-Khonsari (eds.), *Advanced Plasma Technology*, pp. 243–268. Wiley-VCH, Weinheim, Germany (2007).

[137] E. Sardella, P. Favia, R. Gristina, M. Nardulli, and R. d'Agostino, Plasma-aided micro- and nanopatterning processes for biomedical applications, *Plasma Process. Polym.* **3**(6–7), 456–469 (2006).

[138] D. J. Balazs, K. Triandafillu, E. Sardella, G. Iacoviello, P. Favia, R. d'Agostino, H. Harms, and H. J. Mathieu. PECVD modification of medical-grade PVC to inhibit bacterial adhesion: PEO-like and nanocomposite Ag/PEO-like coatings. In R. d'Agostino, P. Favia, C. Oehr, and M. R. Wertheimer (eds.), *Plasma Processes and Polymers*, pp. 351–372. Wiley-VCH, Weinheim, Germany (2005).

[139] E. Sardella, R. Gristina, A. Mangone, M. Casavola, P. Favia, and R. d'Agostino, Plasma deposition of Ag-containing nano-composite bactericidal coatings, *J. Appl. Biomat. Biomech.* **4**(1), 67–67 (2006).

[140] P. Brault, S. Roualdes, A. Caillard, A. L. Thomann, J. Mathias, J. Durand, C. Coutanceau, J. M. Léger, C. Charles, and R. Boswell, Solid polymer fuel cell synthesis by low pressure plasmas: A short review, *Eur. Phys. J. Appl. Phys.* **34**(2), 151–156 (2006).

[141] E. Dilonardo, A. Milella, F. Palumbo, J. Thery, S. Martin, G. Barucca, P. Mengucci, R. d'Agostino, and F. Fracassi, Plasma deposited Pt-containing hydrocarbon thin films as electrocatalysts for PEM fuel cell, *J. Mater. Chem.* **20**(45), 10224–10227 (2006).

[142] E. Dilonardo, A. Milella, P. Cosma, R. d'Agostino, and F. Palumbo, Plasma deposited electrocatalytic films with controlled content of Pt nanoclusters, *Plasma Process. Polym.* **8**(5), 452–458 (2011).

[143] E. Dilonardo, A. Milella, F. Palumbo, G. Capitani, R. d'Agostino, and F. Fracassi, One-step plasma deposition of platinum containing nanocomposite coatings, *Plasma Process. Polym.* **7**(1), 51–58 (2010).

[144] A. Milella, F. Palumbo, E. Dilonardo, G. Barucca, P. Cosma, and F. Fracassi, Single step plasma deposition of platinum-fluorocarbon nanocomposite films as electrocatalysts of interest for micro fuel cells technology, *Plasma Process. Polym.* **11**(11), 1068–1075 (2014).

[145] Z. Machala and D. B. Graves, Frugal biotech applications of low-temperature plasma, *Trends Biotechnol.* **36**(6), 579–581 (2018).

[146] P. Kulkarni, P. A. Baron, and K. Willeke, *Aerosol Measurement: Principles, Techniques, and Applications*, 3rd ed. John Wiley & Sons, New Jersey, USA (2011).

[147] C. Mao, A. Liu, and B. Cao, Virus-based chemical and biological sensing, *Angew. Chem. Int. Ed.* **48**(37), 6790–6810 (2009).

[148] M. Magliulo, A. Mallardi, M. Y. Mulla, S. Cotrone, B. R. Pistillo, P. Favia, I. Vikholm-Lundin, G. Palazzo, and L. Torsi, Electrolyte-gated organic field-effect transistor sensors based on supported biotinylated phospholipid bilayer, *Adv. Mater.* **25**(14), 2090–2094 (2013).

[149] P. Lisboa, M. B. Villiers, C. Brakha, P. N. Marche, A. Valsesia, P. Colpo, and F. Rossi, Fabrication of bio-functionalised polypyrrole nanoarrays for bio-molecular recognition, *Micro and Nanosyst.* **3**(1), 83–89 (2011).

[150] E. M. Hetrick and M. H. Schoenfisch, Reducing implant-related infections: Active release strategies, *Chem. Soc. Rev.* **35**(9), 780–789 (2006).

[151] J. Heuts, J. Salber, A. M. Goldyn, R. Janser, M. Möller, and D. Klee, Bio-functionalized star PEG-coated PVDF surfaces for cytocompatibility-improved implant components, *J. Biomed. Mat. Res. Part A.* **92**(4), 1538–1551 (2010).

[152] M. Mohorčič, I. Jerman, M. Zorko, L. Butinar, B. Orel, R. Jerala, and J. Friedrich, Surface with antimicrobial activity obtained through silane coating with covalently bound polymyxin B, *J. Mater. Sci. Mater. Med.* **21**(10), 2775–2782 (2010).

[153] E. Faure, C. Falentin-Daudré, T. S. Lanero, C. Vreuls, G. Zocchi, C. Van De Weerdt, J. Martial, C. Jérôme, A. S. Duwez, and C. Detrembleur, Functional nanogels as platforms for imparting antibacterial, antibiofilm, and antiadhesion activities to stainless steel, *Adv. Funct. Mater.* **22**(24), 5271–5282 (2012).

[154] C. Vreuls, G. Zocchi, H. Vandegaart, E. Faure, C. Detrembleur, A. S. Duwez, J. Martial, and C. Van de Weerdt, Biomolecule-based antibacterial coating on a stainless steel surface: Multilayer film build-up optimization and stability study, *Biofouling.* **28**(4), 395–404 (2012).

[155] L. A. O'Hare, L. O'Neill, and A. J. Goodwin, Anti-microbial coatings by agent entrapment in coatings deposited via atmospheric pressure plasma liquid deposition, *Surf. Interface Anal.* **38**(11), 1519–1524 (2006).

[156] P. Heyse, A. Van Hoeck, M. B. Roeffaers, J. P. Raffin, A. Steinbüchel, T. Stöveken, J. Lammertyn, P. Verboven, P. A. Jacobs, J. Hofkens, and S. Paulussen, Exploration of atmospheric pressure plasma nanofilm technology for straightforward bio-active coating deposition: Enzymes, plasmas and polymers, an elegant synergy, *Plasma Process. Polym.* **8**(10), 965–974 (2011).

[157] P. Heyse, M. B. Roeffaers, S. Paulussen, J. Hofkens, P. A. Jacobs, and B. F. Sels, Protein immobilization using atmospheric-pressure dielectric-barrier discharges: A route to a straightforward manufacture of bioactive films, *Plasma Process. Polym.* **5**(2), 186–191 (2008).

[158] F. Palumbo, G. Camporeale, Y. W. Yang, J. S. Wu, E. Sardella, G. Dilecce, C. D. Calvano, L. Quintieri, L. Caputo, F. Baruzzi, and P. Favia, Direct plasma deposition of lysozyme-embedded bio-composite thin films, *Plasma Process. Polym.* **12**(11), 1302–1310 (2015).

[159] C. P. Hsiao, C. C. Wu, Y. H. Liu, Y. W. Yang, Y. C. Cheng, F. Palumbo, G. Camporeale, P. Favia, and J. S. Wu, Aerosol-assisted plasma deposition of biocomposite coatings: Investigation of processing conditions on coating properties, *IEEE Trans. Plasma Sci.* **44**(12), 3091–3098 (2016).

[160] Y. H. Liu, C. H. Yang, T. R. Lin, and Y. C. Cheng, Using aerosol-assisted atmospheric-pressure plasma to embed proteins onto a substrate in one step for biosensor fabrication, *Plasma Process. Polym.* **15**(9), 1800001 (2018).

[161] C. Lo Porto, F. Palumbo, A. Treglia, G. Camporeale, and P. Favia, Aerosol assisted atmopheric pressure PECVD of drug containing nano-capsules, *Jpn. J. Appl. Phys.* **59**(SA), SA0801 (2019).

[162] F. Palumbo, A. Treglia, C. Lo Porto, F. Fracassi, F. Baruzzi, G. Frache, D. El Assad, B. R. Pistillo, and P. Favia, Plasma-deposited nanocapsules containing coatings for drug delivery applications, *ACS Appl. Mater. Interfaces.* **10**(41), 35516–35525 (2018).

[163] C. Lo Porto, F. Palumbo, G. Palazzo, and P. Favia, Direct plasma synthesis of nano-capsules loaded with antibiotics, *Polym. Chem.* **8**(11), 1746–1749 (2017).

[164] C. Lo Porto, F. Palumbo, J. Buxadera-Palomero, C. Canal, P. Jelinek, L. Zajickova, and P. Favia, On the plasma deposition of vancomycin-containing nano-capsules for drug-delivery applications, *Plasma Process. Polym.* **15**(5), 1700232 (2018).

[165] C. Lo Porto, F. Palumbo, F. Fracassi, G. Barucca, and P. Favia, On the formation of nanocapsules in aerosol-assisted atmospheric-pressure plasma, *Plasma Process. Polym.* **16**(11), 1900116 (2019).

[166] F. Fanelli and F. Fracassi, Aerosol-assisted atmospheric pressure cold plasma deposition of organic-inorganic nanocomposite coatings, *Plasma Chem. Plasma Process.* **34**(3), 473–487 (2014).

[167] P. Marchand, I. A. Hassan, I. P. Parkin, and C. J. Carmalt, Aerosol-assisted delivery of precursors for chemical vapour deposition: Expanding the

scope of CVD for materials fabrication, *Dalton Trans.* **42**(26), 9406–9422 (2013).

[168] M. Michel, J. Bour, J. Petersen, C. Arnoult, F. Ettingshausen, C. Roth, and D. Ruch, Atmospheric plasma deposition: A new pathway in the design of conducting polymer-based anodes for hydrogen fuel cells, *Fuel Cells.* **10**(6), 932–937 (2010).

[169] F. Fanelli, A. M. Mastrangelo, and F. Fracassi, Aerosol-assisted atmospheric cold plasma deposition and characterization of superhydrophobic organic-inorganic nanocomposite thin films, *Langmuir.* **30**(3), 857–865 (2014).

[170] P. Brunet, R. Rincón, J. M. Martinez, Z. Matouk, F. Fanelli, M. Chaker, and F. Massines, Control of composite thin film made in an Ar/ isopropanol/TiO$_2$ nanoparticles dielectric barrier discharge by the excitation frequency, *Plasma Process. Polym.* **14**(12), 1700049 (2017).

[171] X. Chen, C. Lo Porto, Z. Chen, A. Merenda, F. M. Allioux, R. d'Agostino, K. Magniez, X. J. Dai, F. Palumbo, and L. F. Dumée, Single step synthesis of janus nano-composite membranes by atmospheric aerosol plasma polymerization for solvents separation, *Sci. Total Env.* **645**, 22–33 (2018).

[172] L. Wang, C. Lo Porto, F. Palumbo, M. Modic, U. Cvelbar, R. Ghobeira, N. De Geyter, M. De Vrieze, Š. Kos, G. Serša, and C. Leys, Synthesis of antibacterial composite coating containing nanocapsules in an atmospheric pressure plasma, *Mat. Sci. Eng.: C.* **119**, 111496 (2021).

[173] F. Fanelli, A. M. Mastrangelo, G. Caputo, and F. Fracassi, Tuning the structure and wetting properties of organic-inorganic nanocomposite coatings prepared by aerosol-assisted atmospheric pressure cold plasma deposition, *Surf. Coat. Technol.* **358**, 67–75 (2019).

[174] F. Fanelli, A. M. Mastrangelo, N. De Vietro, and F. Fracassi, Preparation of multifunctional superhydrophobic nanocomposite coatings by aerosol-assisted atmospheric cold plasma deposition, *Nanosci. Nanotechnol. Lett.* **7**(1), 84–88 (2015).

[175] E. M. Liston, L. Martinu, and M. R. Wertheimer, Plasma surface modification of polymers for improved adhesion: A critical review, *J. Adhesion Sci. Technol.* **7**(10), 1091–1127 (1993).

[176] R. H. Hansen and H. Schonhorn, A new technique for preparing low surface energy polymers for adhesive bonding, *J. Polym. Sci. Polym. Lett. Ed.* **4**(3), 203–209 (1966).

[177] R. Wilken, A. Holländer, and J. Behnisch, Surface radical analysis on plasma-treated polymers, *Surf. Coat. Technol.* **116**, 991–995 (1999).

[178] J. E. Klemberg-Sapieha, L. Martinu, O. M. Küttel, and M. R. Wertheimer. Modification of polymer surfaces by dual frequency plasma. In K. L. Mittal (ed.), *Metallized Plastics 2: Fundamental and Applied Aspects*, pp. 315–329. Springer Science, New York, USA (1991).

[179] L. Cop, J. Jordaan, H. P. Schreiber, and M. W. Wertheimer, Process for making a non-polar polymeric material dyeable with an acid dye, US Patent US4744860A (1987).

[180] P. Favia and R. d'Agostino, Plasma treatments and plasma deposition of polymers for biomedical applications, *Surf. Coat. Technol.* **98**(1-3), 1102–1106 (1998).

[181] P. Favia, F. Palumbo, R. d'Agostino, S. Lamponi, A. Magnani, and R. Barbucci, Immobilization of heparin and highly-sulphated hyaluronic acid onto plasma-treated polyethylene, *Plasma Polym.* **3**(2), 77–96 (1998).

[182] G. Camporeale, M. Moreno-Couranjou, S. Bonot, R. Mauchauffé, N. D. Boscher, C. Bebrone, C. Van de Weerdt, H. M. Cauchie, P. Favia, and P. Choquet, Atmospheric-pressure plasma deposited epoxy-rich thin films as platforms for biomolecule immobilization: Application for anti-biofouling and xenobiotic-degrading surfaces, *Plasma Process. Polym.* **12**(11), 1208–1219 (2015).

[183] L. De Bartolo, S. Morelli, L. C. Lopez, L. Giorno, C. Campana, S. Salerno, M. Rende, P. Favia, L. Detomaso, R. Gristina, R. d'Agostino, and E. Drioli, Biotransformation and liver-specific functions of human hepatocytes in culture on RGD-immobilized plasma-processed membranes, *Biomaterials.* **26**(21), 4432–4441 (2005).

[184] F. Garbassi, M. Morra, E. Occhiello, L. Barino, and R. Scordamaglia, Dynamics of macromolecules: A challenge for surface analysis, *Surf. Interf. Anal.* **14**(10), 585–589 (1989).

[185] D. Youxian, H. J. Griesser, A. W. H. Mau, R. Schmidt, and J. Liesegang, Surface modification of poly (tetrafluoroethylene) by gas plasma treatment, *Polymer.* **32**(6), 1126–1130 (1991).

[186] D. J. Wilson, R. L. Williams, and R. C. Pond, Plasma modification of PTFE surfaces. Part i: Surfaces immediately following plasma treatment, *Surf. Interf. Anal.* **31**(5), 385–396 (2001).

[187] R. Lamendola, P. Favia, F. Palumbo, and R. d'Agostino, Plasma-modification of polymers: process control in PECVD of gas-barrier films and plasma processes for immobilizing anti-thrombotic molecules, *European Phys. J. Appl. Phys.* **4**(1), 65–71 (1998).

[188] M. R. Wertheimer and R. Bartnikas. Degradation effects of plasma and corona on polymers. In R. d'Agostino, P. Favia, and F. Fracassi (eds.), *Plasma Processing of Polymers, NATO ASI Series E: Appl. Science, Vol. 346*, pp. 435–450. Kluwer Acad. Publ., Dordrecht, The Netherlands (1997).

[189] F. D. Egitto, Plasma etching and modification of organic polymers, *Pure and Appl. Chem.* **62**(9), 1699–1708 (1990).

[190] P. Favia, F. Palumbo, R. d'Agostino, S. Lamponi, A. Magnani, and R. Barbucci, Immobilization of heparin and highly-sulphated hyaluronic acid onto plasma-treated polyethylene, *Plasmas Polym.* **3**(2), 77–96 (1998).

[191] A. S. Chawla and R. Sipehia, Characterization of plasma polymerized silicone coatings useful as biomaterials, *J. Biomed. Mater. Res.* **18**(5), 537–545 (1984).

[192] M. R. Wertheimer and H. P. Schreiber, Surface property modification of aromatic polyamides by microwave plasmas, *J. Appl. Polym. Sci.* **26**(6), 2087–2096 (1984).

[193] D. L. Cho and H. Yasuda, Influence of geometric factors of the substrate on hydrophilic surface modification of polyurethane sponges by plasma treatment, *J. Vac. Sci. Technol. A.* **4**(5), 2307–2316 (1986).

[194] T. R. Gengenbach, X. Xie, R. C. Chatelier, and H. J. Griesser, Evolution of the surface composition and topography of perfluorinated polymers following ammonia-plasma treatment, *J. Adhesion Sci. Technol.* **8**(4), 305–328 (1994).

[195] P. Favia, M. V. Stendardo, and R. d'Agostino, Selective grafting of amine groups on polyethylene by means of NH_3-H_2 RF glow discharges, *Plasmas Polym.* **1**(2), 91–112 (1996).

[196] P. Favia, F. Palumbo, M. V. Stendardo, and R. d'Agostino. Plasma-treatments of polymers by NH_3-H_2 RF glow discharges: Coupling plasma and surface diagnostics. In B. D. Ratner and D. G. Castner (eds.), *Surface Modification of Polymeric Biomaterials*, pp. 69–77. Springer, Boston, MA (1996).

[197] A. Tressaud, E. Durand, and C. Labrugére, Surface modification of several carbon-based materials: Comparison between CF_4 RF plasma and direct F_2-gas fluorination routes, *J. Fluorine Chem.* **125**(11), 1639–1648 (2004).

[198] S. M. Mukhopadhyay, P. Joshi, S. Datta, J. G. Zhao, and P. France, Plasma assisted hydrophobic coatings on porous materials: Influence of plasma parameters, *J. Phys. D: App. Phys.* **35**(16), 1927 (2002).

[199] R. Barni, C. Riccardi, E. Selli, M. R. Massafra, B. Marcandalli, F. Orsini, G. Poletti, and L. Meda, Wettability and dyeability modulation of poly (ethylene terephthalate) fibers through cold SF_6 plasma treatment, *Plasma Process. Polym.* **2**(1), 64–72 (2005).

[200] A. Tursi, N. De Vietro, A. Beneduci, A. Milella, F. Chidichimo, F. Fracassi, and G. Chidichimo, Low pressure plasma functionalized cellulose fiber for the remediation of petroleum hydrocarbons polluted water, *J. Hazardous Mat.* **373**, 773–782 (2019).

[201] L. Sabbatini, *Polymer Surface Characterization.* De Gruyter, Berlin, Boston (doi.org/10.1515/9783110288117) (2014).

[202] N. Inagaki, K. Narushim, S. Ejima, Y. Ikeda, S. K. Lim, Y. W. Park, and K. Miyazaki, Hydrophobic recovery of plasma-modified film surfaces of ethylene-co-tetrafluoroethylene co-polymer, *J. Adhesion Sci. Technol.* **17**(11), 1457–1475 (2003).

[203] P. Favia, A. Milella, L. Iacobelli, and R. d'Agostino. Plasma pretreatments and treatments on polytetrafluoroethylene for reducing the hydrophobic recovery. In R. d'Agostino, P. Favia, C. Oehr, and M. R. Wertheimer (eds.), *Plasma Processes and Polymers*, pp. 271–280. Wiley-VCH, Weinheim, Germany (2005).

[204] F. Palumbo, R. Di Mundo, D. Cappelluti, and R. d'Agostino, Superhydrophobic and superhydrophilic polycarbonate by tailoring chemistry and nano-texture with plasma processing, *Plasma Process. Polym.* **8**(2), 118–126 (2011).

[205] S. Iijima, Helical microtubules of graphitic carbon, *Nature.* **354**(6348), 56–58 (1991).

[206] M. F. L. De Volder, S. H. Tawfick, R. H. Baughman, and A. J. Hart, Carbon nanotubes: Present and future commercial applications, *Science.* **339**(6119), 535–539 (2013).

[207] V. N. Popov, Carbon nanotubes: properties and application, *Mat. Sci. Eng.: R: Reports.* **43**(3), 61–102 (2004).

[208] P. M. Ajayan and J. M. Tour, Nanotube composites, *Nature.* **447**(7148), 1066–1068 (2007).

[209] A. L. Mohd Tobi, I. Zaman, S. Jamian, and A. E. Ismail, A review on carbon nanotubes reinforced ceramic composite, *ARPN J. Eng. Appl. Sci.* **11**(12), 7406–7414 (2016).

[210] C. Chen, B. Liang, D. Lu, A. Ogino, X. Wang, and M. Nagatsu, Amino group introduction onto multiwall carbon nanotubes by NH_3/Ar plasma treatment, *Carbon.* **48**(4), 939–948 (2010).

[211] D. Tasis, N. Tagmatarchis, A. Bianco, and M. Prato, Chemistry of carbon nanotubes, *Chem. Reviews.* **106**(3), 1105–1136 (2006).

[212] F. Pourfayaz, Y. Mortazavi, A. A. Khodadadi, S. H. Jafari, S. Boroun, and M. V. Naseh, A comparison of effects of plasma and acid functionalizations on structure and electrical property of multi-wall carbon nanotubes, *Appl. Surf. Sci.* **295**, 66–70 (2014).

[213] M. G. Trulli, E. Sardella, F. Palumbo, G. Palazzo, L. C. Giannossa, A. Mangone, R. Comparelli, S. Musso, and P. Favia, Towards highly stable aqueous dispersions of multi-walled carbon nanotubes: The effect of oxygen plasma functionalization, *J. Colloid Interface Sci.* **491**, 255–264 (2017).

[214] M. Thomas and K. L. Mittal, *Atmospheric Pressure Plasma Treatment of Polymers: Relevance to Adhesion.* John Wiley & Sons, Beverly, MA (2013).

[215] B. J. Lee, Y. Kusano, N. Kato, K. Naito, T. Horiuchi, and H. Koinuma, Oxygen plasma treatment of rubber surface by the atmospheric pressure cold plasma torch, *Jpn. J. Appl. Phys.* **36**(5R), 2888 (1997).

[216] M. Noeske, J. Degenhardt, S. Strudthoff, and U. Lommatzsch, Plasma jet treatment of five polymers at atmospheric pressure: Surface modifications and the relevance for adhesion, *Int. J. Adhes. Adhesives.* **24**(2), 171–177 (2004).

[217] S. A. Rich, T. Dufour, P. Leroy, F. Reniers, L. Nittler, and J. J. Pireaux, LDPE surface modifications induced by atmospheric plasma torches with linear and showerhead configurations, *Plasma Process. Polym.* **12**(8), 771–785 (2015).

[218] F. Chen, J. Song, S. Huang, S. Xu, G. Xia, D. Yang, W. Xu, J. Sun, and X. Liu, Simultaneous and long-lasting hydrophilization of inner and outer wall surfaces of polytetrafluoroethylene tubes by transferring atmospheric pressure plasmas, *J. Phys. D: Appl. Phys.* **49**(36), 365202 (2016).

[219] V. Armenise, F. Fanelli, A. Milella, L. D'Accolti, A. Uricchio, and F. Fracassi, Atmospheric pressure plasma treatment of polyurethane foams

with He-O$_2$ fed dielectric barrier discharges, *Surf. Interfaces.* **20**, 100600 (2020).

[220] S. Sun and Y. Qiu, Influence of moisture on wettability and sizing properties of raw cotton yarns treated with He/O$_2$ atmospheric pressure plasma jet, *Surf. Coat. Technol.* **206**(8–9), 2281–2286 (2012).

[221] J. Pawlat, P. Terebun, M. Kwiatkowski, and J. Diatczyk, RF atmospheric plasma jet surface treatment of paper, *J. Phys. D: Appl. Phys.* **49**(37), 374001 (2016).

[222] X. Y. Yu, T. Luo, Y. X. Zhang, Y. Jia, B. J. Zhu, X. C. Fu, J. H. Liu, and X. J. Huang, Adsorption of lead (II) on O$_2$-plasma-oxidized multiwalled carbon nanotubes: Thermodynamics, kinetics, and desorption, *ACS Appl. Mater. Interfaces.* **3**(7), 2585–2593 (2011).

[223] S. Ali, I. A. Shah, A. Ahmad, J. Nawab, and H. Huang, Ar/O$_2$ plasma treatment of carbon nanotube membranes for enhanced removal of zinc from water and wastewater: A dynamic sorption-filtration process, *Sci. Total Env.* **655**, 1270–1278 (2019).

[224] F. Fanelli, F. Fracassi, and R. d'Agostino, Fluorination of polymers by means of He/CF$_4$-fed atmospheric pressure glow dielectric barrier discharges, *Plasma Process. Polym.* **5**(5), 424–432 (2008).

[225] F. D. Egitto, V. Vukanovic, and G. N. Taylor. Plasma etching of organic polymers. In ed. R. d'Agostino, *Plasma Deposition, Treatment and Etching of Polymers*, pp. 321–422. Academic Press, San Diego, USA (1990).

[226] U. Schulz, P. Munzert, R. Leitel, I. Wendling, N. Kaiser, and A. Tünnermann, Antireflection of transparent polymers by advanced plasma etching procedures, *Optics Express.* **15**(20), 13108–13113 (2007).

[227] I. Woodward, W. C. E. Schofield, V. Roucoules, and J. P. S. Badyal, Super-hydrophobic surfaces produced by plasma fluorination of polybutadiene films, *Langmuir.* **19**(8), 3432–3438 (2003).

[228] J. Fresnais, J. P. Chapel, and F. Poncin-Epaillard, Synthesis of transparent superhydrophobic polyethylene surfaces, *Surf. Coat. Technol.* **200**(18–19), 5296–5305 (2006).

[229] A. D. Tserepi, M. E. Vlachopoulou, and E. Gogolides, Nanotexturing of poly (dimethylsiloxane) in plasmas for creating robust super-hydrophobic surfaces, *Nanotechnology.* **17**(15), 3977 (2006).

[230] K. Teshima, H. Sugimura, Y. Inoue, O. Takai, and A. Takano, Ultra-water-repellent poly (ethylene terephthalate) substrates, *Langmuir.* **19**(25), 10624–10627 (2003).

[231] M. Morra, E. Occhiello, and F. Garbassi, Contact angle hysteresis in oxygen plasma treated poly (tetrafluoroethylene), *Langmuir.* **5**(3), 872–876 (1989).

[232] J. P. Youngblood and T. J. McCarthy, Ultrahydrophobic polymer surfaces prepared by simultaneous ablation of polypropylene and sputtering of poly (tetrafluoroethylene) using radio frequency plasma, *Macromolecules.* **32**(20), 6800–6806 (1999).

[233] A. Mukherjee, *Biomimetics: Learning from Nature.* BoD-Books on Demand, London, UK (2010).

[234] M. Yamamoto, N. Nishikawa, H. Mayama, Y. Nonomura, S. Yokojima, S. Nakamura, and K. Uchida, Theoretical explanation of the lotus effect: Superhydrophobic property changes by removal of nanostructures from the surface of a lotus leaf, *Langmuir.* **31**(26), 7355–7363 (2015).

[235] G. D. Bixler and B. Bhushan, Bioinspired rice leaf and butterfly wing surface structures combining shark skin and lotus effects, *Soft Matter.* **8**(44), 11271–11284 (2012).

[236] W. Barthlott, T. Schimmel, S. Wiersch, K. Koch, M. Brede, M. Barczewski, S. Walheim, A. Weis, A. Kaltenmaier, A. Leder, and H. F. Bohn, The Salvinia paradox: Superhydrophobic surfaces with hydrophilic pins for air retention under water, *Adv. Mater.* **22**(21), 2325–2328 (2010).

[237] L. Wen, J. C. Weaver, P. J. Thornycroft, and G. V. Lauder, Hydrodynamic function of biomimetic shark skin: Effect of denticle pattern and spacing, *Bioinspiration Biomimetics.* **10**(6), 066010 (2015).

[238] X. Gao and L. Jiang, Water-repellent legs of water striders, *Nature.* **432**(7013), 36–36 (2004).

[239] L. Feng, Y. Zhang, J. Xi, Y. Zhu, N. Wang, F. Xia, and L. Jiang, Petal effect: A superhydrophobic state with high adhesive force, *Langmuir.* **24**(8), 4114–4119 (2008).

[240] H. Gao, X. Wang, H. Yao, S. Gorb, and E. Arzt, Mechanics of hierarchical adhesion structures of geckos, *Mech. Mater.* **37**(2–3), 275–285 (2005).

[241] K. V. Baryshnikova, A. S. Kadochkin, and A. S. Shalin, Nanostructural antireflecting coatings: Classification analysis (A review), *Opt. Spectrosc.* **119**(3), 343–355 (2015).

[242] J. Sun, B. Bhushan, and J. Tong, Structural coloration in Nature, *RSC Advances.* **3**(35), 14862–14889 (2013).

[243] J. Huang, X. Wang, and Z. L. Wang, Controlled replication of butterfly wings for achieving tunable photonic properties, *Nano Lett.* **6**(10), 2325–2331 (2006).

[244] L. T. Phan, S. M. Yoon, and M. W. Moon, Plasma-based nanostructuring of polymers: A review, *Polymers.* **9**(9), 417 (2017).

[245] In ed. R. d'Agostino, *Plasma Deposition, Treatment and Etching of Polymers.* Academic Press, San Diego, USA (1990).

[246] R. Di Mundo, R. d'Agostino, and F. Palumbo, Long-lasting antifog plasma modification of transparent plastics, *ACS Appl. Mater. Interfaces.* **6**(19), 17059–17066 (2014).

[247] R. Di Mundo, F. Palumbo, and R. d'Agostino, Nanotexturing of polystyrene surface in fluorocarbon plasmas: From sticky to slippery superhydrophobicity, *Langmuir.* **24**(9), 5044–5051 (2008).

[248] R. Di Mundo, F. Palumbo, and R. d'Agostino, Influence of chemistry on wetting dynamics of nanotextured hydrophobic surfaces, *Langmuir.* **26**(7), 5196–5201 (2010).

[249] R. Di Mundo, F. Bottiglione, F. Palumbo, P. Favia, and G. Carbone, Sphere-on-cone microstructures on Teflon surface: Repulsive behavior against impacting water droplets, *Mater. Des.* **92**, 1052–1061 (2016).

[250] R. Di Mundo, F. Bottiglione, F. Palumbo, M. Notarnicola, and G. Carbone, Filamentary superhydrophobic Teflon surfaces: Moderate apparent contact angle but superior air-retaining properties, *J. Colloid Interface Sci.* **482**, 175–182 (2016).

[251] R. Di Mundo, F. Bottiglione, M. Notarnicola, F. Palumbo, and G. Pascazio, Plasma-textured teflon: Repulsion in air of water droplets and drag reduction underwater, *Biomimetics.* **2**(1), 1 (2017).

[252] C. Lo Porto, R. Di Mundo, V. Veronico, I. Trizio, G. Barucca, and F. Palumbo, Easy plasma nano-texturing of PTFE surface: From pyramid to unusual spherules-on-pyramid features, *Appl. Surf. Sci.* **483**, 60–68 (2019).

[253] F. Palumbo, C. Lo Porto, and P. Favia, Plasma nano-texturing of polymers for wettability control: Why, what and how, *Coatings.* **9**(10), 640 (2019).

Chapter 3

The Methods of Streamer Formation in/on Dielectric and Conductive Liquids

Yuri Akishev

SRC RF TRINITI, 108840, Moscow, Troitsk, Pushkovykh Street,
Vladenie 12, Russia, and NRNU MEPhI, 115409,
Moscow, Kashirskoe shosse, 31, Russia
akishev@triniti.ru

At present, numerous methods for liquid treatments, like water cleaning, have been developed. The principles underlying these methods are very diverse and based on the use of various processes — physical, chemical, microbiological, etc. Electro-physical methods occupy an important place among all the other methods used for liquid treatment. They are based on different physical and chemical processes happening in liquid under their impact by the electrical discharge of a specific type and pre-set released power. Atmospheric pressure plasma–liquid systems, forming a non-equilibrium low-temperature plasma (NTP) in liquids, or a thin plasma layer on a liquid surface, are referred to as electro-physical methods of liquid treatment as well. They are of great interest for many scientific, technological and biomedical applications, in civil engineering, environmental protection, etc. A reason is that the gaseous non-thermal plasma provides the enrichment of the liquid by various reactive species (atoms, radicals, UV, excited atoms and molecules, ions) which are able to intensify/initiate the required biochemical processes. Besides, the NTP treatment of liquids is an environmental friendly one. This makes the latter a serious competitive alternative to the traditional biochemical methods using, as a rule, different toxic additives or generating toxic or harmful byproducts.

Nowadays, various approaches are developed for the NTP activation of liquids. Many known plasma–liquid systems are based on the use of NTP formed by transient streamer or spark electrical discharges which are generated directly in/on the liquid to be treated. These discharges exhibit themselves in/on the liquid in the form of bright and numerous short-living, chaotically spreading and branching thin current filaments called streamers. In the case of NTP generation in/on a liquid, streamers are in close contact with the liquid, thus the reactive plasma species are effectively and quickly transferred to the liquid. In fact, both the possibility of creating the streamer discharges in/on liquids and plasma parameters in the formed streamers are significantly dependent on the liquid conductivity. This chapter presents a review of the methods of streamers' formation in/on dielectric and conductive liquids.

Contents

1. Introduction

Modern civilization, for various reasons, is more and more in need of clean water. This need stimulates the research and development of novel approaches designed for the effective cleaning of water and other liquids containing different toxic and hazardous contaminants. At present, numerous methods providing the liquid treatment (in particular, water cleaning) have been developed. The principles underlying these methods are very diverse and based on the use of various processes — physical, chemical, microbiological, etc. The problems discussed in our review are referred to the so-called electro-physical methods which occupy an important place among all the other methods. The electro-physical methods are based on the use of electrical discharges generated in, above, or in contact with a liquid. Besides the high electric field that can be generated in a liquid, different physical and chemical processes in and above the liquid happen either due to the energy released by the electrical discharge itself or by the plasma created by this discharge. Intensity and efficiency of these processes depend on specific type of electric discharge and its pre-set released power.

Note that the high-voltage (HV) electrical discharges in/above a liquid (in particular, water) have been the subjects of intense research for a long time. However, these studies were carried out mainly by the electrical engineering community. This is why these studies were devoted predominantly to developing lightning protection methods, pulsed high-power equipment and HV insulation. In other words, the main attention was paid to the study of the physical and engineering aspects related to the development of HV breakdown in gases and liquids. Plasma-chemical aspects of the processes occurring in gases and liquids under their influence by HV discharges and their plasmas received much less attention. However, the mentioned plasma-chemical processes can play a crucial role in HV discharges and plasmas for the treatment or cleaning of liquids.

At present, many known plasma–liquid systems are based on the use of non-equilibrium low-temperature plasma at atmospheric pressure, frequently referred to as cold or non-thermal plasma (NTP). In fact, the NTP is a quasi-neutral gaseous mixture of neutral and charged (ions and electrons) particles. The electron concentration n_e is much lower compared to the concentration of neutral particles N, i.e., $n_e \ll N$. At the same time, the reduced electric field, E/N, in a cold plasma can be very high. It means, the energy transferred from the electric field to electrons will be high as well. As a result, the average electron energy, T_e, is about (30,000–50,000) K, exceeding approximately by a factor 100 the temperature of ions and neutral particles having a lower temperature, T, close to the room one, i.e., $T_e \gg T$. Such energetic (or overheated) electrons provide very effectively the excitation of electronic states of neutral particles and, if these particles are molecules, the excitation of rotation and vibration states and the dissociation of molecules as well.

Owing to the existence of strongly overheated electrons, the NTP is enriched abundantly with numerous physically and biochemically reactive species. Just these charged and neutral reactive particles are transferred into a liquid and are responsible for the occurrence of the plasma-chemical processes needed to treat the liquid. Because of that, the NTP methods designed for the treatment of a liquid are widely used in many scientific, technological and biomedical applications, in civil engineering, environmental protection, etc.

As mentioned above, the gaseous non-thermal plasma provides the enrichment of a liquid by various reactive species — atoms, radicals, UV radiation, excited atoms and molecules, ions. These species are able to intensify/initiate the necessary biochemical processes in the liquid, for instance, to degrade and convert unwanted persistent molecules present in the waste water. All the enumerated attractive features explain the high interest in NTP methods. Besides, the simultaneous action of all the plasma reactive species in a liquid can provide their synergistic positive effect. In the case of water, it is appropriate to recall that water molecules are strongly polar ones and therefore they interact readily with the positive and negative ions of a salt, making the dissolving process possible.

It is useful to note several additional points about the reactive species generated by the NTP. These species can be divided into two groups. One of them corresponds to the so-called reactive oxygen species (ROS), the most significant of them being OH (hydroxyl radical), O_2^- (superoxide anion radical), HO_2 (hydroperoxyl radical), H_2O_2 (hydrogen peroxide), atomic oxygen O, ozone O_3, etc. Note that ozone is the most stable radical — its life time in ambient air is about 12 hours. The second group corresponds to Reactive Nitrogen Species (RNS), the most significant of them being NO (nitric oxide), NO_2 (nitrogen dioxide), ONOOH (peroxynitrite), $ONOOH^-$, etc. Short-lived RNS formed by plasma activation, such as NO and NO_2, undergo chemical reactions in the presence of water and oxygen, producing nitrite and nitrate anions in the liquid phase. It is well known that all the ROS and RNS have strong biochemical effects in a liquid. Therefore, supply of these reactive species in the liquid to be treated is important. However, plasma-induced biochemical effects in liquids may depend markedly on the generation and penetration depth of these species in the liquid.

Among all the reactive species, the OH hydroxyl radical is the most reactive oxidant (Fig. 1). Hydroxyl radicals may be produced in the NTP due to gas-phase reactions and also through reactions between plasma-generated species in the liquid phase. Under acidic conditions, the most important pathway for OH formation via liquid-phase reactions is the formation of peroxynitrous acid (O=NOOH), through reaction (1) pointed in Fig. 1. Peroxynitrous acid is unstable and decomposes through reactions (2) and (3) (with the indicated branching ratio), resulting in the formation of OH. The lifetime of OH in water is relatively short (about $100\,\mu s$) as it is rapidly consumed by formation of H_2O_2. This is why the penetration depth of gas-phase OH into the liquid does not exceed $10\,\mu m$. In the absence of organic matter, H_2O_2 is relatively stable in water. Superoxide radical anion O_2^- has a relatively long lifetime in solution (about $5\,s$).

There is no common point of view so far on the processes happening during the plasma–liquid interaction including the processes in the NTP, interface plasma–liquid and in bulk liquid. A reason is that transportation of reactive species from the gas phase through the liquid phase into a target is accompanied by complicated chemical reactions among air, H_2O and plasma-generated active species.

Although this review is far from exhaustive, we would like to give readers a broader picture of the existing ideas relevant to this issue. To this end, several hypothetic schemes proposed by different authors, providing insights on the plasma–liquid processes, are presented in Figs. 2–4 (see [2–4]). These schemes are relevant only to pure water, not containing any organic or inorganic impurities and contaminants.

1) $NO_2^- + H_2O_2 + H^+ \longrightarrow O$

2) $O = NOOH \longrightarrow NO_2 + OH$ ~30%

3) $O = NOOH \longrightarrow HNO_3 \longrightarrow NO_3^- + H^+$ ~70%

(a)   (b)

Plasma-liquid interaction

(c)

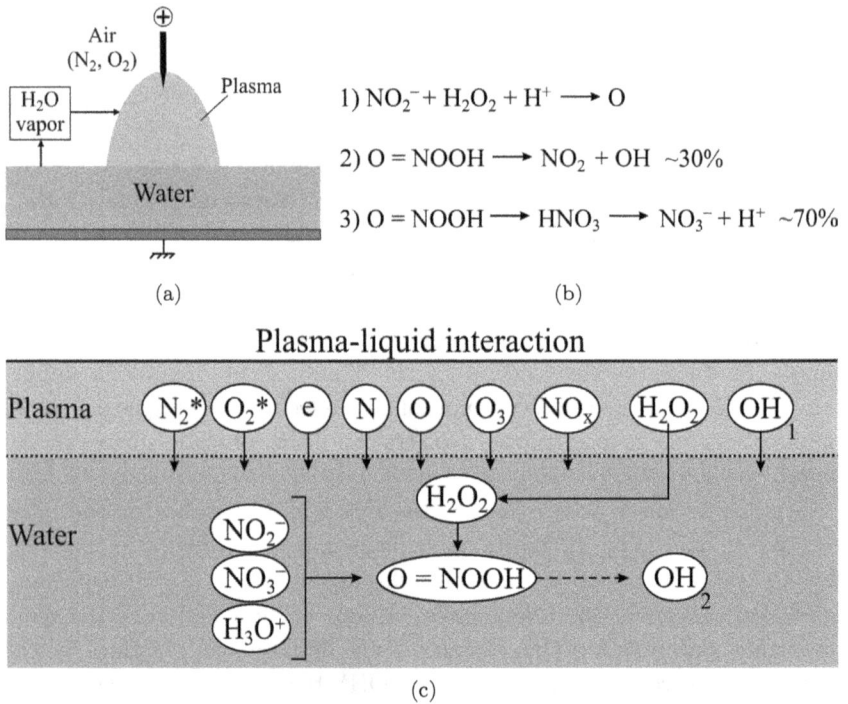

Fig. 1. (a) Sources and pathways of OH produced by positive corona discharge above water, and (b) and (c) the corresponding chemical reactions. Two sources of OH are emphasized: (1) OH generated from plasma-induced reactions at or near the plasma–solution interface, and (2,3) the decomposition of O=NOOH peroxynitrous acid in solution under acidic conditions. Adapted from [1].

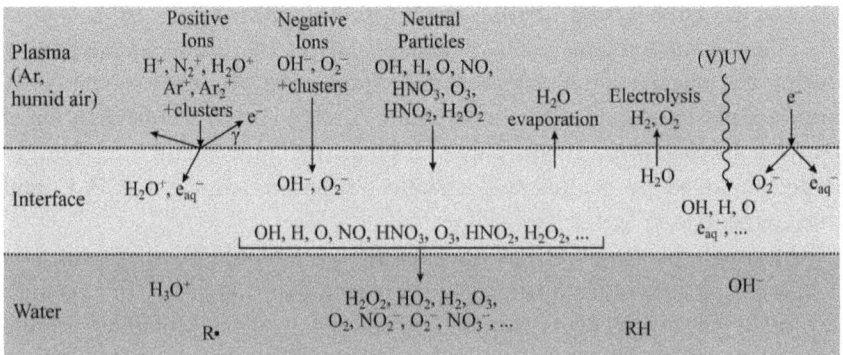

Fig. 2. Schematic diagram of some of the most important species and mechanisms for an argon/humid air plasma in contact with water. Adapted from [2].

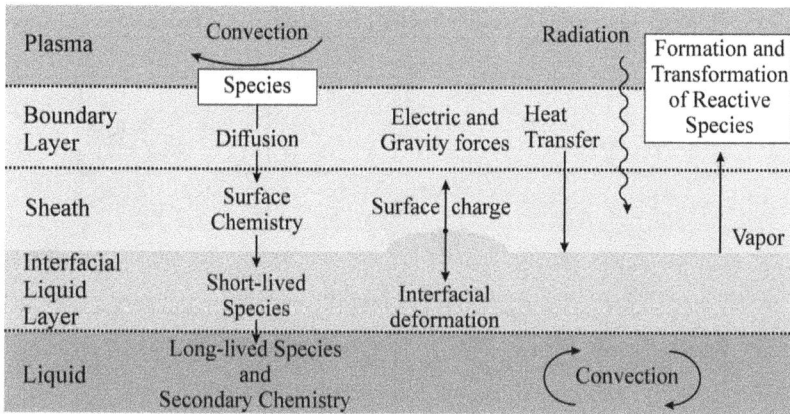

Fig. 3. Schematic representation of the key plasma–liquid interactions. The boundary layer reflects the gradients in temperature, species densities and gas flow velocities. The sheath is present for ionizing plasmas when the liquid surface is charged. Adapted from [3].

If readers wish to acquaint themselves with more comprehensive information on plasma chemistry aspects related to the interaction of different reactive species generated by NTP with the water and pollutants in water, they can find it in the appropriate numerous excellent publications [5–26].

Now we would like to add several points about sources generating the NTP at atmospheric pressure. In most cases, the NTP is created above/in the liquid or due to formation of a thin plasma layer on the liquid surface. The NTP methods are characterized by much lower energy consumption compared to other electro-physical methods based on energetic influence like shock waves in liquids. Increased attention to NTP methods has arisen since the late 1980s. Among the reasons for such close attention to the plasma–liquid systems, one may point not only to their lower energy consumption but to another important advantage of the NTP treatment of liquids — it is an environment friendly method. Because of all the reasons mentioned above, the NTP methods can be referred to as green technology and considered as serious competitive alternatives to the conventional bio-chemical methods, that either use different toxic additives for the liquid treatment or generate toxic or harmful byproducts in the treated water.

So far, numerous approaches have been developed for the NTP activation of liquids. Many known plasma–liquid systems are based on the use of transient streamer or spark electrical discharges which are generated

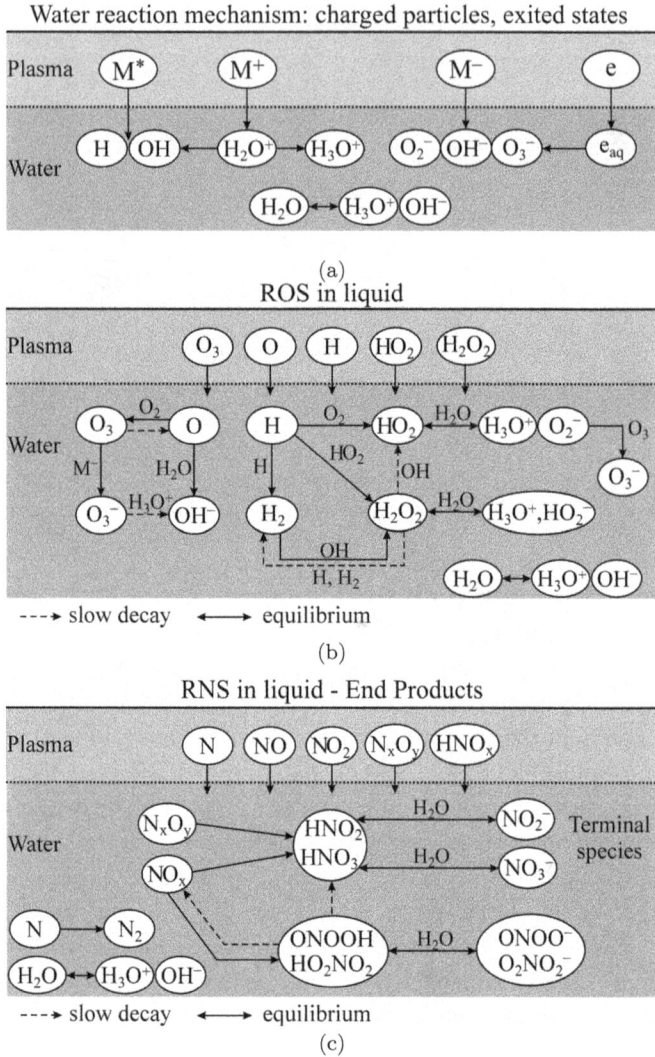

Fig. 4. Schematic diagrams of the most important reactions in liquid activated with reactive plasma species. (a) Ions, excited states in liquid. All charged species (and photons) pass directly into water that is initially at equilibrium: pH=7 and saturated with atmospheric gases. Energetically possible charge exchange and dissociative reactions are shown. (b) ROS in liquid. Neutral species enter the liquid at rates proportional to their Henry's law equilibrium constants. pH-dependent reaction rates are mediated by H_3O^+aq density. (c) RNS in liquid. Hydrolysis of N_xO_y and HNO_x species produce NO_3^- as terminal species on long timescales. Adapted from [4].

Fig. 5. Typical image of the lone streamer formed in plasma–liquid system of a pin-to-plane configuration in the motionless ambient air between the point electrode and surface of tap water. The gap length is 15 mm, and the amplitude of the pulsed voltage $U = 20\,\text{kV}$. With permission from [27]. © 2015 IOP Publishing.

above or directly in/on the liquid to be treated. These atmospheric pressure discharges exhibit themselves, equally above and in/on the liquid, in a form of bright and numerous short-living, chaotically spreading and branching thin current filaments, abundantly enriched with reactive plasma species.

Typical image of the alone streamer being formed between the point electrode of a positive polarity and tap water is shown in Fig. 5. This image was taken at long exposure time corresponding to full completion of the streamer breakdown between the pin and water. The streamer-to-spark transition did not happen because of the electric current limitation by deep water and external circuit. The streamer spatial structure consists of two parts: (a) volume streamer in a form of a single thin current filament shunting the space between the pin and water, and (b) numerous surface streamers contacting the water and spreading on its surface in radial directions from the volume streamer base. Eventually, many surface streamers form the plasma sheet tightly contacting the water.

In most cases, the pulsed electric discharges are used for creation of transient discharges designed for the liquid activation with reactive plasma species. Typical discharge schemes of the developed plasma–liquid systems and the results obtained with their usage are presented in many

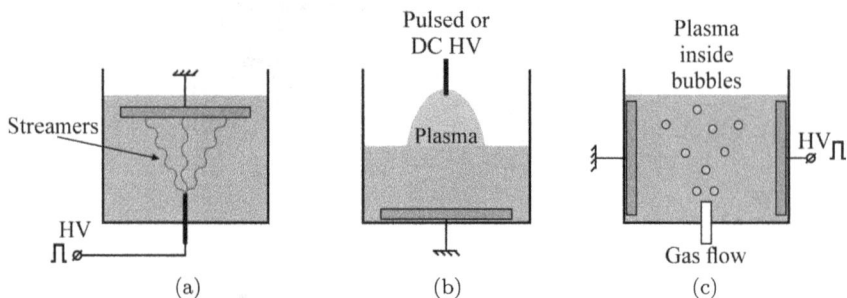

Fig. 6. Schematic depictions of some plasma sources creating the NTP: (a) Underwater; (b) overwater, (c) inside bubbles. Adapted from [5].

reviews and original publications [28–47]. For instance, Fig. 6 presents the sketches illustrating the schemes of plasma sources operating at atmospheric pressure and creating the NTP underwater, overwater and inside bubbles.

All the plasma devices operating at atmospheric pressure in air generate a suite of reactive oxygen and nitrogen species (RONS) that are biochemically active. However, the RONS generation is only the first step in plasma–liquid interaction. Next step is the fast RONS transfer into a liquid to be treated, where the conversion/oxidation of unwanted molecules takes place. The reactive species transportation from the NTP into the liquid is an important process influencing the treatment efficiency of each plasma–liquid system. Indeed, in the case of a discharge created in ambient air above the liquid, the formed radicals are transferred due to diffusion toward the liquid. However, the diffusion transport takes a long time that can exceed the lifetime of some reactive species. Therefore, many reactive species generated in NTP above a liquid disappear long before their arrival to the liquid. This issue will be discussed in detail in Section 3.

In contrast to the discharges created above a liquid, the NTP generation in/on liquid is a more effective approach. In this case, the streamers closely contact a liquid and therefore the formed reactive plasma species are effectively and quickly transferred into the liquid. That is why the main attention in this review will be paid to plasma–liquid systems generating NTP in/on liquid. Note that the possibility of streamer discharges creation in/on liquids and the plasma parameters (n_e and the reduced electric field E/N) in the formed streamers significantly depend on both the solution pH and the liquid conductivity. In particular, the creation of electrical discharges directly in a liquid requires the presence of a sufficiently high conductivity in the solution, and therefore such discharges are characterized by the relatively high electrical energy consumption to produce the reactive species. To clarify in more detail the mentioned discharge features, this chapter presents in what follows a review of the methods of streamers' formation in/on both dielectric and conductive liquids. Preferential attention will be paid to the physical aspects of streamers' formation.

2. Plasma–Liquid Systems for Generation of Streamers Above the Liquids

The simplest electrode configuration used for generation of the pulsed streamer discharges above the liquid is the needle-to-plate, or multiple needles-to-plate, where the liquid serves as an electrode (see Fig. 5). Such electrode systems are generally used for the NTP generation in motionless ambient air. However, due to the reason discussed in the next paragraph, these electrode configurations do not provide fast transfer of the created RONS into liquid. From this point of view, the electrode systems, which are blown through with the plasma forming gas, are more effective for the treatment of the liquid. In this case, the RONS are transferred not due to slow diffusion process but due to fast convection transfer by the gas flow forming plasma jets outside the electrode system.

In Figs. 7(a)–7(c), we present the images of plasma jets formed in N_2 flow that is blown out into free (i.e., without water) space. Plasma jet enters the ambient gas with different concentration of oxygen: (a) pure N_2, (b) 98% $N_2 + 2\%$ O_2, and (c) room air. One can see that the existence of oxygen in the ambience appreciably diminishes the length of N_2 plasma jet. One of the reasons is the fast disappearance of electrons from the plasma jet due to their attachment to electronegative oxygen molecules. The second reason is a high ability of oxygen to quench the electronically excited states of nitrogen. Figure 7(d) shows the image of argon plasma jet entering the ambient air. Figure 7(e) shows the image of argon plasma jet striking the liquid. One can see that high-velocity plasma jet ($V \approx 30\,\text{m/s}$) striking the liquid provides strong hydrodynamic effect leading to deformation of the liquid surface (i.e., formation of a deep recess) and to intensive turbulization of the liquid.

Another approach to treat a liquid by plasma assumes the usage of the flow of aqueous aerosols blown through plasma zone. These aerosols have to have small sizes, i.e., their typical radius R has to be in the range of tens or hundreds micrometers. A reason is that the amount of reactive

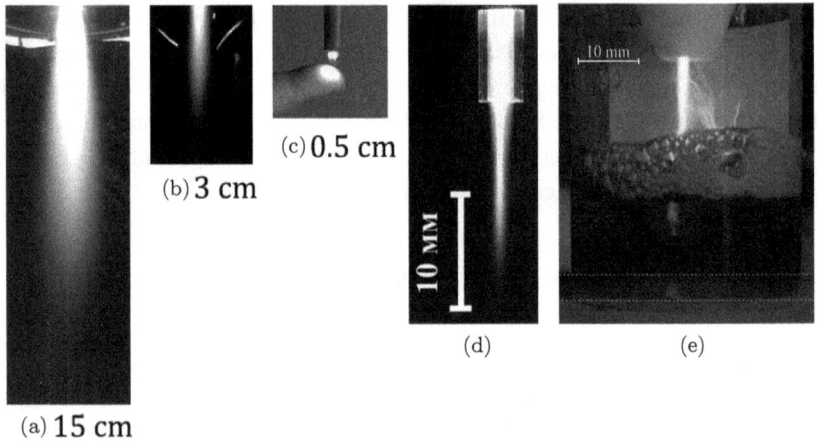

Fig. 7. Images of plasma jets formed by the flow of different gases entering a difference ambience. Here, L is the visual length of plasma jet. (a)–(c) Plasma forming gas in pure N_2 and ambient atmosphere: (a) pure N_2, (b) N_2 +2% O_2, (c) N_2+20% O_2 (air). (d) Image of argon plasma jet entering the ambient air; (e) image of plasma jet processing a liquid. One can see two things: A deep cone cavity in the liquid being formed by plasma jet, and the foam happening due to the liquid turbulization. Images (a)–(c) with permission from [48]. © 2015 De Gruyter. Images (d) and (e) from unpublished results by Akishev.

plasma species entering the droplet is proportional to the droplet surface, i.e., proportional to R^2. However, the amount of harmful contaminants inside the droplet is proportional to the droplet volume, i.e., proportional to R^3. It follows from this that the amount of reactive species per one contaminant in the droplet is proportional to R^{-1}. In other words, the smaller the aqueous droplet, the higher the positive effect from the plasma treatment.

An example of such an approach, based on the usage of corona discharge in the aerosol flow blown through the discharge zone, is shown in Fig. 8.

The images demonstrate the shape evolution of low-velocity air jet, $V \approx$ 1–3 m/s, passing through the negative corona and directed to the liquid. The air jet is enriched with tiny, 20–100 μm, aqueous aerosols that are the objects for charging and activation by the reactive species generated by corona. One can see that corona effectively influences the shape of the aerosol jet. Besides, as it was shown by Akishev *et al.* [50], the plasma destruction rate of bacteria in small water aerosol is very high. Additional contribution to the obtained high destruction rate can be associated with

(a)   (b)   (c)

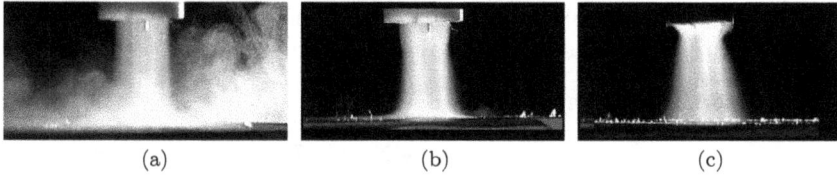

Fig. 8. Photos of the aerosol jet blown through the negative corona. Image of the jet when: (a) Negative corona is switched off; (b) and (c) negative corona is switched on. The gap length is $d_{a-c} = 20\,\text{mm}$ in all cases, while the corona current is: (a) $I = 0\,\mu\text{A}$, (b) $I = 5\,\mu\text{A}$, (c) $I = 220\,\mu\text{A}$. With permission from [49]. © 2009 IEEE.

the charging and strong polarization of small water droplets in the discharge electric field that is strong enough to provide the cell wall electroporation.

3. Efficiency of the Reactive Species Transfer into a Liquid in Different Plasma–Liquid Systems

Before the discussion on the reactive species which can be transferred into a liquid, we estimate the main gas-phase reaction pathways for the appearance of these species. In fact, the effective generation of reactive plasma species in streamers happens predominantly due to collision processes of energetic electrons with the molecules of ambient gas. In the humid air above the water, the main components determining the gas discharge properties and plasma-chemical processes are N_2, O_2 and H_2O. In such a case, the basic processes with participation of electrons promoting the formation of the primary reactive species are the following:

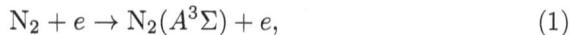

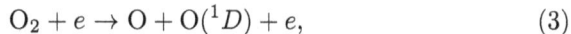

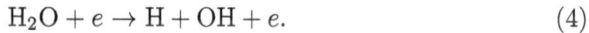

$$N_2 + e \rightarrow N_2(A^3\Sigma) + e, \tag{1}$$

$$O_2 + e \rightarrow O + O + e, \tag{2}$$

$$O_2 + e \rightarrow O + O(^1D) + e, \tag{3}$$

$$H_2O + e \rightarrow H + OH + e. \tag{4}$$

Here, $N_2(A^3\Sigma)$ and $O(^1D)$ are the excited metastable states of molecular nitrogen and atomic oxygen, OH is the hydroxyl radical, O and H are the non-excited atoms of oxygen and hydrogen, respectively. The above-listed species are just the main primary neutral reactive ones created in streamers.

After a fast transient discharge phase, responsible for the generation of primary reactive species, a slow phase of their further chemical transformations sets in. During this post-discharge phase, most of the primary reactive particles are converted into more stable species. The basic gas-phase chemical reactions governing this transformation look as follows:

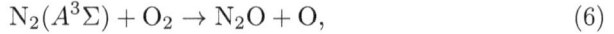

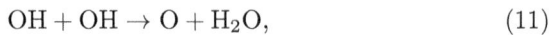

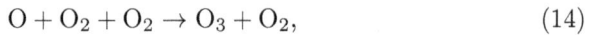

$$N_2(A^3\Sigma) + O_2 \rightarrow N_2 + O + O, \tag{5}$$

$$N_2(A^3\Sigma) + O_2 \rightarrow N_2O + O, \tag{6}$$

$$N_2(A^3\Sigma) + H_2O \rightarrow N_2 + OH + H, \tag{7}$$

$$O(1^D) + O_2 \rightarrow O + O_2, \tag{8}$$

$$O(1^D) + H_2O \rightarrow O + H_2O, \tag{9}$$

$$O(1^D) + H_2O \rightarrow OH + OH, \tag{10}$$

$$OH + OH \rightarrow O + H_2O, \tag{11}$$

$$OH + OH \rightarrow H_2O_2, \tag{12}$$

$$H + O_2 + M \rightarrow H_2O + M, \tag{13}$$

$$O + O_2 + O_2 \rightarrow O_3 + O_2, \tag{14}$$

$$O + O_2 + N_2 \rightarrow O_3 + N_2. \tag{15}$$

Here, M stands for either O_2 or H_2. At first glance, all reactive particles listed in the right-hand sides of the reactions, Eqs. (5–15), have a chance to arrive into the liquid. In fact, a more correct description of plasma chemistry of the secondary reactive species formation reveals that small amounts of such neutral particles, such as NO, HNO_3, HNO_2, etc., can appear as well. In addition, some ions in the plasma can also penetrate into the liquid but only in even smaller amounts. More detailed information on plasma chemistry in the NTP, including the ion reactions, can be found in [1, 26].

Let us compare the activation efficiency of a liquid due to volume and surface streamers. We will characterize this efficiency for each sort j of the reactive plasma particles separately, using the parameter $\theta_j = N^j_{abs}/N^j_{total}$. Here, N^j_{abs} is the number of reactive species j absorbed by the liquid, and N^j_{total} is a total number of reactive particles j produced by the volume or surface streamers.

We notice in advance that the efficiency of plasma activation by the volume and surface streamers cannot be the same. Two fundamental circumstances determine this difference. First, each reactive species formed by the plasma has its own characteristic lifetime, τ_j^*, determined mainly by its collision quenching or by the plasma-chemical reactions converting primary species into something else. Second, it is necessary to take into account that neutral reactive particles travel due to diffusion in any direction from the place of their appearance, in particular, toward the surface of the liquid to be treated as well. This means that there is a maximum distance R_j that can be traversed by the j-particle before its disappearance. The maximum distance can be estimated using the Einstein–Smoluchowski equation for the 3D case, in the form $R_j \approx \sqrt{6D_j\tau_j^*}$, where D_j is the diffusion coefficient of the j-particles. One can prove for the case of volume streamers of length L that the activation efficiency θ_j of j-particles may be estimated as

$$\theta_j \approx \frac{1}{2}\left(1 - \frac{L}{2R_j}\right), \quad \text{if } L \leq R_j, \tag{16}$$

$$\theta_j \approx \frac{1}{4}\frac{R_j}{L}, \qquad \text{if } L \geq R_j. \tag{17}$$

Expressions (16) and (17) show that the activation efficiency of any reactive species being produced by a volume streamer diminishes with increasing streamer length, i.e., with the increase of a distance between the liquid and the high-voltage pin.

One may also conclude from these expressions that the total number of j-particles absorbed by the liquid cannot increase further if the streamer length L exceeds the maximum distance R_j. In other words, the maximum amount of reactive species j transferred from the volume streamer into the liquid occurs when $L = R_j$. However, even in this case, the activation efficiency θ_j will be only about 25%. In contrast to volume streamers, the activation efficiency of surface streamers is larger and approximately equal to 50%. This is irrespective of their length, because they get much closer to the liquid than volume streamers do. As a result, the plasma sheet formed by surface streamers on the liquid is very thin, and therefore the reactive plasma species (even primary reactive particles) can be quickly transferred into the liquid.

Thus, based on the above-simplified analysis, one can state that the use of surface streamers is preferential for plasma activation of liquids as compared to volume streamers. This is the reason why, in this chapter, attention is drawn chiefly to the streamer discharges being formed in/on the liquid to be treated. However, before going on, we will discuss if there is the possibility to further increase the efficiency of a streamer discharge being formed above a liquid.

4. Increase of the Reactive Species Transfer into a Liquid by its Mechanical Agitation

One approach to increase the transfer of reactive species into the liquid consists in applying strong wave-like disturbances to the liquid surface by using, for instance, intense mechanical vibrations. We will demonstrate this effect in the case of streamer discharges above the agitated tap water. The general sketch of the plasma–liquid system with multi-pin AC streamer discharges, in ambient air, above the disturbed tap water is shown in Fig. 9.

Each pin-like electrode was stressed by a sinusoidal high-voltage with amplitude of about 10 kV and frequency of 50 Hz, loaded with an individual capacitor of 110 pF. The HV pin-like electrodes were disposed at a distance of 30 mm from each other. Tap water serves as the grounded electrode, and the water conductivity was $\sigma = 700 \, \mu\text{S cm}^{-1}$. The average distance between pins and water surface is 5 mm.

The water was agitated by intensive mechanical vibrations which were created by an horizontal back-and-forth movement of the dielectric vibrator. To increase the mechanical interaction between water and vibrator, the latter has a periodical structure made of spikes 6 mm in height. The periodical structure on the vibrator influences the landscape of the disturbed water surface as well.

As it turned out, both the magnitude of the disturbances on water surface and the image of surface landscape of the disturbed water depend

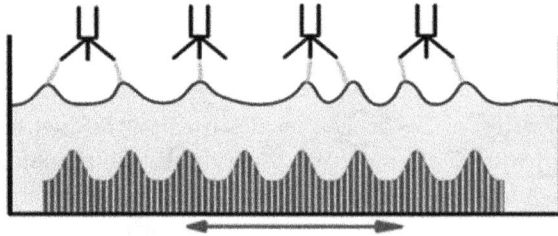

Fig. 9. Sketch of the setup of AC multi-pin streamer discharge in air above tap water agitated by strong mechanical vibrations in the horizontal direction. The vibrator is pictured in green. The bidirectional arrow shows the direction of horizontal periodical displacements of the vibrator with amplitude of 25 mm. Thickness of a water layer over the vibrator is approximately equal to the height of its spikes. Streamers (in magenta color) "feel" the water landscape and choose the shortest path between HV needles and the nearest hump on the waved water surface. With permission from [51]. © 2015 Research Group for Industrial Applications of Plasmas.

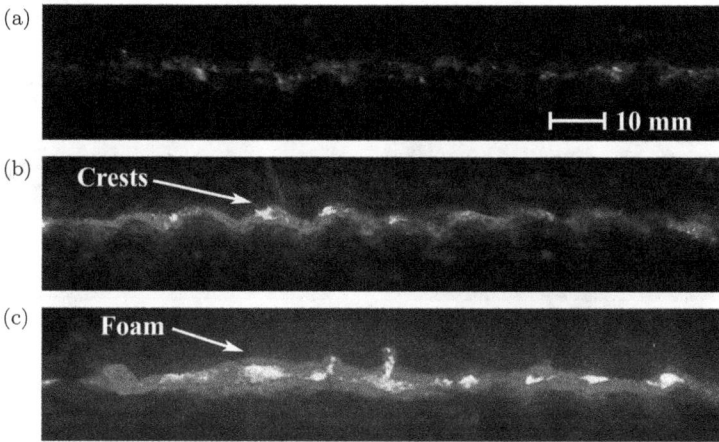

Fig. 10. Images of the surface landscape of the agitated water under different frequencies of the vibrator. (a) $F = 3.5\,\text{Hz}$, the surface disturbance amplitude is low, the regular surface structure is practically absent. (b) $F = 9\,\text{Hz}$, the regular surface wave amplitude is high, and the spatial period (12 mm) of the surface wave is equal to that of the vibrator structure. (c) $F = 12\,\text{Hz}$, the regular structure with a period of 12 mm disappears, but sometimes small droplets arise above water and a foam composed of small bubbles, of size 5–10 mm, appears abundantly on the water surface. With permission from [51]. © 2015 Research Group for Industrial Applications of Plasmas.

on the mechanical vibration frequency because the frequency determines the intensity of mechanical disturbances. The experimental data proving this statement are presented in Fig. 10.

Since the amplitude of the vibrator displacement is kept fixed, a higher vibrational frequency leads to an increase in the speed of the vibrator motion. One may see in Fig. 10(a) that the low-frequency mechanical vibration (i.e., slow moving vibrator) influences the water surface negligibly. The increase in frequency up to 9 Hz (Fig. 10(b)) changes the water landscape, leading to the appearance of regular wave-like structures on the water surface. The spatial period of this structure is 12 mm, that is equal to the spatial period of the spike vibrator structure. The strongest disturbance of the water surface was observed at a frequency of 12 Hz (Fig. 10(c)). In this case, the agitated water is turbulent and its surface looks similar to the surface of a sea during a strong storm. High-frequency agitation of the water leads to practical disappearance of the regular surface structure. Furthermore, small droplets above the water arise and a foam

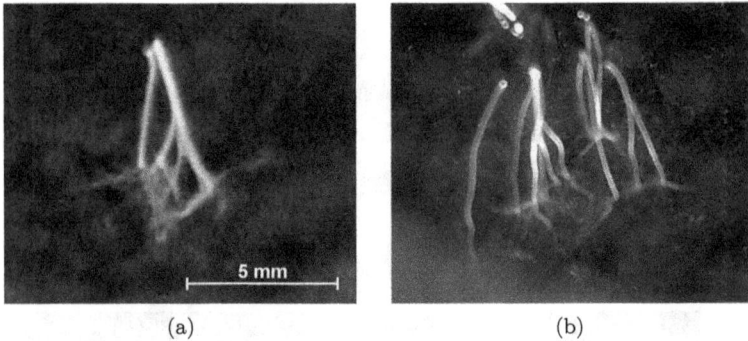

Fig. 11. Images of AC streamer discharges above the strongly agitated water surface, for the applied vibrator frequency $F = 12\,\text{Hz}$. (a) One may see the numerous surface streamers extending from the tops of irregular waves toward the bottom of valleys between them. (b) Streamers were formed simultaneously from different HV needles. With permission from [51]. © 2015 Research Group for Industrial Applications of Plasmas.

of small bubbles (diameter ranging from $3\,\text{mm}$ to $8\,\text{mm}$) appears on the water surface. It turns out that this regime is the most effective for plasma activation of liquid by streamers. The image of streamer discharges above the strongly agitated tap water is shown in Fig. 11.

The comments to Fig. 11 are as follows. Volume streamers "feel" the instant water landscape and choose the shortest path between each HV pin and the waved water surface. That is why they predominantly strike the wave tops but not the wave valleys. Contrarily, surface streamers broadly branch from the tops of irregular waves toward the bottom of valleys between them. If one compares Figs. 5 and 11, one may clearly realize that the area occupied by surface streamers, spreading from the base of the volume streamer on the agitated water (Fig. 11), is larger than that on the tranquil water (Fig. 5). Moreover, due to the existence of high crests and large bubbles on the surface of strongly agitated water, volume streamers are a bit shorter compared to those in a streamer discharge above the tranquil water. All the features mentioned above lead to a higher efficiency in plasma activation by streamer discharges above a strongly agitated water.

Another feature of this regime is that the tops and valleys on the disturbed water are distributed chaotically in space and time. This leads to the fact that the HV pins do not always generate streamers simultaneously (Fig. 11). This feature is illustrated schematically in Fig. 9. Nevertheless, the experiments showed that the average electric power consumed by the

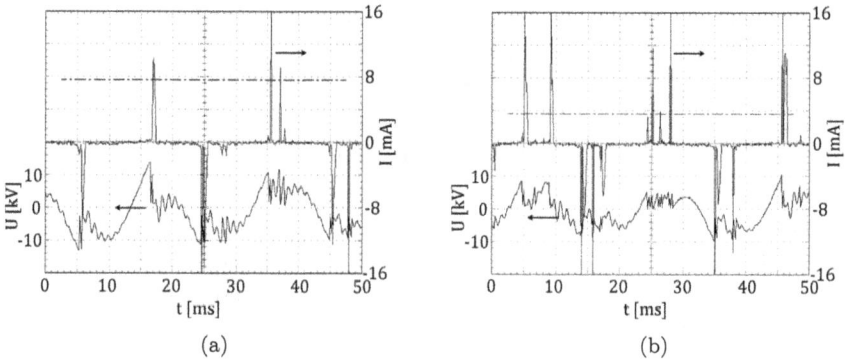

Fig. 12. Current I (upper panels), and voltage U waveforms (lower panels) of the AC streamer discharge: (a) Without mechanical vibration of the water. (b) With mechanical vibration of the water. Axes Scales: Time $[t] = 5\,\text{ms/div}$, current $[I] = 4\,\text{mA/div}$ and applied potential $[U] = 10\,\text{kV/div}$. The current pulses amplitudes in (b) are higher, and happen more frequently, than those in (a). With permission from [51]. © 2015 Research Group for Industrial Applications of Plasmas.

AC streamer discharge above strongly agitated water is approximately twice as higher as in the case of the tranquil water. A reason is that the amplitudes of current pulses in the case of the agitated water are higher and happen more frequently than in the tranquil water case (see Fig. 12).

5. Interaction of Streamers with Foam of Water Bubbles Filled with Air

Why is the knowledge about the interaction of streamers with foam of bubbles of great practical interest? By now the appropriate experiments were performed [21], which confirmed that the plasma–activation of liquid based on atmospheric pressure gas discharge in foam is indeed a very promising approach. The experiments proved definitively that the plasma–foam system is the most efficient one for hydrogen peroxide production in liquid phase. However, the physical reason providing a higher efficiency of reactive species generation by electric discharge in foam compared with other non-foam plasma–liquid systems has not been practically investigated. This is the first motivation for the following discussion on the interaction of streamers with bubble foams.

The second motivation sounds as follows. In many industrial processes based on the use of a liquid, foams appear. However, the presence of foams in products or processes is not always desirable. One of the reasons is that foaming reduces the productive volume and can even block the outlet of an installation. In spite of the great importance of foam control, it still remains an empirical art. The most commonly used methods for the foam destruction are the chemical ones based on the addition of antifoam agents. However, the chemical agents can result in significant unwanted side effects, in particular, deterioration in the quality of the final product. Physical methods for foam control are free of such effects. Physical methods include the use of ultrasound, thermal and electrical treatments. Electrical foam breakers are based on passing an electric discharge through the foamy region to break up the foam.

One of the first studies on the control of the water foam treatment by high-voltage discharge pulses was done in [52]. The authors claimed that pulsed discharges lead to foam destruction due to the bubbles' explosion initiated by streamers penetrating through the bubble walls. In fact, there were no experimental observations in [52] proving such streamer–bubble interaction. Because of the importance in understanding how the streamer interacts with a bubble, more detailed information on the bubble burst induced by streamer strike is presented in the following.

To clarify the role of bubbles in plasma activation of the agitated water by streamers, the issue of streamers' interaction with a foam consisting of small (average diameter was ≤2 mm) and large (average diameter was about 5 mm) bubbles, or a mixture of small and large bubbles, has been studied [27]. The size of large bubbles corresponds to that of bubbles created by mechanical vibrations. The experiments have revealed that large bubbles are more sensitive against the streamer strike compared to small ones (Fig. 13).

In the case of a foam consisting of a mixture of small and large bubbles, the streamer strike destroys predominantly the large bubbles, but the small bubbles practically survive altogether (Fig. 14).

It is interesting to note that surface streamers spread mainly on the surface of large bubbles rather than on small bubbles. Besides, a single streamer activates many bubbles in a foam at the same time. Despite the fact that the bubble is destroyed after the streamer strike, the reactive species, that were effectively transferred from the streamer into the bubble thin wall, will not diffuse to the ambient air: They remain in the liquid of the destroyed bubble and therefore increase total water activation. Hence, strong mechanical agitation of water increases the efficiency of its plasma activation by the streamer discharge above the water.

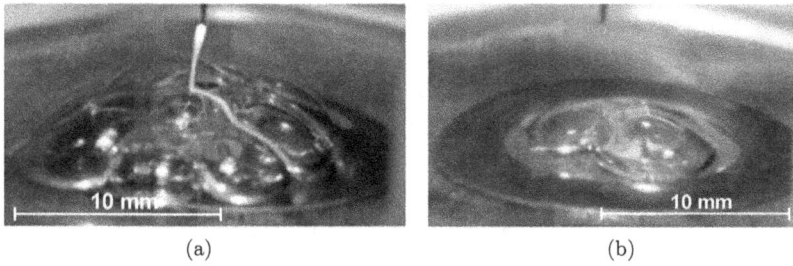

(a)　　　　　　　　　　　　　　(b)

Fig. 13.　Image of a foam consisting predominantly of large bubbles: (a) Under streamer striking; (b) after streamer striking. The foam is being formed on thin water layer above the grounded metallic disc. In (b), most of the large bubbles are destroyed, and the foam area occupied by large bubbles gets drastically reduced. With permission from [27]. © 2015 IOP Publishing.

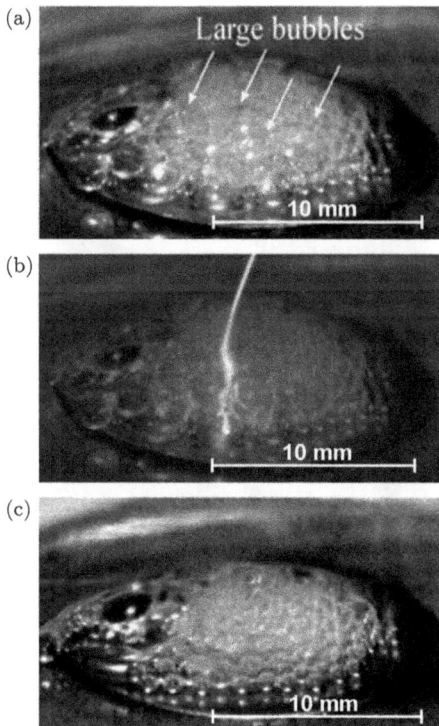

Fig. 14.　Image of the foam consisting of small and large bubbles: (a) Before streamer striking; (b) under streamer striking; (c) after streamer striking. Average diameter of large bubbles is about 5 mm. After the streamer strikes, most of the large bubbles are destroyed, but the amount of small bubbles remains practically the same. With permission from [27]. © 2015 IOP Publishing.

In fact, in a real plasma–foam system, each streamer interacts simultaneously with a huge number of bubbles of a foam. This circumstance leads to the overlapping in time of many individual streamer–bubble interactions and strongly complicates a possible clarification of the important question: How does the streamer influence a foam? In order to avoid the mentioned problem, the experimental conditions for streamer–foam interaction were simplified, firstly by dealing with an isolated bubble [27]. To imitate a foam, a bubble was formed on a liquid yielding a floating bubble. Secondly, a pulsed discharge was used to generate a single streamer, which always originates from a fixed space point above the bubble.

We note that numerical calculations on the interaction of a streamer with a bubble on a liquid were done by Babaeva and Kushner in [53, 54], but they dealt with a tiny liquid bubble, the diameter of which ($80\,\mu$m) was smaller compared to the streamer diameter. However, in [27] the opposite situation was obtained in which the bubble diameter (≈ 1 cm) drastically exceeded the streamer diameter ($\approx 100\,\mu$m).

A general scheme of the experimental set-up used in [27] is shown in Fig. 15. The pin-to-plane electrode geometry was used to study the interaction of the single streamer with the floating bubble, artificially formed on the liquid surface. The high-voltage needle electrode was a stainless steel wire of 1 mm diameter with a sharpened tip. The grounded electrode was an aluminum plate 35 mm in diameter. This electrode was covered with a thin layer, about 2–3 mm, of liquid. Plasma-forming gas was ambient air at atmospheric pressure. An isolated bubble floating on a liquid was formed by a slow injection of a small amount of air through a narrow nozzle at the center of the aluminum plate. To do that, a syringe

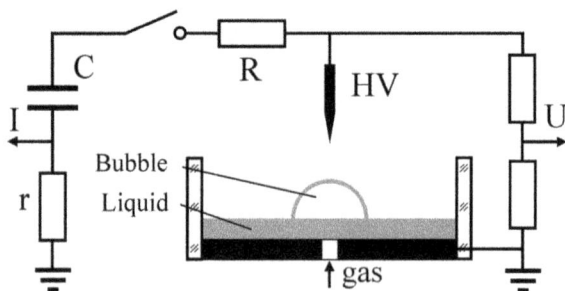

Fig. 15.   Schematic representation of the set-up used for searching for a streamer–bubble interaction. With permission from [27]. © 2015 IOP Publishing.

was used. The diameter of a bubble basis ranged from 5 mm to 15 mm. The length of the inter-electrode gap could be tuned in the experiment, and the distance between the point electrode and the bubble top varied in the range 2–10 mm.

The experiments were made with the use of tap water and transformer oil. The chemical reason why tap water and transformer oil were used for investigation is as follows. In many cases, liquids which are to be treated in plasma-foam systems contain oil or fat as additives or contaminants, which float on the water. A physical reason why these liquids were chosen for study was that they have completely different electrical properties. The electrical conductivity σ and dielectric permittivity ε of tap water and transformer oil are: $\sigma_{water} \simeq 7.5 \times 10^{-4} \, \mathrm{S\,cm^{-1}}$ and $\sigma_{oil} \simeq 1.5 \times 10^{-13} \, \mathrm{S\,cm^{-1}}$, $\varepsilon_{water} \simeq 82$ and $\varepsilon_{oil} \simeq 2.2$. The typical values of surface tension γ for tap water and transformer oil are also quite different: $\gamma_{water} \simeq 73 \, \mathrm{mN\,m^{-1}}$ and $\gamma_{oil} \simeq 46 \, \mathrm{mN\,m^{-1}}$. The threshold electric field strength for breakdown of a fresh transformer oil is about $105 \, \mathrm{V\,cm^{-1}}$ and diminishes approximately by half for oil with water vapor contaminants.

The laboratory-self-made DC high-voltage power supply was used to charge a capacitor C, of capacitance $C = 500 \, \mathrm{pF}$. This power supply provided an output voltage of a positive polarity and magnitude up to 20 kV. The capacitor high voltage was quickly transferred to the point electrode through the thyratron switch. Delivering of a positive HV pulse to the needle leads to streamer breakdown in a pin-to-plane gap. The discharge current was controlled by a ballast resistor with a variable resistance of $R = (0\text{--}5) \, \mathrm{k\Omega}$. The pictures illustrating a bubble–streamer interaction were taken by a Canon EOS 550D digital camera.

6. Interaction of an Isolated Positive Streamer with a Single Water Bubble Floating on Water

Experimental results on the interaction of an isolated positive streamer with a single water bubble floating on the water are presented in Figs. 16 and 17. The following conclusions can be drawn.

First, the data in Fig. 16 clearly show that the streamer striking on a water bubble does not burn/perforate instantly a thin bubble wall at the striking point. This is the reason why streamer slides down to the water over the external bubble side, but does not propagate through the bubble along the shortest path from the needle to water. It is interesting to note that a streamer does not strike a bubble perpendicularly to its surface.

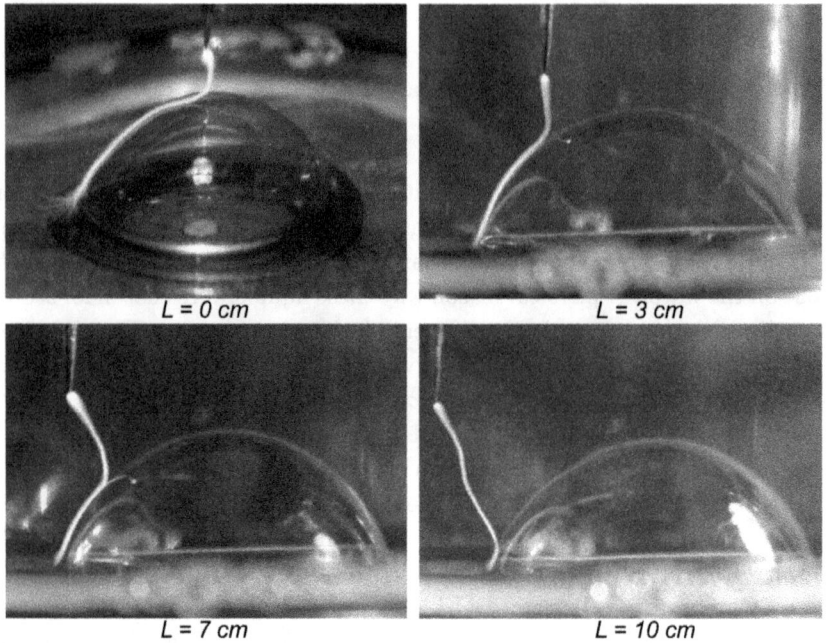

Fig. 16. Set of streamer images at different distances L of the needle from the bubble top. Diameter of the bubble base is 15 mm. The bubble is filled with ambient air. Water depth under the bubble is 3 mm. The 'image' of short and thin streamer offshoots can be seen on all the photos, in fact, it is the reflection of the volume streamer from the bubble surface. The gap between the HV needle and the water surface is 9 mm, the applied voltage $U = 15.2$ kV, ballast resistor $R = 5$ kΩ. With permission from [27]. © 2015 IOP Publishing.

A possible reason is that the water bubble is not an ideal electrical conductor. Nevertheless, due to the finite conductivity of the water, the bubble wall enables to create the electrostatic image of the charged streamer head. The charge of this electrostatic image has an opposite sign compared to the streamer head charge. Therefore, the streamer head is 'attracted' by its electrostatic image to the bubble even if the HV needle is far away from the bubble.

Second, the pictures in Fig. 16 show that water bubbles provide a transformation of volume streamers into surface streamers when the volume streamers reach the bubble wall. Indeed, one can see in Fig. 16 that the streamer striking into the bubble top fully becomes a surface streamer sliding on the external surface of the bubble wall.

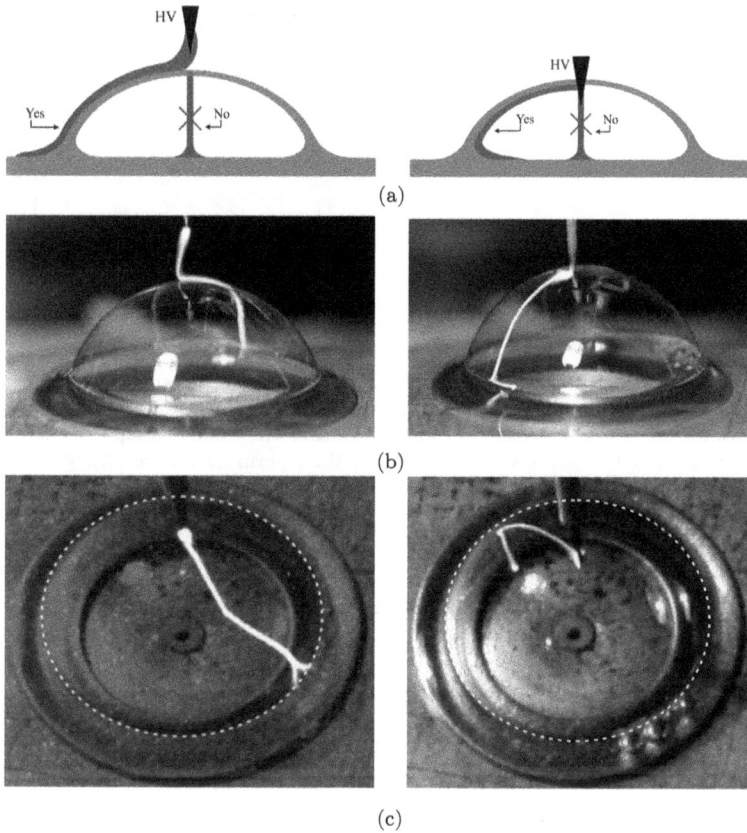

Fig. 17. Images illustrating the interaction of a water bubble filled with air with positive streamer created outside the bubble (left column) and inside the bubble (right column). (a) $U = 15.2\,\text{kV}$ sketch elucidating a general idea; (b) side view; (c) top view (the dotted circle shows the bubble boundary). Left column: the HV needle is above the bubble; the gap between the needle and water surface is 9 mm; diameter of the bubble base is 10 mm. Right column: the HV needle is immersed into the bubble; diameter of the bubble base is 15 mm; the needle–water gap is 7 mm. With permission from [27]. © 2015 IOP Publishing.

Third, the streamer created outside the water bubble does not penetrate into the bubble. The same behavior demonstrates the streamer created inside a bubble, i.e., this streamer cannot come out of a bubble. In other words, the externally created streamer never enters in water bubble, and the internally created streamer never comes out of the water bubble. Even more surprisingly, the internally created streamer does not propagate along the

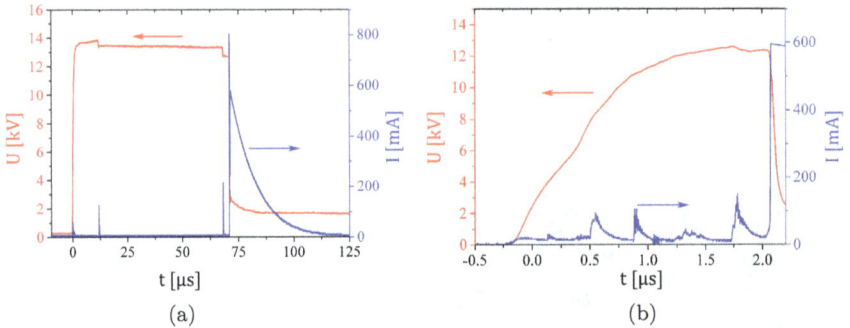

Fig. 18. The voltage and current waveforms of a transient streamer discharge in the cases (a) Without a bubble and (b) with a bubble floating on the water. In both cases, the distance between the needle and water surface is the same and equal to 9 mm; the diameter of the bubble base is equal to 15 mm. With permission from [27]. © 2015 IOP Publishing.

shortest path, through gas from the HV needle to the water surface. Rather, it propagates down to the water along the inner bubble surface. Both the experimental results proving these statements and sketches elucidating the stated general idea are presented in Fig. 17.

Fourth, in their turn, the floating water bubbles influence the electrical breakdown characteristics of a streamer pin-to-plane discharge. Figure 18 shows the voltage and current waveforms of the transient streamer discharge with and without a bubble floating on the water. One can see in this figure, the existence of the water bubble in the pin-to-plane gap drastically accelerates the appearance of streamer breakdown. Due to the occurrence of many pre-breakdown low-current pulses, a time lag for an appearance of full breakdown drastically diminishes from 70 μs down to 2 μs. This circumstance has a great importance for real plasma-foam systems because the breakdown delay decrease allows both: An increase in the repetition frequency of the pulsed voltage feeding the discharge in the presence of foam, and in the total efficiency of the plasma-foam system.

7. Positive Streamer Interaction with an Isolated Oil Bubble Floating on Oil

The existence of an oil film and oil bubbles on a liquid to be treated is a typical situation for industrial liquid waste. Therefore, there is a need to find out the behavior of the streamer initiated above the oil bubble.

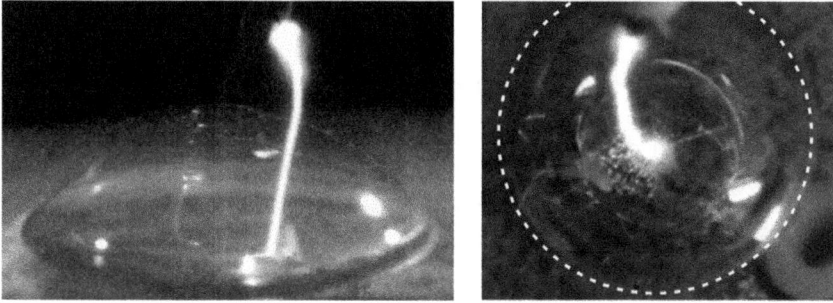

Fig. 19. Images illustrating the interaction of oil bubble with a positive streamer created outside the bubble. The bubble is filled with air and floats on transformer oil with a depth of 2 mm. The left picture is the side view; the right picture is the top view. The dotted circle shows a bubble base boundary. The diameter of the basis is 10 mm; the height of the bubble is 3.4 mm. The air gap between the HV needle and the oil surface is 5 mm. The applied voltage is $U = 13\,\mathrm{kV}$. With permission from [27]. © 2015 IOP Publishing.

As it turned out, the streamer breakdown of a pin-to-plane air gap, with an isolated oil bubble floating on a thin layer of oil, exhibits a different behavior compared to that in the presence of a water bubble. In this case, the streamer does not slide on the bubble surface but quickly perforates it, penetrates into the oil bubble and completes a full pin-to-plane streamer breakdown already inside the bubble. The streamer breakdown completion is followed by a spark. The images in Fig. 19 illustrate this feature of the streamer–oil bubble interaction. One can see in the right picture of this figure (top view) that the cathode end of the streamer definitely has struck the oil surface inside the bubble.

Based on the experimental data, one can state that the pin-to-plane breakdown in the presence of an oil bubble can be divided into three stages. The first stage corresponds to fast partial breakdown between HV needle and a bubble surface. During the second stage, the perforation of the bubble wall happens. After that, the third stage sets in, at the very beginning of which the fast streamer breakdown inside the oil bubble occurs. Simultaneously, the electric breakdown of a thin oil film on the grounded electrode happens as well. The end of the third stage corresponds to the transition of the streamer to spark, which completes the full pin-to-plane breakdown. However, the duration of each stage is not fixed but exhibits stochastic scatter at the same applied voltage and the bubble diameter. This scatter is shown in Fig. 20. The time spent by the streamer for the perforation of the oil bubble fluctuates a bit at the given applied voltage.

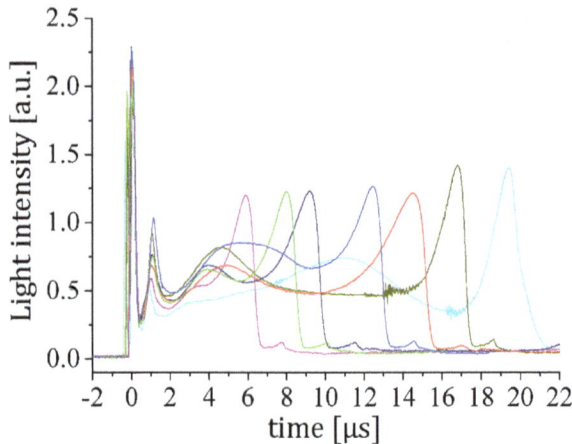

Fig. 20. The temporal behavior of the total light emitted in the course of pin-to-plane breakdown in the presence of oil bubble situated under the HV needle. Every color curve corresponds to separate breakdown triggering and has three well-pronounced sharp peaks. The oil bubble is filled with the air and floats on transformer oil with a depth of 2 mm. The bubble base diameter and the bubble height are 10 mm and 3.4 mm, respectively. The air gap between the HV needle and oil surface is 5 mm, and $U = 13$ kV. With permission from [27]. © 2015 IOP Publishing.

The duration of the second stage decreases from 500 ns to 50 ns if the applied voltage increases. The same effect occurs if the ballast resistance decreases. As for the third stage, its duration fluctuates markedly within the limits of $(5–20)\,\mu$s even at the fixed applied voltage.

Figure 20 helps to elucidate some details of the three-stage breakdown process. Indeed, it shows the temporal behavior of total light emission recorded by a photomultiplier in the course of the three-stage breakdown. Each color curve corresponds to separate breakdown triggering and has three well-pronounced sharp peaks. The first splash of light is connected with the first stage. The completion of the second stage and the very beginning of the third stage correlates with the second light splash. The full breakdown completion correlates with the last splash of light corresponding to a bright spark. The full breakdown leads to a dramatic drop down of the discharge voltage to the level of (300 ± 50) V.

8. Dynamics of the Water and Oil Bubble Destruction by Positive Streamers

A streamer striking into bubbles leads to their breaking. The information on dynamics of the bubble destruction induced by streamer striking has a great importance for the development and optimization of the electric methods for the foam control. The proper experimental data are presented in Fig. 21. Results in Fig. 21(a) are for reference [55], while Figs. 21(b) and 21(c) display the size of the water and oil bubbles, respectively, during their decay.

All of the pictures of the collapsing water bubble were taken with exposure time of about 0.2 ms. Shooting was carried out with an external bright illumination provided by a blue laser. The opaque screen in the form of a bubble cross-section, but a bit bigger in size, was used to diminish the parasitic light collected by the camera. The arrows in Fig. 21(b) show the instant boundary of the rest of the water bubble in the course of its destruction.

The images from [55] demonstrate the glycerol–water bubble collapse induced by the mechanical perforation of a bubble at its top. This example shows the distinction in the dynamics of the spontaneous bubble destruction induced by different methods. One can see in Fig. 21(a) that the initial small hole grows spontaneously in size, by increasing its diameter in an axially symmetric way. The disappearance of the thin wall of a bubble transforms itself into a liquid ring which thickens and slides down to the water.

A water bubble destruction induced by streamer happens in a different way. The duration of the streamer–water bubble interaction does not exceed 30 μs, but the total time for the bubble destruction is about 2 ms. It means that the water bubble collapse, induced by the streamer, occurs further without the influence of any other streamer. Such a scenario is analogous to that shown in Fig. 21(a).

However, there is an essential dissimilarity between these scenarios due to a strong distinction in the geometrical form of the initial damage caused to the bubble by the streamer. In the case of Fig. 21(a), the initial damage is a dot prick at the bubble top. In the case of Fig. 21(b), the initial damage is more serious: It is a meridional rip along the full length of line where the streamer touches the bubble. As a result, the water bubble destruction has: First, no axial symmetry and, second, it happens a bit faster. This statement is proved in Fig. 21(b), showing the development in time of the bubble destruction induced by a streamer.

The sequence of images in Fig. 21(c) illustrate the development of the oil bubble destruction induced by high-current streamer (spark) strike. As a result, the destruction of the oil bubble really looks like a burst or explosion, where the typical time for destruction is practically reduced by a factor of 100 compared to that for a water bubble. The main reason of the explosion-like destruction of an oil bubble is a fast streamer-to-spark transition that

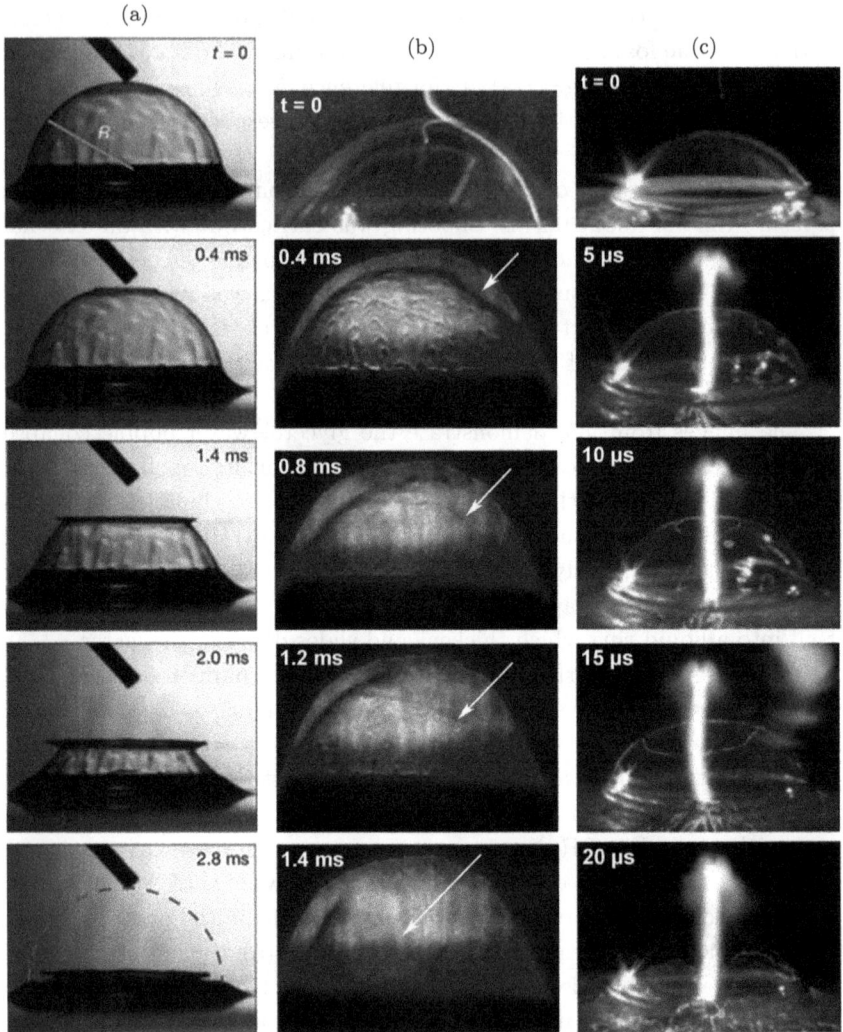

Fig. 21. Images (disposed in columns) illustrating the induced collapse of different bubbles. (a) The glycerol-water bubble filled with air, where an axially symmetrical collapse is induced by mechanical perforation of a bubble at its top. With the kind permission from [55]. © 2010 Springer Nature. (b) The water bubble filled with air, where an asymmetrical collapse is induced by a side striking of a positive streamer. (c) The oil bubble filled with air, where an almost axially symmetric collapse is induced by high-current streamer/spark. Shooting of an oil bubble collapse was carried out without external illumination because sparks produce their own bright light. The experimental conditions in (b) are the same as in Fig. 16, while those in (c) are the same as in Fig. 20. With permission from [27]. © 2015 IOP Publishing.

provides a hydrodynamic effect, i.e., the fast rise of gas pressure inside the oil bubble caused by strongly overheated spark, where the gas temperature in the spark reaches 5000–6000 K [56, 57]. It leads to the fact that the bubble becomes swollen and the size of the hole in the bubble quickly increases with time until full bubble burst occurs.

9. Possible Mechanism of Bubble Erosion by a Streamer

In fact, the destruction of a water bubble by a surface streamer is not a burst or explosion. This is a slow process of the bubble wall shrinkage which is controlled by the forces of the surface tension, but not by the higher gas pressure inside the bubble. Indeed, the pressure drop ΔP across a bubble wall is estimated to be $\Delta P \approx \gamma_{\text{water}}/R$, where γ_{water} and R are the surface tension and radius of the water bubble, respectively. In our case, $R \approx 0.5$–0.75 cm, which leads to the magnitude of $\Delta P \approx 10 \, \text{N m}^{-2}$. Such a pressure drop is very low compared to the ambient atmospheric pressure $P \approx 105 \, \text{N m}^{-2}$, and it cannot therefore provide any essential gas flow from a bubble in the course of its slow destruction.

Let us discuss the possible mechanism providing the perforation of the water bubble by a positive streamer. The slow energy release provided by the streamer, with a characteristic time of about 30 μs, cannot lead to the formation of shock waves from the streamer. Therefore, there are no mechanical forces which can influence the bubble up to its local perforation. However, the streamer is characterized by a high gas temperature of about 1500±300 K [58, 59]. This circumstance leads to the reasonable conclusion that the streamer sliding on a water bubble causes a local heating of a thin bubble wall up to its evaporation. Indeed, the processing of I-V characteristics shows that the amount of energy ε_{SS} released by the surface streamer is around 20 mJ. Approximately half of this energy can be dissipated to the heating and evaporation of the narrow strip in the bubble wall adjoining the surface streamer,

$$\chi_w \, \rho_w \, \ell_w \, d_w \, \Delta_w \leq \frac{1}{2}\varepsilon_{\text{SS}}, \tag{18}$$

where $\chi_w = 2256 \, \text{J g}^{-1}$ is the specific heat of water vaporization, ρ_w is the mass density of water ($\rho_w = 1 \, \text{g cm}^{-3}$), ℓ_w is the water strip length (in the experiments, this length for the isolated bubble does not exceed 1 cm), d_w is the width of the water strip (this width is approximately equal to the streamer diameter, i.e., $d_w \approx 10^{-2}$ cm) and Δ_w is the thickness of a water bubble wall.

From inequality Eq. (18) it follows that the bubble wall thickness does not exceed $5\,\mu$m. This thickness agrees with an estimation of depth $\delta \simeq 4\vartheta\tau \leq 5\,\mu$m to which the temperature propagates into the water wall during its slow heating (here $\vartheta \approx 1.7\,10^{-7}\,\mathrm{m^2\ s^{-1}}$ is the thermal diffusivity of water at $100°$C, $\tau \approx 30\,\mu$s is the duration of streamer energy release). The estimated bubble thickness is also not far from the value $\Delta = 11\,\mu$m, reported in [55]. Inequality in Eq. (18) allows us also to understand why the small bubbles in foam are more stable against streamer striking. A reason is that the surface streamer simultaneously covers several small bubbles. In this case, the total length of the water strip adjoining the streamer is longer compared to that for the large isolated bubble. Consequently, it leads to an increase in the amount of water adjoining the streamer, and therefore the energy released by the streamer will not be enough to evaporate this liquid. The higher stability of small bubbles against streamer striking is the unwanted effect for the streamer control of foam, but it does not influence fast transfer of reactive species from the surface streamer into the water.

The elapsed time τ for the axially symmetric bubble destruction induced by the small initial perforation can be estimated as [55],

$$\tau \approx \sqrt{\rho R^2 \Delta / \gamma}, \qquad (19)$$

where ρ and γ are the density and surface tension of a liquid, R and Δ are the radius and thickness of a bubble wall. The elapsed time for water bubble ripped by streamer (see Fig. 21) is about $1.4\,$ms, which is a bit lower compared to that (about $2\,$ms) estimated by the relation Eq. (19), with $R \approx 1\,$cm and $\Delta \approx 5\,\mu$m. This distinction can be attributed to both a large size of the asymmetrical initial streamer erosion of the bubble (in the form of the meridian rip) and the thinner bubble wall.

10. Numerical Calculations of the Surface Streamer Propagation on Bubbles

In principle, the surface streamer can slowly propagate on the water if the parameters of streamer and liquid meet this condition,

$$E_{\mathrm{sh}} \approx E_{\mathrm{sb}} \frac{\sigma_s}{\sigma_\ell} \geq E_{\mathrm{s}}^*, \qquad (20)$$

where E_{sh} is the electric field strength at the streamer head, E_{sb}, the one in the streamer body, and E_{s}^*, the critical electric field magnitude at the streamer head providing its propagation. Here, σ_s and σ_ℓ are the conductivities of the streamer and liquid, respectively. An estimation of E_{sh} can be derived from the continuity of the conductive current around the streamer head.

Actually, the magnitudes of E_{sb}, E_{s}^* and σ_s do not depend strongly on the experimental conditions, as compared to the liquid conductivity σ_ℓ, which can vary over a broad range. As a rule, $E_{\mathrm{sb}} \ll E_{\mathrm{s}}^* \approx 2 \cdot 10^5 \,\mathrm{V}$ cm^{-1} [55], therefore it is not always possible to satisfy the strong condition of Eq. (20). However, the necessary condition for streamer propagation is significantly weakened for the case of streamer propagation on a thin liquid layer. In such a case, this condition looks as follows:

$$E_{\mathrm{sh}} \approx E_{\mathrm{sb}} \frac{\sigma_s}{\sigma_\ell} \frac{d_s}{\Delta} \geq E_{\mathrm{s}}^*, \tag{21}$$

where d_s is the streamer diameter and Δ, the thickness of a liquid layer. For the case of water bubbles, the inequality $d_s \gg \Delta$ is fulfilled. This is a reason why the propagation of surface streamers on water bubbles is easily realizable.

The calculated electric current and electric field distributions around the surface streamer moving on the water bubble are shown in Fig. 22. In this case, the resistance of surface streamer per unit length is lower compared

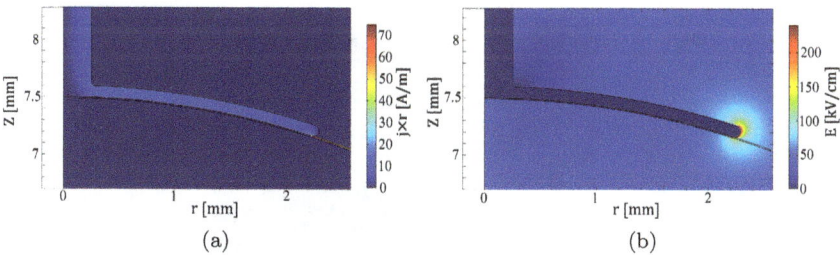

(a)   (b)

Fig. 22. The calculated distributions for: (a) The reduced electric current ($\mathbf{j} \cdot \mathbf{r}$), and (b) electric field. The distributions are obtained in the liquid and around the moving streamer on the bubble at the time $t = 0.3\,\mu s$, after applying the external voltage of amplitude $20\,\mathrm{kV}$. The surface streamer resistance per unit length is lower than that of the bubble wall adjoining the streamer. The electric current flows within the streamer body, and enters the liquid in the vicinity of its head (a). This circumstance leads to a high concentration of the current at the head and to strong electric field creation in this region (b), providing the streamer propagation. Coordinates z and r are measured from the bubble base and symmetry axis. With permission from [27]. © 2015 IOP Publishing.

to that of the bubble wall adjoining the streamer. The pictures correspond to the time $t = 0.3 \, \mu s$ after applying the voltage of $20 \, kV$. One can see in Fig. 22(a) that the conductive electric current flows from the HV pin within the streamer body but not within a bubble wall. A reason is that the electric field in a bubble wall under the streamer body is very low (see Fig. 22(b)). This is why the electrical breakdown of the water bubble by the streamer is absent. The current enters the liquid bubble wall in the vicinity of the streamer head and flows further toward the ground within the thin bubble wall. The strong concentration of the electric current at the streamer head leads to formation of the high electric field $E > E_s^*$ around the head that provides the streamer propagation. The image in Fig. 23 shows the calculated distribution of electric potential around the surface streamer shortly after the volume streamer strikes a water bubble.

The interaction of streamer with an oil bubble happens in another scenario. In contrast to the water bubble, a strong electric field across the oil bubble wall evolves rapidly. As a result, the critical magnitude E_{oil}^*, needed for the electrical breakdown of the bubble wall, is reached before the electric field at the streamer head can reach the critical value E_{str}^* required for streamer propagation. This is the reason why a surface streamer does not propagate on the oil bubble after it's striking by the volume streamer. The electrical breakdown of the bubble wall will occur first, then followed by the perforation of a bubble and penetration of the streamer inside the

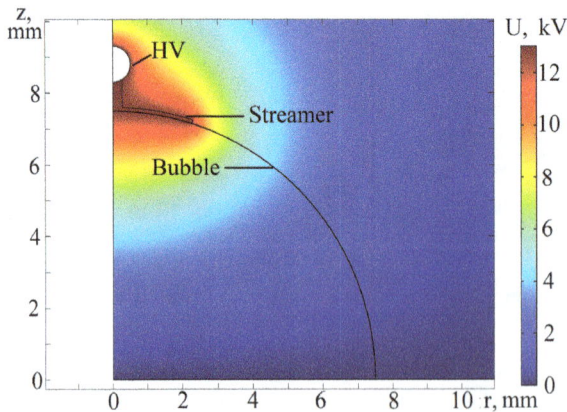

Fig. 23. The calculated electric potential distribution around a streamer after it strikes the water bubble. Coordinates z and r are measured from the bubble base and symmetry axis. With permission from [27]. © 2015 IOP Publishing.

bubble. Only then, a fast streamer transition into spark occurs, which is followed by the bubble burst.

11. Composition of Active Plasma Species Generated by a Surface Streamer on a Water Bubble

Let us discuss the efficiency of plasma activation of water bubbles by surface streamers. If we take into account both the high gas temperature in the surface streamer and high water vapor concentration inside the streamer due to evaporation of water from the bubble wall, the production of ozone by the streamer will be appreciably supressed. One can therefore state that the O, OH, HO_2 radicals and stable H_2O_2 hydrogen peroxide prevail in the composition of the primary reactive species produced by a streamer during the discharge phase. Hereby, in the case of high concentration of water vapor surrounding the streamer, the leading reactions forming the secondary reactive species in the post-streamer plasma will be as follows:

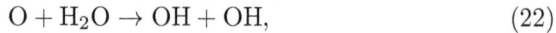

$$O + H_2O \rightarrow OH + OH, \tag{22}$$

$$OH + OH \rightarrow H_2O_2, \tag{23}$$

$$HO_2 + H_2O \rightarrow H_2O_2 + OH. \tag{24}$$

The final product of these reactions is a hydrogen peroxide H_2O_2 with a (up to a factor three) higher concentration compared to that in post-streamer plasma with a low concentration of water vapor. Due to the very long lifetime of H_2O_2, these species have a chance to travel from post-streamer plasma to the liquid and be absorbed. This statement was proved in [27] by numerical calculations on chemistry kinetics of the reactive species in post-streamer plasma rapidly enriched with a water vapor at high concentration. The set of chemical reactions is based on those listed above. The results obtained are presented in Fig. 24.

One can see in Fig. 24 that indeed the final product in post-streamer plasma is H_2O_2, the concentration of which exceeds its initial concentration by factor of 3. The offered mechanism of the efficiency increasing is based on fast evaporation of a narrow water strip of the bubble wall tightly contacting the surface streamer. This evaporation leads to the diffusion of the water vapor into the body of surface streamer. Larger (by a factor of 10 or more) and faster enrichment of the streamer plasma with water vapor leads to the occurrence of all the processes mentioned above. This mechanism can be

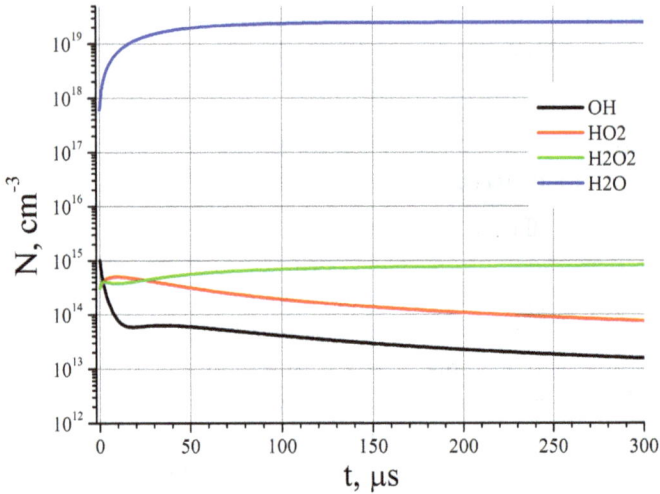

Fig. 24. Time behavior of the reactive species in a post-streamer plasma rapidly enriched with water vapor at high concentration. The enrichment occurs due to fast evaporation of the narrow water strip of the bubble wall tightly contacting the surface streamer. With permission from [27]. © 2015 IOP Publishing.

applied for the explanation of the higher efficacy of plasma–foam systems in the activation of a liquid reported in [21].

12. Plasma–Liquid Systems for Generation of Streamers in/on the Liquids

Other approaches for activation of liquids assume the use of transient streamers or spark discharges generated within the liquid. In this case, gaseous bubbles play a crucial role in the generation of such discharges. One can divide these discharges into two groups:

First group: It consists of transient discharges which form tiny bubbles themselves. A reason is that bubbles can pre-exist in the near electrode region due to dissolved gases even for nanoseconds voltage pulse widths. For large pulse widths (from sub-microseconds to DC), especially in high conductive water solutions, the breakdown process is preceded by the vapor formation. A reason is that a local heating occurs in the near electrode region by the pre-breakdown current in the liquid. Note that a typical order of magnitude of the local electrical breakdown field of water is 10^6 V/cm (in the case of microsecond pulsed breakdown). This magnitude exceeds by at least a factor 30 the breakdown electrical field of atmospheric pressure in air. One of the reasons is that free electrons are generally absent in water because even if they are present, they are quickly solvated within 1 ps time scales.

Second group: It involves the transient discharges in/on large gaseous bubbles, which are either injected in the liquid, or formed/existing in it permanently due to its bubbling. In this case, streamers tightly slide along the gas–liquid boundary inside the bubble and reactive species can therefore be quickly transferred from streamers into liquid as well. Besides, the voltage required for initiation of streamers in gaseous bubbles is much lower compared to that for the electric breakdown in water.

In fact, by now numerous and various configurations of the transient discharges belonging to the first group have been developed and there is no possibility to review them all. Therefore, only sketches of the most widespread bubble–plasma systems, which help to illustrate the main ideas discussed here, are shown in Fig. 25.

The electrode configurations in Figs. 25(a) and 25(b) are related to the first group of the transient discharges in liquid. Microbubbles appear in the course of the pre-breakdown stage at the tip of a high voltage metal needle (or sharp protrusions on the flat surface of the electrode) or in/top of tiny

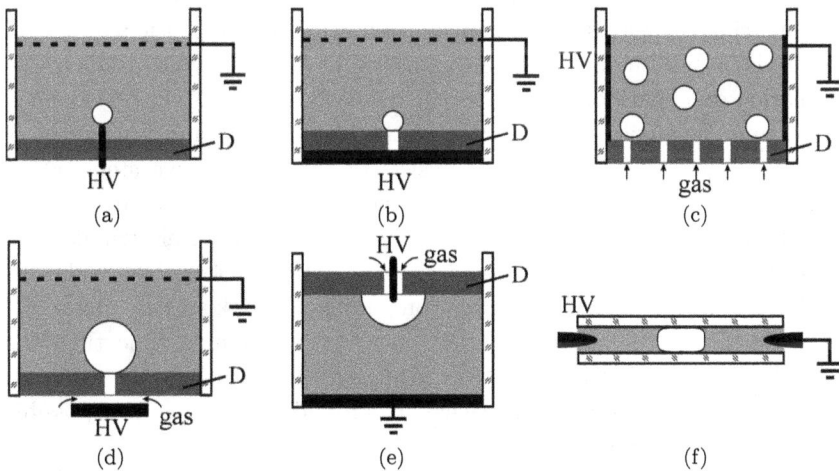

Fig. 25. The sketches illustrating the creation of bubbles in liquids by different ways: (a) and (b) Tiny bubbles filled with water vapor (corresponding to the first group) occur in the course of the low-current pre-breakdown stage. (c)–(f) Large gaseous bubbles (corresponding to the second group) are introduced in liquids in advance by an external gas bubbling. With permission from [27]. © 2015 IOP Publishing.

hole in a dielectric barrier covering the electrode. This happens due to the high concentration of the pre-breakdown current at these small areas resulting in strong overheating of liquid followed by its fast evaporation [11, 14].

In the second group (Figs. 25(c)–25(f)), bubbles are much larger and introduced in liquids by an external gas bubbling [60–65]. It should be noted that in the first group of plasma-liquid systems, the bubbles are predominantly filled with water vapors (or dissolved gases), but in the second group, the bubbles are filled with air or other gases.

13. Transient Discharges with Bubble Formation at a HV Metallic Electrode

A sketch of the typical electrode configuration relevant to such transient discharges (i.e., discharges of the first group) is shown in Fig. 25(a). A specific spatial and temporal behavior of these transient streamer-spark discharges depends on the polarity of the applied high-voltage pulse. However, a sequence of the events taking place in the streamer-spark discharges is qualitatively similar for both polarities.

Based on the works in [66–72], and the literature cited therein, one can point to the main steps in the development of these discharges. Full electrical breakdown is generally defined as the moment when a conductive plasma channel forms an electrical connection between the two metal electrodes inside the liquid. The completion of full breakdown leads to the formation of a spark or arc.

A time lag between application of the high voltage and breakdown is always observed. This time lag consists of three successive steps: (1) a non-luminous initiation phase or streamer inception associated with the bubble formation at the tip of an electrode with a strong electric field, in which impact ionization occurs, leading to the initiation of electron avalanche in the bubble; (2) streamer propagation phase; (3) spark and arc phase.

According to [69], the initiation phase takes a longer time than the streamer propagation and presents a huge jitter. Although the streamer transforms to a spark when it reaches the grounded electrode, it is possible to stop the streamer without a spark by adjusting the experimental parameters such as the amplitude of applied pulsed voltage or its pulse width, distance between the electrodes, charging power pulse energy, parameters of the external circuit and electric conductivity of the liquid.

Positive streamers: A general impression about the underwater positive streamer-spark discharges can be obtained from looking at their images

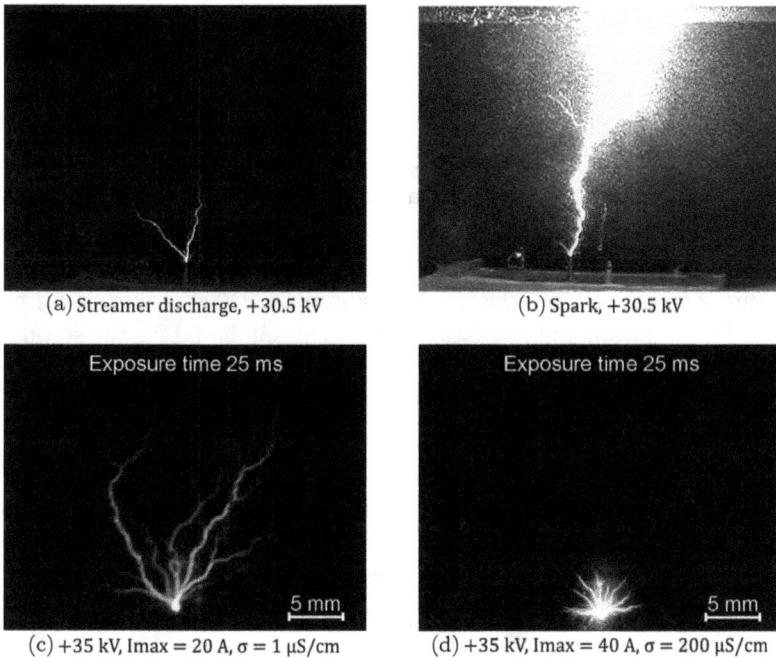

(a) Streamer discharge, +30.5 kV

(b) Spark, +30.5 kV

(c) +35 kV, Imax = 20 A, σ = 1 μS/cm

(d) +35 kV, Imax = 40 A, σ = 200 μS/cm

Fig. 26. Two modes of the pulsed electric discharge in water under voltage of a positive polarity: (a), (c) and (d) Streamers; (b) spark ($U = 30.5\,\text{kV}$). In (c) and (d) $h = 55\,\text{mm}$. Streamer mode corresponds to the not completed (or partial) discharge when streamers do not arrive at the opposite electrode. With permission from [73]. © 2000 Springer Nature.

shown in Fig. 26. Based on publications [66–72], and the papers cited therein, one can summarize some features of positive streamers commonly observed and described in the literature. Depending on streamer propagation structure, the positive streamers in a dielectric liquid are classified as subsonic bush-like streamers or supersonic filamentary streamers. In other words, streamer discharges at positive polarity have two propagation modes: A primary streamer and a secondary streamer.

Primary streamer: It appears at lower applied voltages with a propagation speed of (23) km/s, having a semi-spherical bush-like structure ($\sim 500\,\mu\text{m}$) composed of many filaments. All but a few filaments disappear within 400 ns, resulting in a tree-like structure. The primary streamer propagates during the flow of repetitive pulsed currents.

Secondary streamer: Above a certain threshold voltage, the secondary streamer propagates with a speed typically larger than 10 km/s, which can even reach 30 km/s. The transition between these modes occurs abruptly. In all cases the streamer channel is of gaseous nature. In the case of low conductive water, the secondary streamer propagation is followed by re-illuminations, which are restarts of the discharge in the formed gas channel, and show a stepwise propagation. The secondary streamer and the re-illuminations are characterized by a continuous current with a duration of around 100 ns and subsequent spike currents. The pressure of shock waves launched by secondary streamer propagation reaches 3 GPa. Such a high pressure leads to a local electric field above 20 MV/cm, suggesting field-induced dissociation and ionization of molecules in liquid for the secondary streamer inception.

The current flowing during the first streamer mode is very small and typically less than 1 mA. During the second mode of streamer propagation, the current flows intermittently. The intermittent frequency depends on the liquid conductivity. The current may become continuous at sufficiently high conductivity. Correspondingly, the channel length increases stepwise and the channel lights up only during current flow. Not all channels created in the course of streamer propagation reignite. The propagation velocity of secondary streamers is constant over a wide voltage range, while the number of streamer branches increases with voltage. A general scheme clearly elucidating the streamer-to-spark transition process, by the example of the breakdown of distilled water in the needle-to-plane gap, is shown in Fig. 27.

The set of excellent continuous images showing the transient streamer-to-spark process in its dynamics, starting from formation of bubble cluster at electrode tip to initiation and growth of primary streamer, and the next set of continuous images of transition process from propagation of underwater positive streamer to spark discharge followed by bubble expansion around the spark are presented in Figs. 28 and 29.

Figure 28 shows continuous images of the process, from formation of the bubble cluster at the electrode tip, to initiation of the underwater positive primary streamer. The time interval between the images is 100 ns, and the horizontal width of the images is 690 μm. Initially, as it is shown by the arrows, small bubbles form on the electrode surface, and after 100 ns, those bubbles grow to about 10 μm. Next, with bubble growth, multiple bubbles begin to form and grow from different locations on the electrode surface. After about 1 μs, contact occurs between the bubbles that have grown at different points on the electrode surface, a bubble cluster is formed, and growth continues.

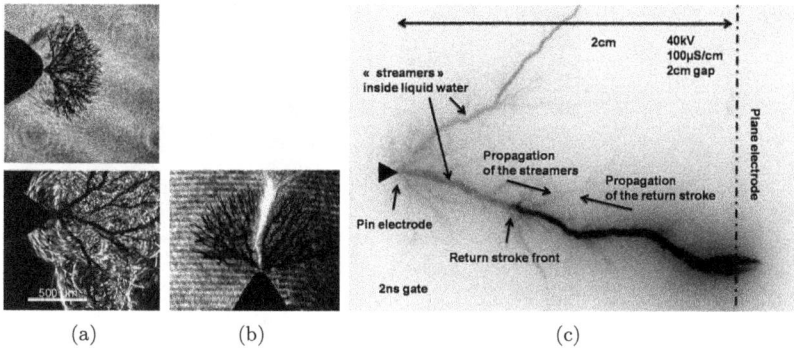

(a)   (b)   (c)

Fig. 27. (a) Bushlike (top) and filamentary (bottom) streamers in water obtained by shadow graph and Schlieren techniques. (b) Primary-second streamer transition. With permission from [68]. © 2007 AIP Publishing. (c) Scheme of streamer to spark transition process: When the first of the streamers reaches the opposite plane electrode, a thermalization return stroke propagates back to the pin electrode. With permission from [69]. © 2010 IOP Publishing.

After about $2\,\mu$s, further local bubble growth from the unified bubble cluster begins (indicated as local bubble growth in Fig. 28). Approximately $0.5\,\mu$s later, the tip of this local bubble changes to a protruding shape, and a streamer initiates from the tip of the protruding bubble (indicated as bubble extension in Fig. 28). After the streamer propagates in a semi-spherical shape, the streamer channel expands due to heat. In the conventional method, it was necessary to record sequential images of multiple discharges while shifting the recording time of each discharge. On the contrary, successful visualization of the continuous changes happening during one discharge process was possible, as shown by the series of images in Fig. 28.

Figure 29 shows continuous images of the process until the underwater positive primary streamer propagates, reaching spark discharge and the subsequent condition of bubble expansion. The horizontal image width is $3.08\,$mm, and the time is indicated in the figure. Inception of the primary streamer occurs at $3.2\,\mu$s. Streamer propagation slows down at $3.6\,\mu$s, and it restarts at $6.9\,\mu$s. The streamer reaches the earth electrode at $7.8\,\mu$s. Spark discharge occurs simultaneously in the two streamer channels at $7.9\,\mu$s. The light emission by the spark discharge of the finer streamer channel disappears at $8.0\,\mu$s, but the emission from the thicker streamer channel continues until $9.4\,\mu$s. Thereafter, the thicker streamer channel begins to expand due to the heat of the spark discharge, and at $25\,\mu$s, the channel diameter is approximately 3 times larger than its original size.

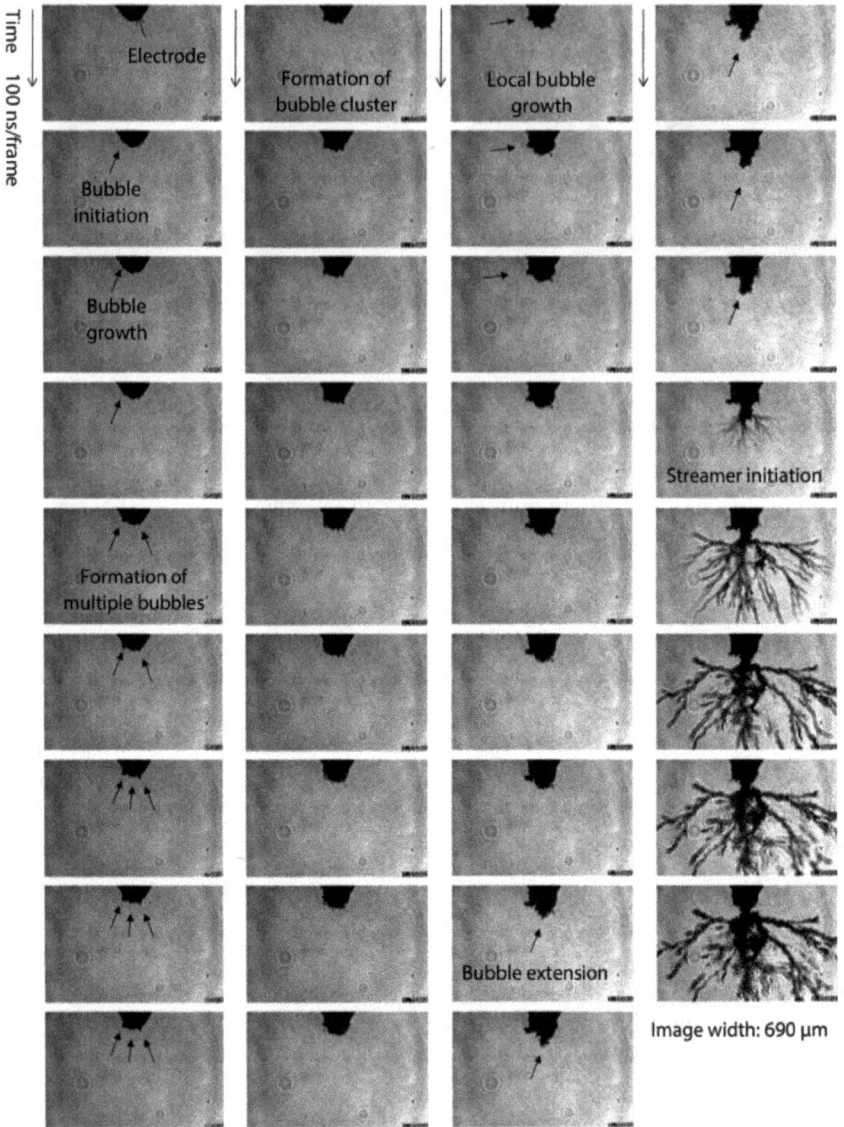

Fig. 28. Continuous images from formation of bubble cluster at the electrode tip to initiation and growth of the primary streamer. For discharge, ultrapure water was used. An inter-electrode gap distance 2 mm is used, and discharge was initiated by applying a high voltage of +14 kV to the needle. With permission from [74]. © 2018 Shimadzu.

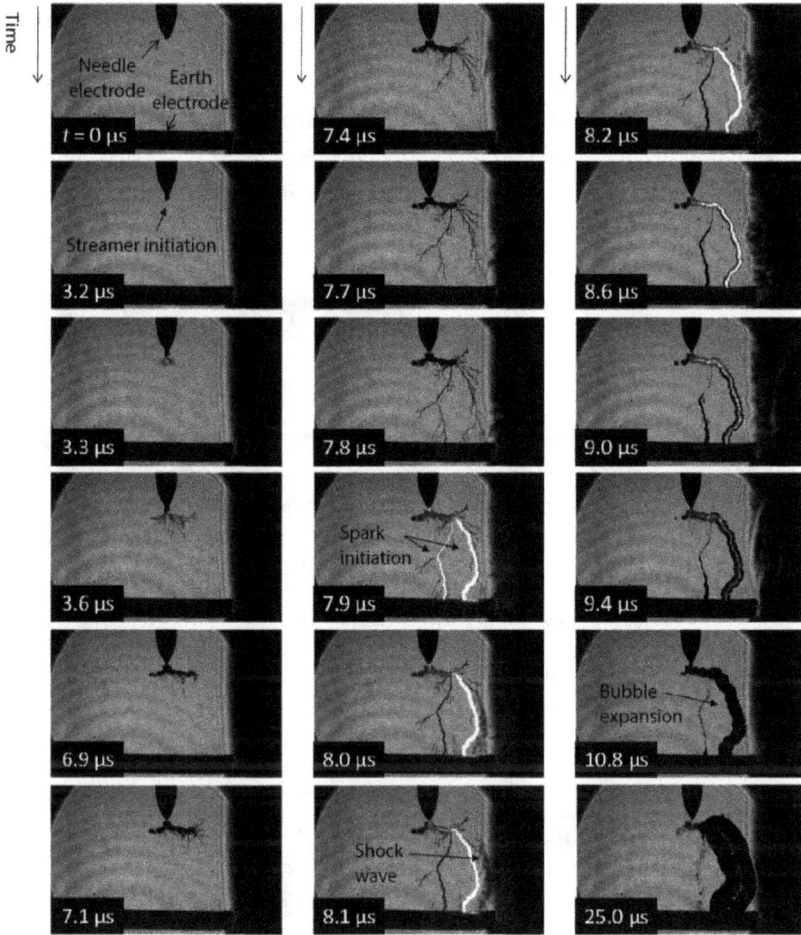

Fig. 29. Continuous images of a transition process from propagation of an underwater positive streamer to spark discharge and process of bubble expansion around the spark. With permission from [74]. © 2018 Shimadzu.

Negative streamers: The discharge phenomena for negative streamers differ significantly from those of their positive counterparts under the same experimental conditions such as the shape of the electrode, the gap length, the amplitude of the applied pulsed voltage and the type of liquid [75].

The propagation speeds of tree-like and filament-like negative streamers are several kilometers per second, whereas the positive streamers are more than ten times faster. The positive streamer emits light strongly, whereas

the negative streamer illumines only a microscopic region at the needle for
the same applied voltage. The maximum length of the positive streamer
propagation is much in excess of that for the negative streamer. Since the
light emission of these streamers is weak, they are generally imaged using
the Schlieren or shadowgraph method that measures the density change as
a refractive index change. Some results illustrating the behavior of negative
streamers are presented in Figs. 30 and 31.

Fig. 30.   Series of shadowgraphs of negative streamer propagation from a needle electrode
in ultrapure water. The exposure time is 50 ns at 10 Mfps. The applied voltage is $V =$
20 kV having a rise time of 100 ns. The gap length is 2 mm. With permission from [75].
© 2018 AIP Publishing.

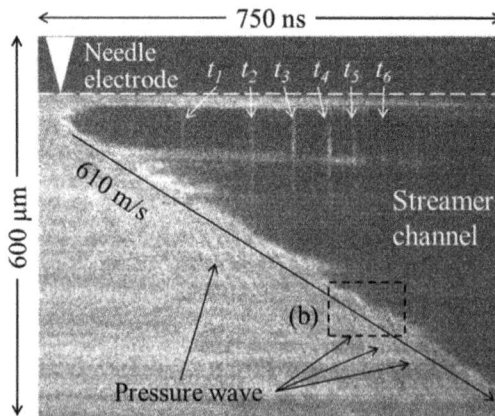

Fig. 31.   Streak image of negative streamer propagation obtained using a laser backlight
with a sweep speed of 50 ns/mm at an applied voltage $V = 21$ kV. Negative streamers
have a speed of (820 ± 50) m/s in the presence of pulsed currents, and (100 ± 20) m/s
when pulsed currents were absent. It was also observed that 1480 m/s pressure waves
were generated during streamer propagation. With permission from [75]. © 2018 AIP
Publishing.

14. Transient Discharges with Bubble Formation in/at Narrow Hole(s) in a Dielectric Barrier

Sketch of the electrode configuration relevant to these discharges is shown in Fig. 25(b). There are two types of such liquid discharges that differ with the location of a dielectric barrier: (1) porous dielectric barrier covers one of the metallic electrodes; (2) dielectric barrier with hole divides two volumes with liquid (diaphragm discharge). As a rule, these discharges exhibit a low-frequency pulsed-periodical regime while they are powered by steady-state DC voltage. Existence of hole(s) in a dielectric barrier leads to strong concentration of the electric current in hole that, in its turn, provides the energy injection resulting in fast heating and evaporation of the liquid contained in the hole. Eventually, a gas bubble appears in/at the hole and streamer breakdown inside a bubble happens. Further, the bubble grows in size due to continuing gas heating and liquid evaporation by the energy released from streamers. However, the bubble growth leads to a situation in which the voltage drop on the elongated bubble wont be enough to sustain gas discharge inside the bubble. After the discharge stops, gas inside the bubble begins to cool, vapors will condensate and the bubble shrinks and disappears. However, after some interval of time, all the above events repeat again and again.

The set of video images presented in Fig. 32(a) nicely illustrates the described dynamics of the isolated microbubble in DC low-current and

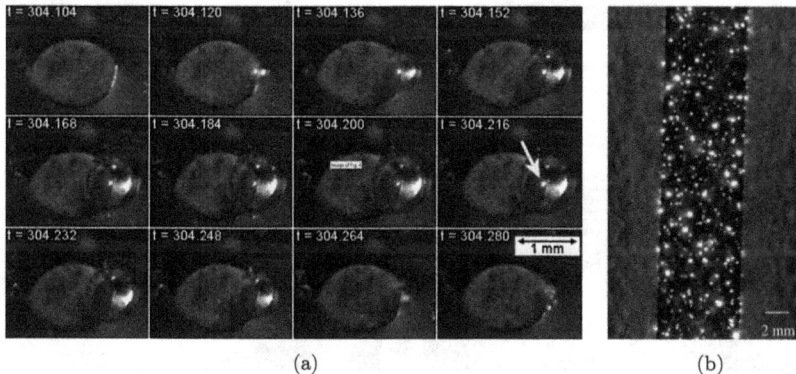

Fig. 32. (a) Sequence of video images showing a small area of a metallic electrode covered by a thin dielectric layer with a narrow hole. This set covers the lifetime of a single discharge event from the birth to death of a lone bubble in/at the hole. Times are given in ms. With permission from [76]. © 2015 Elsevier. (b) Large-scale image of many unsteady microdischarges that chaotically scintillate at the holes. With permission from [77]. © 2007 Elsevier.

low-voltage electrolyte discharge with an electrode covered with a dielectric barrier having very narrow pores [76, 77]. The applied voltage of low amplitude does not allow long streamers' formation. Figure 32(b) presents the total image of the electrolyte discharge with numerous short-length streamers formed inside microbubbles initiated in narrow pores.

This type of discharge with microbubbles in narrow holes can be easily scaled to much greater sizes. One of the pulsed labscale versions of such a discharge is elaborated in [78] and shown in Fig. 33. In this case, the HV electrode was covered with the ceramic coating, the parameters of which are the following: Layer thickness of 0.2–0.3 mm, open porosity of (25)% and pore dimension of 1–10 μm. This low-current streamer discharge his also of practical interest for many biomedical applications. As it turned out, the number of microdischarges depends appreciably on the water

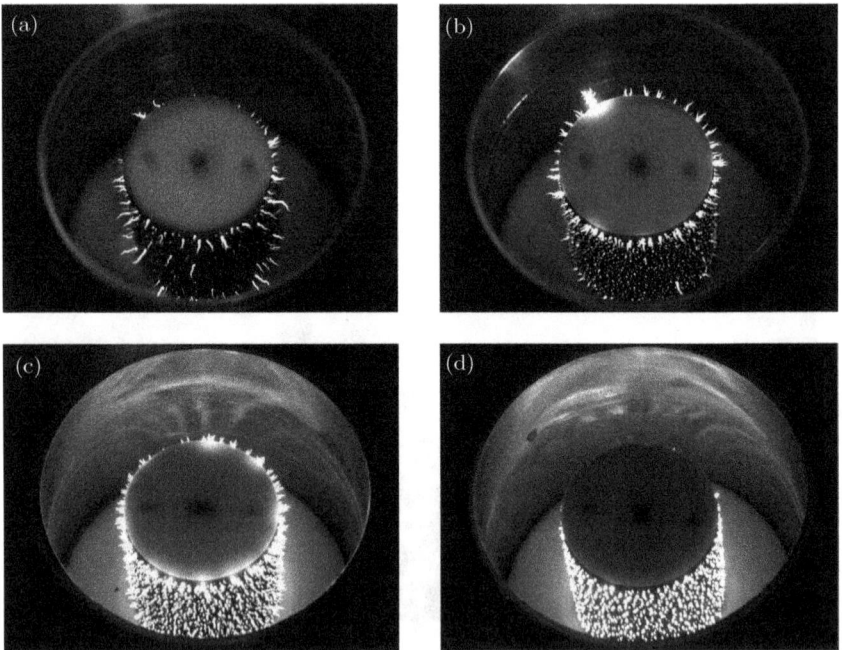

Fig. 33. Multichannel pulsed electrical discharge in water generated by using porous-ceramic-coated metallic electrodes. Effect of water conductivity on electrical discharge characteristics: (a) 600 μS/cm; (b) 1.5 mS/cm; (c) 6 mS/cm and (d) 15 mS/cm. Here, $U = +28$ kV, $R_{in} = 60$ mm, $R_{out} = 120$ mm. With permission from [78]. © 2008 IEEE.

conductivity — the higher the conductivity, the greater the number of microdischarges that appear.

15. DC Diaphragm Discharge in Capillary Tube Between Two Separated Volumes of Liquid

The idea of this discharge is also based on forcible and strong electric current concentrations in a narrow dielectric tube as described by the electrode configuration shown in Fig. 25(b). The schematic sketch of the pulsed DC diaphragm discharge, its image averaged over many cycles, and the current and voltage waveforms are presented in Fig. 34.

In this discharge set up, both a diameter and length of the capillary hole exceed appreciably those parameters of pores in a dielectric barrier. It gives a possibility to increase markedly the electric power that can be input in the diaphragm discharge. Formation of the vapor bubble inside the connecting hole allowed the electrical breakdown of the bubble and

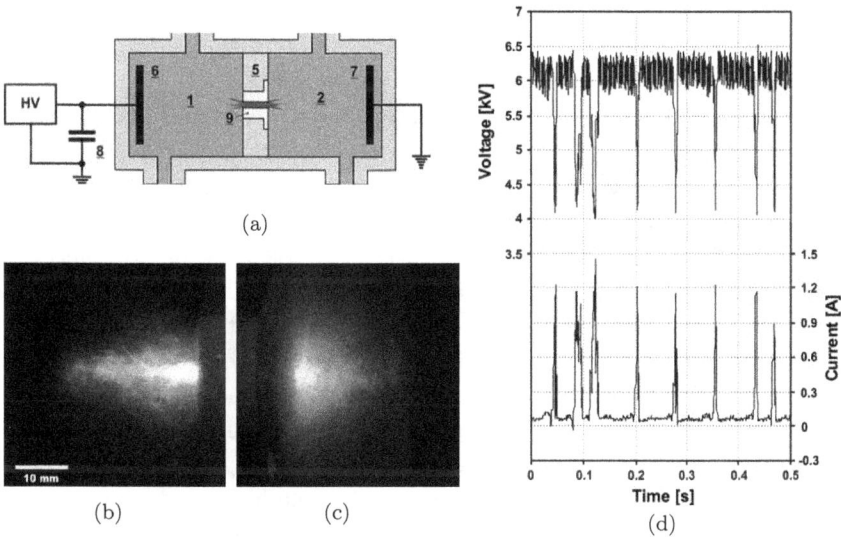

Fig. 34. (a) Sketch of a diaphragm discharge set-up. Two chambers (1, 2) are filled with a solution and separated from each other by a dielectric barrier (5) having a capillary hole (9); an HV capacitor (8) of 200 nF is connected in parallel to the DC power supply. Photograph of underwater DC diaphragm discharge. (b) View from the positive polarity side, and (c) from the negative one. The power input is 2 kW, the solution conductivity is 500 μS/cm. (d) Sequence of voltage and current pulses produced by DC diaphragm discharge in water. With permission from [79]. © 2013 Springer Nature.

the initiation of a streamer discharge, which strongly expanded into the relatively large volume of surrounding water on both sides of the connecting hole. In other words, the streamer discharge of the positive and the negative polarity is generated simultaneously. Expansion of the discharge leads to the termination of the conditions required for the existence of the discharge and to the interruption of the discharge current between the electrodes. After the end of this process, the connecting hole begins to fill back with water and the whole process is repeated.

Thus, the apparatus is operated at the pulse regime although it is charged by a DC power supply. By proper choosing of the connecting hole dimensions (length and diameter) and the capacitance of the high-voltage capacitor (8) (shown in Fig. 34), it is possible to establish in the connecting hole an electrical current of sufficient density needed for the evaporation of a small amount of liquid, and thus, the formation of the vapor bubble inside the capillary and its subsequent electrical breakdown.

16. Transient Discharges with Bubble(s) Inserted into a Liquid in Advance

General schemes of these discharges are shown in the sketches presented in Figs. 25(c)–25(f). By now, there are a lot of papers devoted to theoretical and experimental investigations of the streamer discharges with bubbles immersed in a liquid [48, 80–88]. Excellent contribution to this issue was made by Kushner and coworkers [81–83].

For the case of bubbles filled with air, the following picture emerges. The dielectric constant and conductivity of the liquid largely determine the path and pattern of positive streamer propagation. Streamers in bubbles, immersed in liquids with a high permittivity, preferentially propagate along the surface of the bubble. This indirect path is associated with electric field refraction at the boundary between materials with dissimilar dielectric constants. Liquids having a low permittivity produce streamers that propagate along the axis of the bubble. The higher the conductivity of the liquid, the smaller the permittivity at which this transition occurs. The particular pattern of the streamer path is also a function of the bubble size, applied voltage and its polarity. The behavior of streamers in bubbles is similar to that of surface dielectric-barrier discharges where the liquid serves as a lossy dielectric electrode. Note the streamer propagation in the water bubble not immersed in the water but floating on the water (see Fig. 17) presents exactly the same behavior.

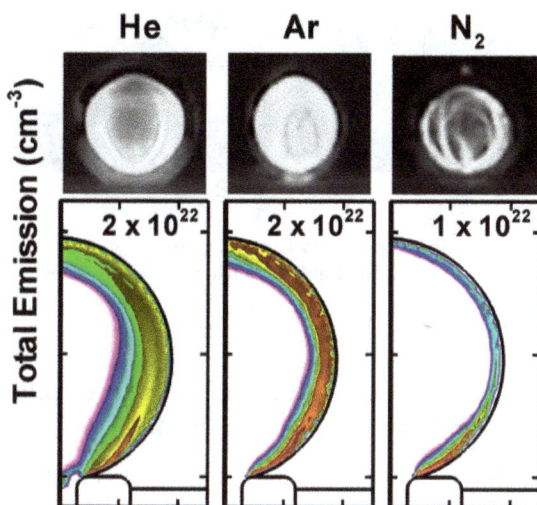

Fig. 35. Experimental and calculated time integrated total emission intensity from discharges sustained in He, Ar and N_2 bubbles. With permission from [82]. © 2014 IOP Publishing.

Some experimental and numerical results on the streamer propagation along the surface of the bubble immersed in the water are presented in Fig. 35. These pictures show the time integrated total emission intensity from discharges sustained in He, Ar and N_2 bubbles immersed in water. One may see that the discharges typically propagate along the surface of the bubble where the gradient in dielectric constant is largest, producing electric field enhancement, therefore discharge plasma (adjoins) concentrates at the gas–water interface.

The predicted total emission is most uniform in the He bubble and most filamentary in the N_2 bubble. A filamentary discharge sticking to the inner surface of the bubble was predominantly seen in molecular gases, such as N_2 and O_2, which caused a wrinkled disturbance on the bubble surface after the discharge. In contrast, in rare gases, such as He, Ne and Ar, a more uniform volumetric discharge was seen in addition to the filamentary one. When the input power increases, it leads to the collapse of the original bubble after the discharge due to the pressure produced from the streamer discharge in the bubble. One of the scenarios of this collapse is shown in Fig. 36. In this case, the collapse happens due to instability of the bubble surface shape under increasing gas pressure.

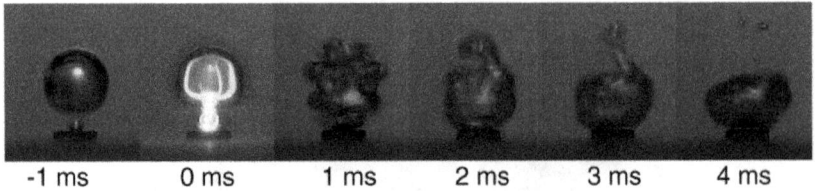

-1 ms   0 ms   1 ms   2 ms   3 ms   4 ms

Fig. 36. Different appearances of the discharge after-effect observed at $30\,kV$ of applied voltage and $d = 0.3\,mm$ for a N_2 bubble. With permission from [84]. © 2011 IOP Publishing.

17. Streamer Behavior in a Single Bubble in Capillary Tube Filled with an Aqueous Solution

Another method for the creation of the streamer discharge in gas bubble(s) consists of inserting in advance the bubble(s) in a capillary dielectric tube that either is filled with the water solution or connects two large volumes with the liquid (see the set-up sketched in Fig. 25(f)). Such a configuration was investigated in [61, 62]. At first glance, this configuration is similar to the diaphragm discharge. However, there is an essential difference between them. In the case of diaphragm discharge, the bubble does not pre-exist in the capillary hole, but it is periodically created due to heating and evaporation of the water in the hole. Because of the large heat of vaporization of water, this process consumes a huge amount of energy, leading to the low energy efficiency in the reactive species production by this discharge, and reduces the operating lifetime of the dielectric capillary hole working at high temperature due to the strong mechanical stress caused periodically by the pressure shock. In the case of the configuration shown in Fig. 25(f), in contrast to the diaphragm discharge set-up, there is no need to spend energy for the bubble generation — the bubble is prepared in advance in the capillary hole and can be introduced in the hole periodically by external bubbling.

The above-mentioned features can result in a strong reduction of energy costs of the reactive species generated by the streamer discharge in the configuration of Fig. 25(f). Besides, this configuration enables generating the streamer discharge in long bubbles, a result that is practically impossible to achieve in the case of plasma–liquid systems by bubbling large volumes of water solution. A reason is that the liquid can electrically shunt the bubble and the needed voltage drop along the bubble, required to provide the necessary condition for streamer breakdown in it, cannot be reached.

The conducting/insulating behavior of a liquid is determined by the Maxwellian relaxation time, $\tau_M \simeq \varepsilon \varepsilon_0 / \sigma$, where ε and ε_0 are the dielectric permittivity of the liquid and vacuum, respectively, and σ is the electrical conductivity of the liquid. For pure water, τ_M is of the order of few microseconds. If a liquid is exposed to a long duration electric pulse, $\tau \gg \tau_M$, water behaves as a resistive medium whose resistance depends on its conductivity σ. For much shorter times, $\tau \ll \tau_M$, water behaves as a dielectric medium with $\varepsilon \approx 80-82$, and can sustain a high electric field until a breakdown threshold is reached. This is why the electrical breakdown formation in a bubble, fully surrounded by a conductive liquid, requires the use of HV pulses with a short rise-time, $\tau \ll \tau_M$. Only if this strong condition is satisfied (although it is especially difficult to achieve for high-conductive liquids), one can reach the critical electric field strength required to provide gas breakdown inside a bubble. Additionally, the larger the bubble diameter, the higher the amplitude of the applied HV pulse must be.

One can conclude therefore that plasma–liquid systems designed for treatment of conductive liquids and based on the bubbling and the use of pulsed HV generators with short rise-time can face serious technical problems when the sizes of plasma reactors grow up to real industrial scales. The practical problems mentioned above explain the increased interest in detailed studies of streamer discharges in the configuration shown in Fig. 25(f). Besides, there is an interest from a physical point of view in this plasma-liquid system. A reason is that the streamer breakdown in the bubble, inserted in a dielectric tube filled with a liquid, occurs in a medium made of three phases: Gas, liquid and dielectric solid. In this case, it would be reasonable to expect that the transient streamer-spark discharge will have some new features compared to those typical of a two-phase system (gas bubble fully surrounded by liquid) and one-phase system (needle-plane configuration in air).

Studies of the streamer breakdown in the air bubble, inserted in a quartz capillary tube of 100 mm length and 2.5 mm inner diameter, were performed by Akishev *et al.* [57]. The length of the bubbles and their position in the tube were varied. A first series of experiments was aimed at studying the slow evolution of the bubble shape when the HV electrode of the capillary tube was stressed by steady-state DC voltage of various amplitudes. Two regimes were revealed in the evolution of the bubble shape depending on how much the applied voltage exceeds the breakdown voltage of the bubble of pre-set length. If the applied voltage slightly exceeds the threshold voltage, a quasi-periodical regime of the current pulsations was

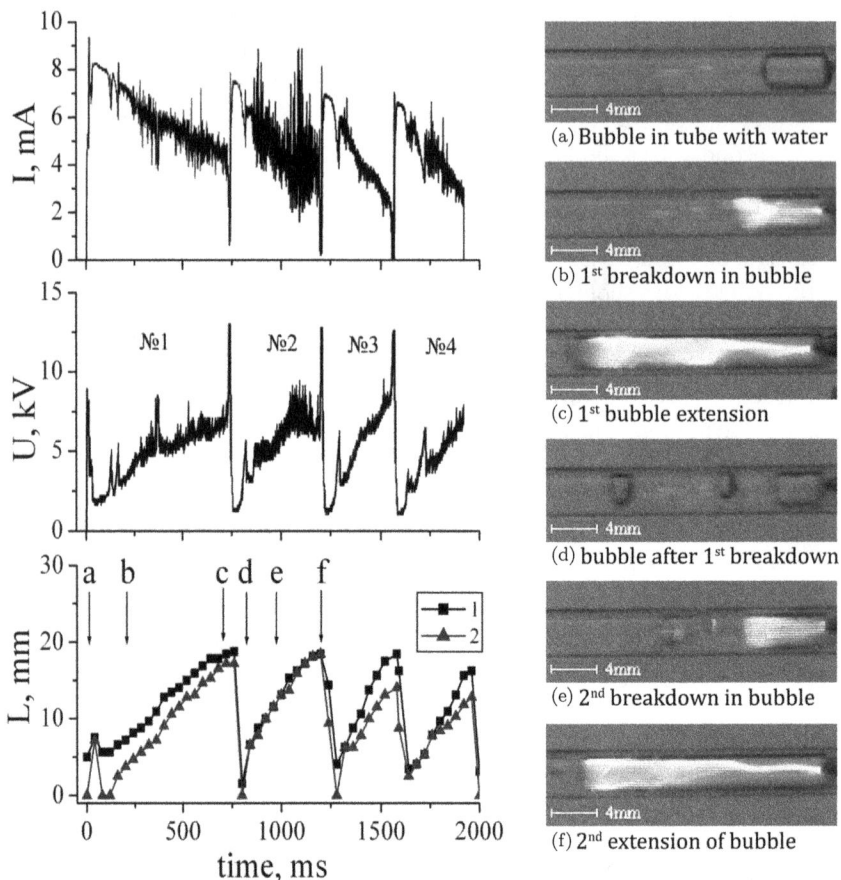

Fig. 37. Electrical characteristics and images of streamer breakdown in a single bubble inserted in a dielectric tube with tap water. Applied voltage is $U_0 = 20\,\text{kV}$. Exposure time of each shot is 20 ms. (Left side) Simulateneous temporal behavior of current I, voltage U and lengths L of: Gas bubble (full squares) and gas discharge column (full triangles), under periodical electric breakdown. (Right side) Snapshots of bubble and gas discharges under periodical electric breakdown; the letters (a–f) correspond to the times of the shots indicated in the left side. With permission from [61]. © 2008 INOE Publishing House — Integra Natura Omnia Aeterna.

observed with typical period of several hundreds of milliseconds. The volt-current parameters of this regime correlated with images of the bubbles and streamers inside the bubble are presented in Fig. 37. The streamers are non-stationary in space and time, twist like snakes and non-uniformly occupy the bulk of bubble. The high-frequency noisy component of the electric current that is seen in Fig. 37 is associated with the non-stationarity of the streamers' behavior in the air bubble.

One can see from the figure that after the streamer breakdown occurs, the current jumps to its maximum magnitude, and the streamers start slowly increasing their length. This is not due to their penetration into the water through the wall of initial bubble, rather their length is equal to the bubble length and increases due to the bubble elongation in which they propagate (see Fig. 37). During the bubble elongation, the average current of the plasma-liquid system slowly decreases, while the voltage across the bubble increases.

After reaching the maximum bubble length (corresponding to the time at which the current diminishes by roughly a factor two compared to its initial magnitude), the current abruptly drops down to zero, and streamers in the bubble disappear as well. Despite the fact that the applied voltage quickly returns to its initial magnitude, the next streamer breakdown does not occur immediately after the previous one, but with a delay of about 10 ms. In the course of this time, the elongated bubble shrinks practically to its initial size, but sometimes the shrinking bubble disintegrates into several smaller bubbles. However, after the next breakdown happens in the bubble nearest to the HV electrode, this elongating bubble absorbs all the previous small bubbles which it meets along its path. In fact, such a quasi-periodical regime lasts not more than 10 cycles, after that the elongating bubble with streamers reaches the opposite electrode, and transition into spark followed by strong increase in the current amplitude and light radiation from the plasma channel takes place.

In the second regime, when the applied voltage markedly exceeds the threshold breakdown voltage, the quasi-periodical regime of the current pulsations accompanied with regular elongation followed by shrinking of the bubble is absent. In this case, both the magnitude of current pulses and frequency of these pulsations are much higher than those in the first regime. The streamers corresponding to these current pulses follow one another, but pauses between the neighbor streamer splashes are much shorter than in the first regime. As a result, the extended bubble has no time to shrink before the next streamer breakdown. In such a case, the bubble elongates practically continuously to the opposite electrode that shortly results in the spark formation.

The described results above correspond to the self-pulsing regime that is characterized by long-term dynamics owing to the slow elongation–shrinking bubbles process. Therefore these results were obtained with a low time resolution. However, there is interest in clarifying in detail the streamers' behavior preceding this regime, i.e., over short periods before starting the bubble elongation. This was done with the use of a high-speed camera. The results are shown in Fig. 38. To avoid the problem associated with a slow bubble elongation, the experiments were performed with the applied high voltage pulse having the rise-time not larger than 100 ns. Besides, the metal needle electrodes were removed from the ends of the tube and large volumes of liquid solution at both sides of the tube were used (see Fig. 38). The tube was filled with an aqueous solution with variable conductivity, that is, either distilled water with a low conductivity $\sigma = 20 \, \mu S/cm$, or a saline with a high conductivity $\sigma = 12 \, mS/cm$. The bubble was filled with air, nitrogen and helium.

Figure 38 shows the waveforms of the discharge current and light radiation from the bubble during the streamer breakdown development in bubble when a voltage pulse of a positive polarity is applied. Close examination of the current waveform leads to the conclusion that the development of a discharge in a bubble before its elongation involves three stages, the duration of which has a stochastic character.

The first stage exhibits the pulsed regime that corresponds to a stepwise breakdown in the bubble and lasts about 15–30 μs.

The second stage corresponds to a slightly increasing quasistationary discharge current and lasts about 30–50 μs at the threshold applied voltage. The quasistationary stage duration shrinks with increasing of the applied

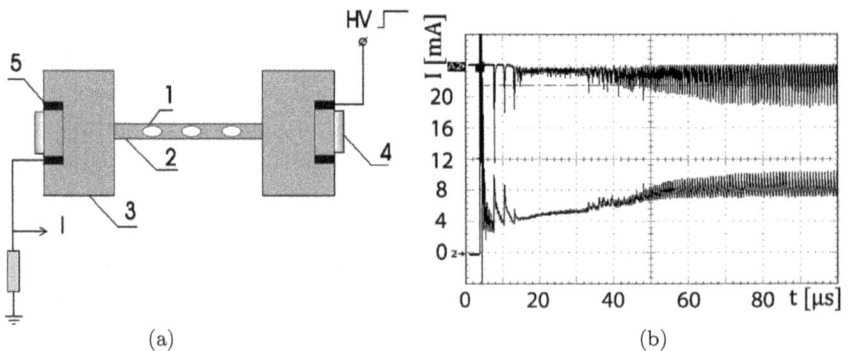

Fig. 38. (a) Sketch of the experimental set-up: 1-gaseous bubbles; 2-quartz tube; 3-volumes with aqueous solution; 4-quartz windows; 5-ring electrodes. (b) The typical waveforms of the electric current (Ch2) and the photomultiplier tube (PMT) signal (A2). Axes Scales: [t]=10 μs/div, [I] = 4 mA/div. Distilled water; $U = +10 \, kV$; an isolated air bubble of 4 mm in length is situated in the tube center. With permission from [89]. © 2007 ICPIG.

voltage amplitude up to its full disappearance. The stationary current magnitude is then determined by the voltage pulse amplitude, the total resistance of the water column in the tube and the ballast resistor of the external circuit.

The third stage is characterized by the presence of regular current pulses with a high frequency exceeding 1 MHz and lasts until the bubble begins to elongate. A more detailed development of streamer breakdown in the air bubble is presented in Fig. 39. Figure 39(a) is related to the first stage and shows the dynamics of incomplete bubble breakdown corresponding to the initial current pulse after applying a voltage pulse. Figure 39(b) shows the image of the completed breakdown at the second stage with a quasistationary current.

It follows from data in Fig. 39 that the air bubble breakdown in distilled water, completed by shunting two opposite bubble sides by streamers, is not a continuous one-step process. In fact, it consists of a sequence of incomplete stepwise breakdowns, each of them elongating the streamer

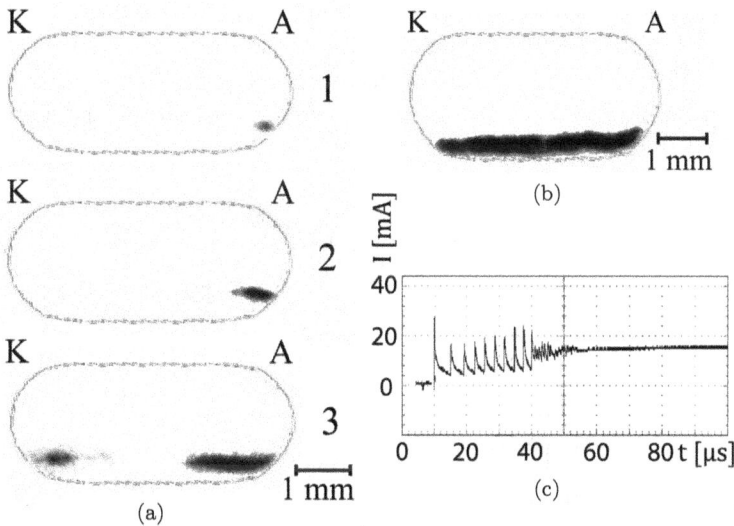

Fig. 39. The set of data characterizing the breakdown development in an isolated air bubble at positive polarity of the applied voltage. Letters K and A denote the bubble sides facing the cathode and the anode, respectively. (a) Sequence of streamer images on not completed breakdown during the initial current pulse after applying the voltage; exposure time is 100 ns; time interval between neighbor images is 100 ns. (b) The streamer image related to the completed breakdown in the second stage; exposure time is 100 ns. (c) The discharge current waveform related to the first and second stages. With permission from [89]. © 2007 ICPIG.

length a but, and it is correlated with the appropriate splash of the current shown in Fig. 39(c). Eventually, the stepwise breakdowns' completion is followed by the full bubble breakdown characterized by quasistationary current, which in its turn transits into a non-stationary stage with regularly repeating streamer breakdowns. Seemingly, an appearance of the third stage is associated with an increase of the water vapor concentration in a bubble happening due to energy injection produced by the two preceding stages. A reason is that the H_2O molecule is an electronegative one and therefore increases the losses of electrons in the streamer channel due to their attachment to H_2O. This process results in instability of the formed streamer up to its annihilation. After that, a new streamer will form again.

In the case of a negative polarity of the pulse voltage, the breakdown process is significantly accelerated. The first stage, with many not completed bubble breakdowns, disappears because even at the initial current pulse, full breakdown of the bubble already occurs. Here, the cathode and (one or more) anode spots are quickly formed at the opposite bubble ends, which are connected by the plasma channel (see proper information in Fig. 40). However, the formation of a quasistationary stage does not happen immediately after full bubble breakdown, as it is the case of positive polarity. Instead of that, many regular current pulses corresponding to repetition of the full bubble breakdowns take place before a quasistationary stage. The

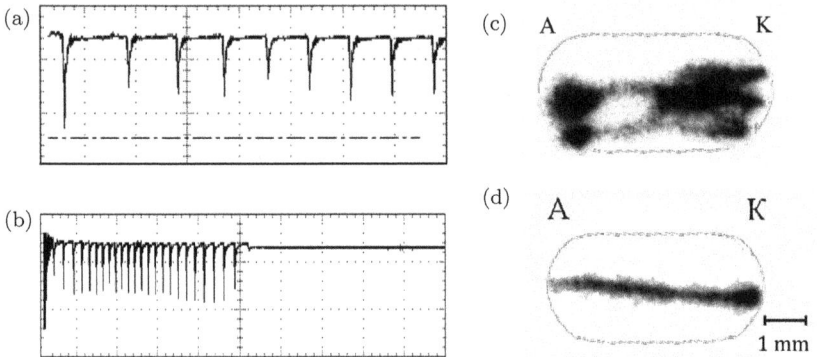

Fig. 40. The breakdown in distilled water under a voltage of negative polarity. Letters K and A denote the bubble sides facing the cathode and the anode, respectively. (a) The current waveforms for the first stage, and (b) for the second one. Axes Scales in: (a) $[t]$ = 1 μs/div, $[I]$ = 4 mA/div; (b) $[t]$ = 5 μs/div, $[I]$ = 4 mA/div. (c) the discharge image for the first stage, and (d) for the second one. Here, the exposure time is 100 ns. With permission from [89]. © 2007 ICPIG.

frequency and quantity of regular negative pulses exceed approximately three times those observed in the first stage at positive polarity. However, the negative pulses' magnitude is approximately three times lower compared to the positive pulses' one. One may conclude from this that the transition to the quasistationary current regime requires the same energy injection into the bubble regardless of the polarity of the applied voltage.

Finally, let us discuss the breakdown behavior in the air bubble inserted in the capillary tube filled with a saline solution which has a higher electrical conductivity compared to that of tap water. The proper experimental results illustrating this behavior are presented in Fig. 41.

The high conductivity of the aqueous solution significantly (qualitatively and quantitatively) changes the breakdown dynamics in a bubble. In saline, the low-current discharge stage corresponding to the streamer breakdown completion always quickly transits into a high-current spark. Contrarily, in distilled water, the completed streamer breakdown never ends with a spark. Moreover, during the applied voltage pulse exposed to saline, a whole series of spark breakdowns is observed with the current amplitude decreasing from one breakdown to another but with increasing the breakdown frequency (see Fig. 41(b)). Each of the first few high-current spark breakdowns (up to about 10) is always preceded by a sequence of streamer breakdowns with current pulses of growing amplitude up to 20 mA. This streamer stage is

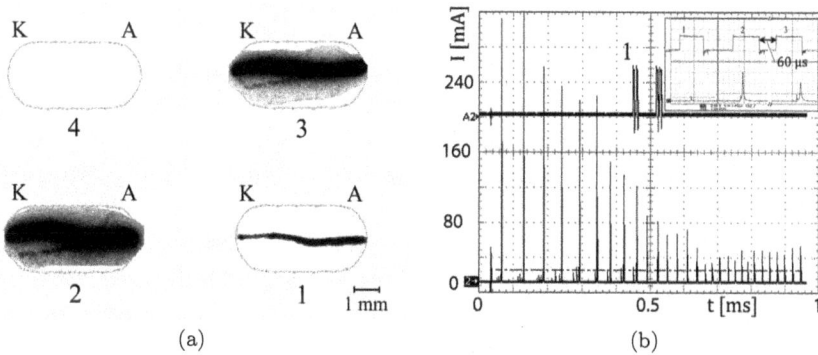

Fig. 41. The breakdown in the air bubble in saline at positive polarity of the applied voltage. Letters K and A denote the bubble sides facing the cathode and the anode, respectively. (a) The snapshots of the discharge in the bubble at streamer-spark breakdowns; each image corresponds to the proper current pulse shown in the inset in (b). (b) The waveform of the current pulses sequence caused by streamer-spark breakdowns, the bubble length is 5 mm, $U = +16$ kV. Axes Scales: $[t] = 100\,\mu s$/div, $[I] = 40$ mA/div. The exposure time of each snapshot is $5\,\mu s$. With permission from [89]. © 2007 ICPIG.

similar to a stepwise streamer breakdown of the air bubble in distilled water. With an increase in the number of spark breakdowns, the stepwise streamer breakdown stage gradually disappears, and sparks arise immediately after the breakdown voltage is reached across the bubble. Note that in saline, in order to reach the regime with the bubble length oscillations, it is necessary to diminish the tube diameter.

18. Streamers Behavior in Two Bubbles in Capillary Tube Filled with an Aqueous Solution

In the streamer regime with the bubble pulsations, the elongating bubble absorbs all other bubbles in its way before the streamer breakdown happens in them. In this connection, there is a question: Is it possible to form the streamer breakdowns in the chain of bubbles from one bubble to the other without their absorption by each other? This question his not only of scientific interest but a practical one as well. Indeed, a positive answer to this question opens the way to develop plasma–liquid systems which can activate many gaseous bubbles simultaneously, which can increase the efficiency of such a system.

An answer to the above question can be obtained in the situation when the bubbles do not move or elongate, that is, under the application of voltage with a fast rise time. In this case, the total energy injection into each gas bubble has to be low, in order that both the bubble sizes and their locations are practically not changed during the breakdown process [89].

The experiments were performed with two air bubbles of 4–5 mm in length. High-voltage pulses up to 20 kV in the amplitude were used. Typical rise-time and duration of the pulses are 100 ns and 12 ms, respectively. It was found that the streamer breakdown in tube with two bubbles takes a bit longer than that with a single bubble of the same size. Besides, the partial and full streamer breakdowns do not occur simultaneously along the whole gas bubble chain. The breakdown process looks like a relay race of the streamer breakdowns successively transferred from the bubble situated closer to HV-electrode toward the bubble located closer to the grounded electrode. In total, the breakdown process in air bubbles in tube with water can be divided into three main stages followed by the final stage #4 characterized by fully completed electric breakdown in all bubbles and steady-state current (see the proper information in Fig. 42). The detailed description of the events taking place at each one of the three stages is given in what follows.

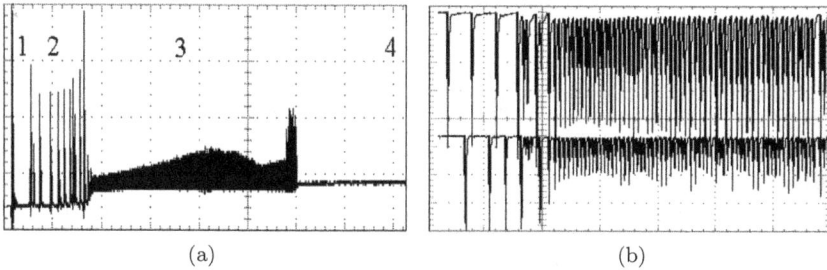

(a)   (b)

Fig. 42. (a) Typical shape of the current waveform related to the breakdown formation in tube with two air bubbles. Distilled water; bubbles dispose 5 mm from each other at the middle of quartz tube; bubbles #1 and #2 are located closer to the anode and the cathode, respectively. $U = +16$ kV. Axes Scales: $[t] = 50\,\mu s$/div, $[I] = 3$ mA/div. Sections (1-4) in the plot mark the stages of the breakdown process. (b) Photomultiplier tube signals of illumination from bubbles #1 (upper curve) and #2 (bottom curve); seldom and frequent pulses correspond to the second and third stages, respectively. In the second stage, one can see the delay in between the PMT signals from bubbles #1 and #2. $U = +13$ kV. Axes Scales: $[t] = 10\,\mu s$/div, $[I] = 3$ mA/div. With permission from [89]. © 2007 ICPIG.

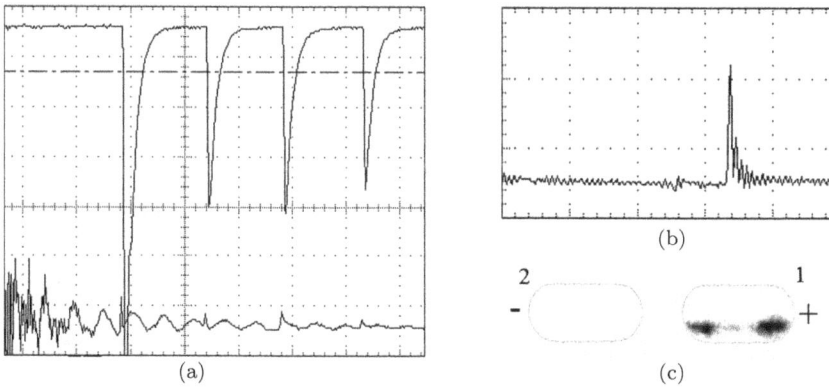

(a)   (c)

Fig. 43. The data set illustrating the events at the first stage of breakdown with a chain of two air bubbles. (a) The PMT signal (upper panel) of luminosity from bubble #1; small current pulses (lower panel) correlate with luminosity from this bubble; $U = +10.5$ kV. Axes Scales: $[t] = 1\,\mu s$/div, $[I] = 1$ mA/div. (b) An example of small current pulse correlated with the discharge inside bubble #1. Axes Scales: $[I] = 1$ mA/div, $[t] = 500$ ns/div. (c) Two bubbles' snapshots at the moment of the current pulse shown in (b). $U = 17$ kV and exposure time 100 ns. With permission from [89]. © 2007 ICPIG.

First stage: There is some delay in starting the breakdown process after applying the HV voltage. This delay is clearly seen in Fig. 43(a) from the PMT signal showing the light emission from bubble #1. After delay, short and low amplitude current pulses appear. Despite their low amplitude, bubble #1 emits high intensity luminosity that proves the existence of a high-strength electric field inside bubble #1 at the first stage. The low-current pulses correspond to the partial breakdowns only in the bubble #1 that is located closer to HV electrode. This statement is proved in Fig. 43c. This figure shows snapshots of two bubbles taken simultaneously with the current pulse presented in Fig. 43(b). One may clearly see that this current peak exactly corresponds to the partial breakdown only in bubble #1.

Second stage: At this stage, the periodical electric breakdowns happen in both bubbles, though not simultaneously, but they are coupled to each other. At the very beginning, the breakdown happens in bubble #1. After a time delay, this breakdown promotes the breakdown in bubble #2. The information on this relay race scenario of the breakdowns in two bubbles is presented in Fig. 44. Notice the fact that the current pulses related to the breakdowns in bubble #2 are higher in the amplitude compared to those in bubble #1 (see Fig. 44(a)). A reason is that an air bubble without plasma has high impedance restricting the electric current. Therefore, if bubble #1 is filled with plasma created by breakdown, it diminishes its impedance and allows one to increase markedly the current through bubble #2 during its breakdown.

Note that the luminosity from the bubble during its low-current partial breakdown is created mostly by the ionization wave front characterized by a strong electric field. The ionization wave propagates only for a short time, and after the wave stops, the luminosity appreciably diminishes despite the existence of the plasma created by the ionization wave. The mentioned

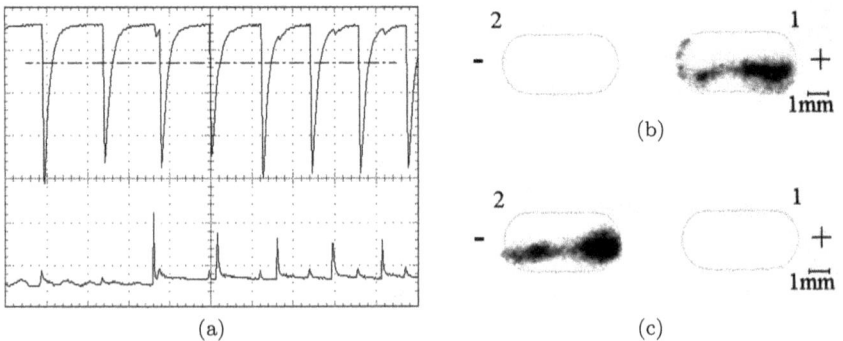

Fig. 44. Set of the experimental data illustrating the events at the second stage. (a) PMT signal from bubble #1 (upper panel) and the electric current (lower panel); $U = +10.5\,\text{kV}$. Axes Scales: $[t] = 1\,\mu/\text{div}$, $[I] = 1\,\text{mA/div}$. (b) Snapshots of discharges in bubble #1, and (c) in bubble #2. With permission from [89]. © 2007 ICPIG.

fact explains why the luminosity from bubble #1 practically disappears at the moment of the delayed breakdown in bubble #2 (see Figs. 44(b) and (c)). The narrowness of the PMT signals, the width of which is less than the delay between the current pulses, corresponding to the breakdowns in bubbles #1 and #2 (see Fig. 43(a)), agrees with this statement. The delay in between the breakdowns in bubbles #1 and #2 gradually diminishes by the time of the termination of the second stage.

Third stage: At this stage, discharge current does not fall down to zero over the whole period between two neighboring current pulses (Figs. 42(a) and 45). It means there is a quasi-steady-state component in the electric current. The magnitude of this component is sufficient to support permanently the conductivity in bubble #1 but not enough to provide the plasma stability in bubble #2. The snapshots of bubbles at the third state presented in Fig. 45 prove this statement as well: Bubble #1 permanently emits the light but bubble #2 emits the light only during the current pulses. The same behavior in the bubble light emission is demonstrated also in Fig. 42(b), where the PMT signal corresponding to bubble #1 does not fall down to zero over the whole period between two neighboring pulses. So, one can state that the breakdowns in bubble #2 are predominantly responsible for the regular current pulses at the third stage. The frequency of the regular current pulses exceeds 1 MHz, which is close to the frequency of the regular pulses at the third stage of the breakdown in the tube with a single bubble. Full breakdown of bubble #2 is completed to the end of the third stage. It means both bubbles are connected with a steady-state plasma channels, and the fourth stage with a steady-state discharge current is established.

Breakdown delay: The first breakdowns occur after the voltage pulse is applied. The time lag depends on bubble location, and polarity and

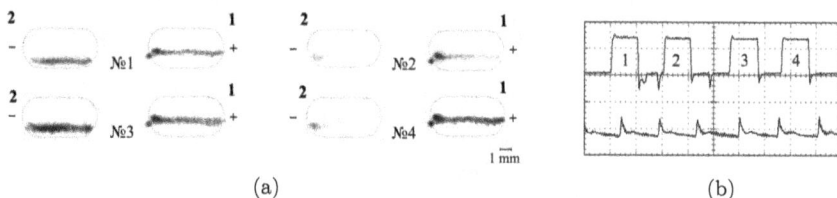

(a)   (b)

Fig. 45. Set of the experimental data illustrating the events at the third stage. (a) Snapshots of the bubbles at moments (1–4) corresponding to those in (b). Images 1 and 2 enumerate the bubbles. (b) The current waveform at the third stage (bottom). Axes Scales: $[t] = 0.5\,\mu s/\text{div}$, $[I] = 3\,\text{mA/div}$. The synchronized pulses showing the moments (1–4) for the snapshots (top). $U = +16\,\text{kV}$. With permission from [89]. © 2007 ICPIG.

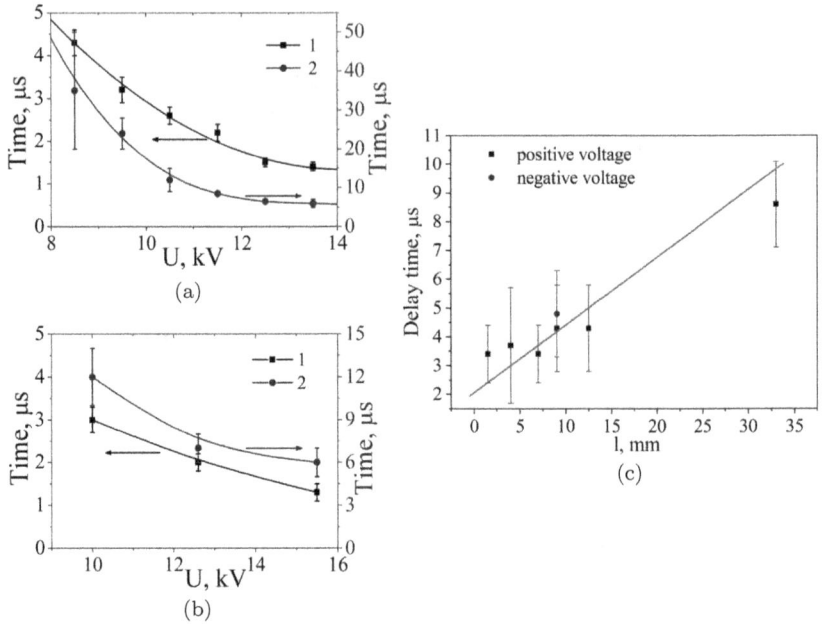

Fig. 46. (a) and (b) Dependence of the time lag for breakdowns in bubbles #1 (full squares) and #2 (full circles) of 4 mm in length vs applied voltage U [kV]. The water gap in between bubbles is 5 mm in (a), and 9 mm in (b), for a positive polarity HV pulse. (c) Time delay between the first partial breakdowns in bubbles #1 and #2 vs gap between the bubbles. $U = +15.5$ kV. With permission from [89]. © 2007 ICPIG.

amplitude of HV pulse. In general, the closer to the HV electrode and the higher the HV amplitude, the shorter the lag (see Figs. 46(a) and 46(b)). These figures also show that in the breakdown within neighboring bubbles, the time lag strongly depends on the polarity of the pulse, diminishing with the amplitude of the HV pulse, and growing with distance between bubbles (Fig. 46(c)).

19. Slow-Developing Streamer Discharges on Liquids

Many of the plasma–liquid systems are based on the use of transient streamer or spark discharges which are generated directly on the surface of a liquid. In general, the specifics of plasma–liquid interaction depend strongly on the magnitude of the electric current supporting the surface plasma. In turn, this magnitude depends on the regime of surface discharge, either completed (full) or not completed (partial) discharges. For instance, pulsed

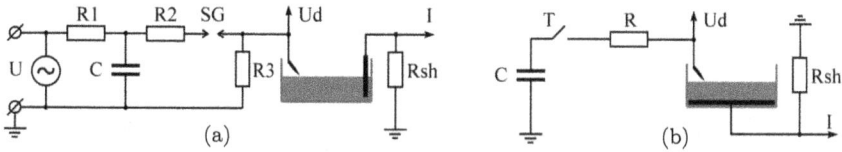

Fig. 47. General sketch of two experimental set-ups with different locations of the grounded electrode: (a) At the opposite side of the cuvette, and (b) at the bottom of the cuvette. HV needle is a sharpened wire of 0.5 mm in diameter. The width and deepness of the water basin are variable parameters; length of the basin is 150 mm; the water basin is made of acrylic plastic. U is a sinus voltage of 50 Hz in frequency; U_d is the voltage applied to the discharge; $C = 12.5$ nF or 50 nF; SG is the spark gap; T is a thyratron; $R = 400\,\Omega$ or $1\,\mathrm{k}\Omega$; $R_1 = 100\,\mathrm{k}\Omega$; $R_2 = 400\,\Omega$; $R_3 = 1\,\mathrm{k}\Omega$; $R_{sh} = 1\,\Omega$. With permission from [47]. © 2017 Springer Nature.

high-current completed discharges on a water surface have been investigated [90–92]. These discharges can only be generated using the electrode system set-up shown in Fig. 47(a).

The completed discharges exist in the form of a bright current filament (transient spark) completely shunting the inter-electrode gap and having a high gas temperature ($\approx$5000 K). As a result, a thin layer of hot steam appears between the high-current filament and water, elevating the filament above the water surface. Thus, the transport of reactive species from the current filament into the water will be restricted.

Our main interest lies in not-completed surface discharges which are restricted in their duration by the duration of the applied HV voltage pulse. In this case, the NTP appears on a liquid in the form of numerous short-living, chaotically spreading and readily branching thin current filaments, i.e., streamers which tightly adjoin the liquid. Due to that, the reactive plasma species can be quickly and effectively transferred into the liquid. At first glance, the streamer discharge on liquid looks like the surface dielectric barrier discharge (SDBD) on solid dielectric barriers. Indeed, in some cases streamer discharge on a liquid can be considered as an SDBD with a liquid dielectric barrier. However, it is necessary to take into consideration the following things.

As mentioned above (see Section 17), the conducting/insulating behavior of a liquid is determined by the Maxwellian relaxation time, $\tau_M \simeq \varepsilon\varepsilon_0/\sigma$. For pure water, τ_M is of the order of a few microseconds. For much shorter times, that is, $\tau \ll \tau_M$, water behaves as a dielectric medium with dielectric constant $\varepsilon \approx 80 - 82$, and surface streamers develop fast on the "dielectric" water. If the liquid is exposed to an electric pulse with a long duration,

$\tau \gg \tau_M$, the water behaves as a resistive medium whose resistance depends on its conductivity σ.

In this case, surface streamers develop slowly because the conductive current in liquid bulk plays a dominant role compared to the displacement current. Moreover, an essential increase in liquid conductivity can lead to the impossibility of streamer propagation because of the shunting of the surface streamer by liquid. However, the formation of slowly developing streamers is a simple task because it does not require the use of sophisticated and expensive HV nanosecond generators. This circumstance determines the attractiveness of slowly developing streamers on water from a practical point of view.

Next, we present the results of surface streamers' behavior slowly propagating on a tap and distilled water. The sketches of the used plasma–liquid systems are those shown in Fig. 47. The streamers were initiated predominantly by an HV needle disposed 1 mm above the water. The grounded electrode is placed at the bottom of a water basin (see Fig. 47(b)).

The integrated images of positive surface streamers propagating on a shallow and deep open tap water are shown in Fig. 48. In these figures, the applied voltage amplitude U is a variable parameter. Close examination of the streamers' structure reveals that the main thick streamers are always surrounded with numerous thin, short and dim streamers. Besides, there is a diffuse glow plasma homogeneously filling in all space in between thin streamers. Therefore, the surface streamer discharge enters in contact with water and with the diffuse glow plasma as well. Total water area occupied by streamers and glow plasma increases with the applied voltage amplitude and water depth. Maximum length of surface streamers that freely propagate in all the radial directions from the initiating HV needle increases with the voltage amplitude, but are not longer than 30–35 mm.

A further improvement of the plasma-liquid system is the development of a sectioned plasma system, where every streamer is ballasted by an individual resistor. The obtained experimental results are presented in Fig. 49. Comparison of the streamers' length, shown in Figs. 48 and 49, regarding the areas of the liquid covered by them, allows one to draw the following conclusion: The approach based on the use of the sectioned and ballasted electrode system activating a larger liquid area is a very promising one.

Of course, in order to get maximum efficiency of such a plasma–liquid system, it is necessary to take into consideration also the h/σ ratio, where h is the depth of a liquid measured from the plane grounded electrode, and σ, the electrical conductivity of the liquid to be treated. As it turned out,

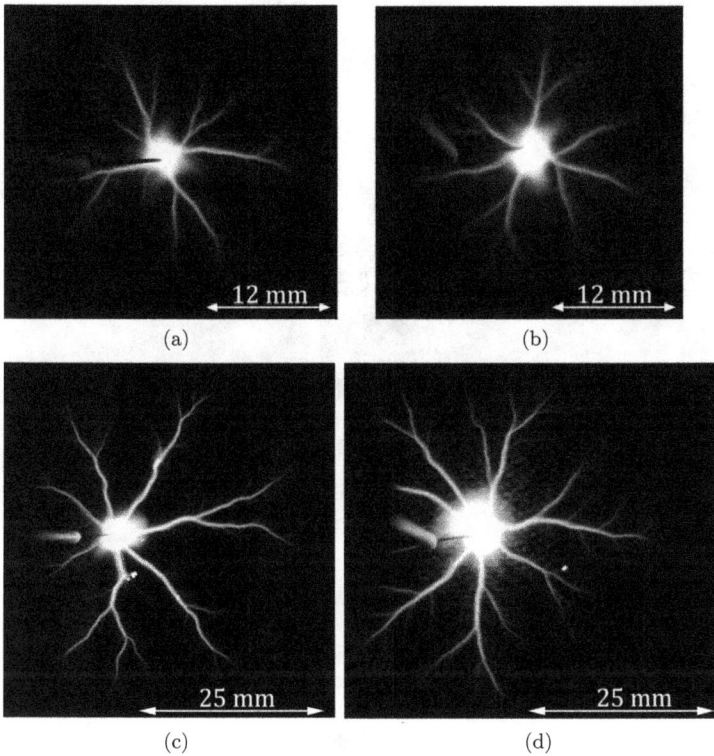

Fig. 48. Top view of positive streamers on open surface of a deep and shallow tap water ($\sigma \approx 700\,\mu\text{S/cm}$). The photos are taken with a long exposure time. U is the initial voltage on the capacitor C; h is the water depth. (a) $h = 65\,\text{mm}$; $U = 7\,\text{kV}$. (b) $h = 5\,\text{mm}$; $U = 7\,\text{kV}$. (c) $h = 65\,\text{mm}$; $U = 23.5\,\text{kV}$. (d) $h = 5\,\text{mm}$; $U = 23.5\,\text{kV}$. The scale is the same for all the pictures. With permission from [93]. © 2017 IOP Publishing.

the higher this ratio, the larger the area of liquid that is covered by surface streamers.

Strong influence of the aqueous solution conductivity on the appearance of streamers and their length in the sectioned system can be seen in Fig. 50. The surface streamers on distilled water are generated much easier, that is, at lower applied voltage and have much longer length compared to those on tap water. Their brightness and branching are pronounced to a lesser degree in comparison with streamers on tap water. Note that a distance in between needles to initiate the streamers is 12 mm, that is two times longer compared to the one in Fig. 49, and, in the event, it is enough to diminish the streamers competition for the space.

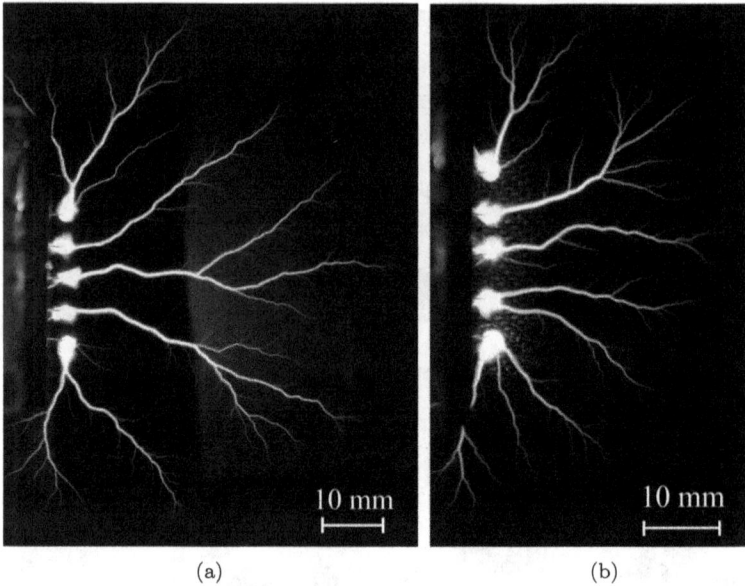

Fig. 49. The images of the sectioned streamers initiated on open tap water. In all cases, the applied voltage amplitude is 23.5 kV. (a) The water depth is 65 mm; length of the longest streamer is 50–55 mm. (b) The water depth is 5 mm; length of the longest streamer is 25–30 mm. The distance in between needles initiating streamers is 6 mm, that is, streamers are close to each other and compete for the space on the water "repulsing" each other. Therefore, their paths are not in parallel but diverge markedly forming fan-shaped configuration. With permission from [93]. © 2017 IOP Publishing.

There is another approach that allows one to provide further increasing of the streamer length. This can be done by confining the streamer propagation within a long and narrow dielectric slit placed on water. The proper data for different applied voltages and slit widths are given in Fig. 51. It turns out that there is a close correlation between the slit width and streamer branching, i.e., the narrower the slit, the smaller the streamer branching, and the longer the streamer length. Maximum streamer length of 110 mm has been reached at the applied voltage of 23.5 kV and slit width of 2.6 mm. The appropriate experimental results are presented in Fig. 51.

What may happen if many surface streamers are initiated simultaneously inside separate and parallel narrow and long dielectric channels placed on water? The experimental results with the sectioned plasma-liquid system are presented in Fig. 52. Here, all the surface streamers were excited simultaneously, and each streamer propagates inside its own narrow and long dielectric channel of 2.5 mm width. These results show that despite

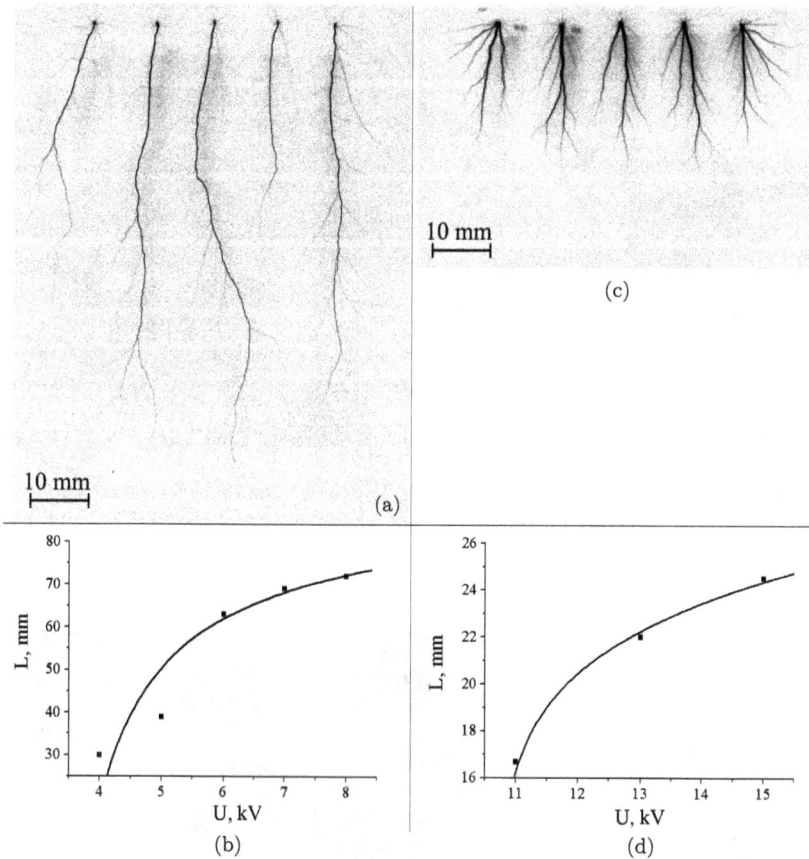

Fig. 50. (a) and (c) Negative images of the surface streamers generated by the sectioned discharge on the open water, and (b) and (d) the dependence of the maximum streamer length L [mm] vs the applied voltage U [kV]. The water depth is 13 mm. Distance in between needles initiating streamers is 12 mm. (a,b) Distilled water, for $U = 8$ kV. (c,d) Tap water, for $U = 15$ kV. With permission from [93]. © 2017 IOP Publishing.

the isolation of streamers within individual channels, they "feel" their neighbors, yielding a competition between them.

This competition exhibits itself in the different lengths of neighboring streamers, and is more pronounced the deeper the water beneath the streamers. In summary, a general trend for streamers' behavior on water can be formulated as follows: Maximum streamer length increases with a water depth in all configurations of the surface streamer discharges on a liquid.

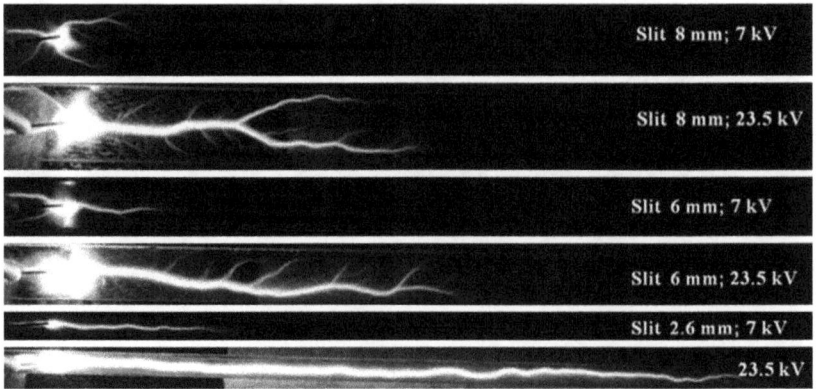

Fig. 51. Top views of the isolated positive streamers on shallow tap water (the depth $h = 5$ mm). Each streamer was excited separately from others in water basin and propagated within long, straight and narrow dielectric channels placed on surface of the water. As to the scale in the photos: The longest streamer length is 110 mm. Such a streamer was obtained in a channel with a slit width of 2.6 mm. With permission from [93]. © 2017 IOP Publishing.

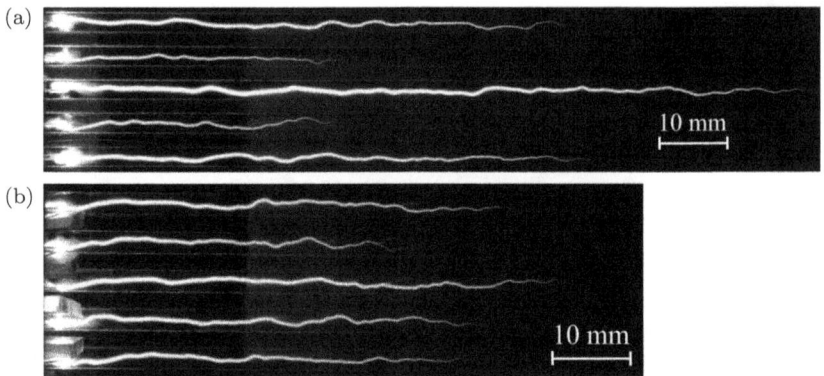

Fig. 52. The images of the sectioned streamers initiated in the separated dielectric channels placed on water. These images show the competition between streamers in the course of their propagation. In all cases, the amplitude of applied voltage is 23.5 kV and slit width is 2.5 mm. (a) The water depth is 65 mm; length of the longest streamer is 115 mm. (b) The water depth is 5 mm; length of the longest streamer is 65 mm. With permission from [93]. © 2017 IOP Publishing.

As shown above, narrow and straight-line slit in a dielectric channel placed on water strongly influences the streamers resulting in their possibility to extend to long distances. What about existence of such influence in the case of slits of a non-straight-line configuration?

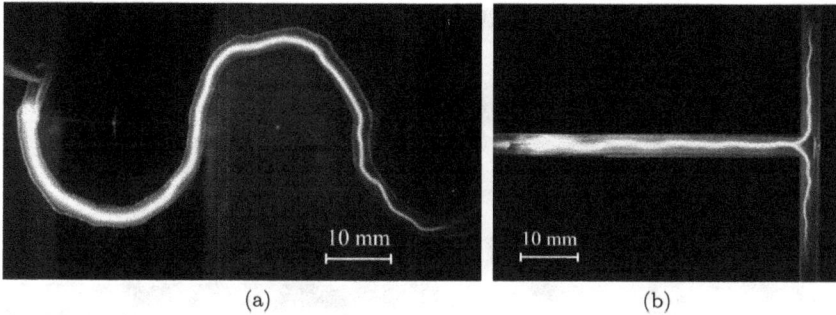

Fig. 53. Control of the streamer configuration by the shape of slit in a dielectric plate placed on water. Water depth is 65 mm; slit width is 2.5 mm; the applied voltage is $U = 23.5$ kV. (a) Sinusoidal configuration of the slit; (b) T-form configuration of the slit. With permission from [93]. © 2017 IOP Publishing.

Experiments were performed using dielectric plates having the slits of different non-straight-line configurations. As it turned out, it is quite possible to control the streamers' trajectory and even forcibly to split them. Results are presented in Fig. 53. Note the double splitting (bifurcation) of the streamer in the channel, leading to a T-shaped form, occurring before reaching the perpendicular wall, thus avoiding a collision with it.

Another interesting question related to the streamer guided by a narrow dielectric channel sounds as follows: is it possible for a surface streamer to penetrate through a thin dielectric film placed on its way perpendicularly to its propagation? If yes, what is the mechanism for the current transfer at this place: The displacement current through solid films or the conductivity current through film pre-broken by the streamer? The answer to this question is not so evident. Indeed, it has been shown [94] that plasma jet can pass through a thin dielectric film due to the re-ignition of a discharge behind a dielectric film, that is, without breakdown of the film. However, in [93] it is shown that the volume positive streamer passes through a thin film of transformer oil only after the film breakdown.

To answer the question stated above, the experiment with a long isolated streamer initiated by voltage of 23.5 kV was performed. The streamer propagates in a narrow slit between two dielectric walls placed on the water surface. The walls have 10 mm thickness and 4.5 mm height; width of the slit formed by walls is 2.8 mm. A thin film of mica with thickness of 10 μm, height of 10 mm and dielectric permittivity $\varepsilon = 6$ was placed across the slit at a distance of 25 mm away from the tip of the HV needle.

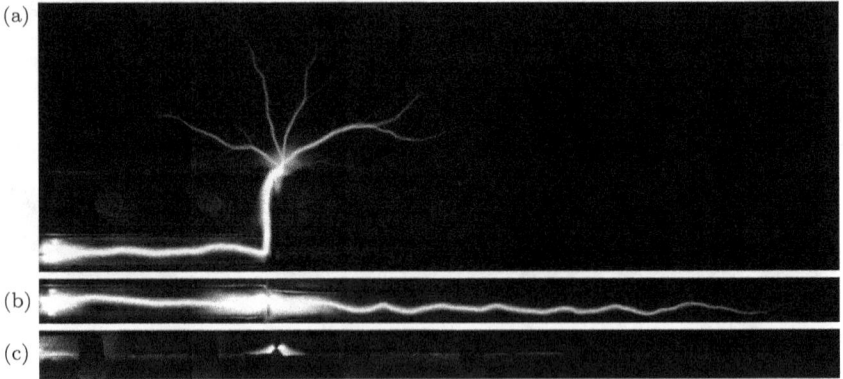

Fig. 54. About streamer possibility to propagate through a thin mica film placed across narrow slit. (a) The image (top view) of the streamer formed by the first voltage pulse: streamer does not penetrate through mica film due to displacement current but goes aside and jumps over a thick dielectric wall to the water. (b) The image (top view) of the streamer formed by the second voltage pulse: streamer eventually induces the electrical breakdown of the film followed by its perforation, after that penetrates through a small hole $(30\,\mu m)$ and propagates through a slit to the other side of the film. (c) The image (side view) of the streamer formed by the third voltage pulse: Streamer freely penetrates through a small hole created by the preceding streamer; the hole was created a little above the water surface due to the formation of meniscus at both sides of the thin film. With permission from [93]. © 2017 IOP Publishing.

The obtained answer summarily sounds as follows. A single surface streamer is not able to penetrate through a thin dielectric film, but a sequence of two streamers can do that due to perforation of the film by its electrical breakdown being performed by the second streamer. The relevant visual information clarifying this process is presented in Fig. 54.

Based on the results presented above, one can state that: Placing a dielectric plate on tap water, with long and narrow slits of the needed configuration, enables one to increase appreciably the surface streamer length, control its trajectory, forcibly to split the streamer and even to destroy the remote thin dielectric film once reached. The above considerations look very attractive from a practical point of view and can be used in different applications.

To complete the physical picture about the development of streamers on the water surface, additional information about the behavior of the streamer speed and its length is presented in Fig. 55. One can see that the instant streamer head speed appreciably diminishes with an increase of the streamer length. Nevertheless, the streamer extends practically up to

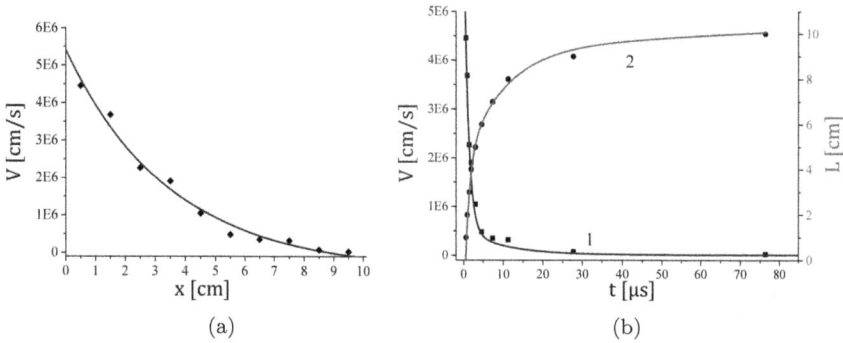

Fig. 55. The streamer parameters characterizing its propagation on tap water in narrow dielectric channel of 2.8 mm width. The applied voltage is $U = 23.5\,$kV. (a) The streamer head velocity vs instant streamer length. (b) The streamer head velocity (curve 1) and instant streamer length (curve 2) vs time. Markers are the experimental data; solid curves are the smooth approximations of the experiment. With permission from [93]. © 2017 IOP Publishing.

90% of its maximum length very quickly (approximately in $20\,\mu$s). After that, due to the decreasing discharge current, the streamer continues the extension but with an essentially lower speed, to stop when its maximum length is reached. After stopping, the surface streamer does not disappear simultaneously at once all over its length. This happens due to the slowly diminishing streamer length toward the HV electrode.

20. Spatial Memory of Surface Streamers Periodically Generated on Water

There is another important question relevant to plasma–liquid systems that is based on the use of surface streamers periodically forming on the liquid to be treated. This question can be formulateds as follows: Is there any spatial memory for such streamers? In other words, does the subsequent surface streamer follow along the same trajectory that was opened by the preceding streamer? It is well known that spatial memory is observed for streamers in a dielectric barrier discharge [95,96]. One of the reasons is that the dielectric barrier keeps the surface charge deposited by the streamer on the solid barrier at the preceding half-cycle till the subsequent half-cycle, and this charge promotes the same localization of the subsequent streamer. However, this mechanism does not work in the case of water, because water has some conductivity, and the electric charge, therefore, cannot be accumulated on

(a)

(b)

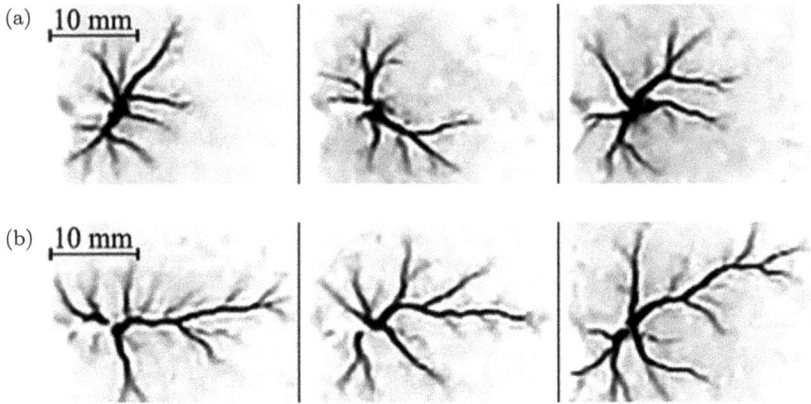

Fig. 56. The typical sequent of negative images of streamers periodically sliding on tap water. Three upper and bottom images correspond to the sequence of three successive current pulses in: (a) Negative half-cycle, and (b) positive one. The exposure time of each picture is $800\,\mu s$. Time interval in between the neighbor pictures is $833\,\mu s$, *i.e.*, each frame corresponds only to a single HV pulse. The pictures prove that spatial memory of streamers on water is absent — the configurations (or trajectories) of streamers formed one after another do not repeat one another. With permission from [47]. © 2017 Springer Nature.

the water surface. This is why the streamers on water will have no spatial memory in positive and negative half-cycles of a periodical applied voltage.

This statement is proved by the images presented in Fig. 56. These images show the sequences of negative and positive streamers generated on tap water. The surface streamers were initiated with a sharpened thin metallic needle placed above the water and stressed with a pulsed-periodical high voltage with amplitude $U = 16.5\,\text{kV}$. The images in Fig. 56 are typical ones and were taken with a fast frame camera.

The three upper and bottom images correspond to the streamers initiated by three successive HV pulses in negative (a) and positive (b) half-cycles, respectively. One can clearly see in these pictures that there is no similarity of the streamer images shown in the neighboring photos. In other words, there is no spatial memory for surface streamers periodically sliding on the water surface. In general, the memory can be absent also for surface streamers sliding on a solid barrier having some finite conductivity. In fact, the absence of spatial memory of streamers sliding on liquid is a positive effect because it allows one to evenly activate the water surface. The results presented in this paragraph can be of interest to scientists and to engineers dealing with the development of plasma–liquid systems for various purposes.

21. Conclusions

The review of the plasma–liquid systems based on the use of different types of moderate power streamer discharges IN/ON aqueous solutions is done. Attention has been primarily paid to the physical properties of these discharges and their electrode configurations, rather than to chemical processes happening in liquids after their activation by reactive plasma species generated by the streamer discharges. Preferentially, the pulsed slow-developing streamer discharges in/on water under conditions when the displacement current plays a negligible role in the establishment of these discharges were the subjects of this review. A reason is that these discharges can be driven by commercially available HV power suppliers generating electrical pulses with a relatively low rise-time of microsecond or sub-microsecond range. In contrast, fast-developing streamer discharges require the usage of sophisticated and expensive HV-pulsed generators of nanosecond or sub-nanosecond range, which can be an obstacle when scaling nanosecond plasma–liquid systems to an industrial level. The information presented in this review can be useful for scientists studying the physics and chemistry of electrical discharges in/on liquid and for engineers dealing with both the elaboration of plasma–liquid systems and the using of non-thermal plasma for various applications [49, 51].

Acknowledgements

I would like to express my sincere gratitude to my wife Lidia Akisheva and Dr. Alexander Petryakov for their time and valuable help in preparing the manuscript.

References

[1] C. E. Anderson, N. R. Cha, A. D. Lindsay, D. S. Clark, and D. B. Graves, The Role of Interfacial Reactions in Determining Plasma–Liquid Chemistry, *Plasma Chem. Plasma Process.* **36**(6), 1393–1415 (2016).
[2] S. Samukawa, M. Hori, S. Rauf, K. Tachibana, P. Bruggeman, G. Kroesen, J. C. Whitehead, A. B. Murphy, A. F. Gutsol, S. Starikovskaia, *et al.*, The 2012 plasma roadmap, *J. Phys. D: App. Phys.* **45**(25), 253001 (2012).
[3] I. Adamovich, S. D. Baalrud, A. Bogaerts, P. J. Bruggeman, M. Cappelli, V. Colombo, U. Czarnetzki, U. Ebert, J. G. Eden, P. Favia, *et al.*, The 2017

plasma roadmap: Low temperature plasma science and technology, *J. Phys. D: App. Phys.* **50**(32), 323001 (2017).

[4] J. Kruszelnicki, A. M. Lietz, and M. J. Kushner. Interaction between atmospheric pressure plasmas and liquid micro-droplets. In *Proceedings of the International Conference on Plasmas with Liquids (ICPL 2017)*, pp. 5–9. Prague, Czech Republic (March, 2017).

[5] P. Bruggeman and C. Leys, Non-thermal plasmas in and in contact with liquids, *J. Phys. D: App. Phys.* **42**(5), 053001 (2009).

[6] J. Hoigne and H. Bader, The role of hydroxyl radical reactions in ozonation processes in aqueous solutions, *Water Res.* **10**(5), 377–386 (1976).

[7] W. H. Glaze, J.-W. Kang, and D. H. Chapin, The chemistry of water treatment processes involving ozone, hydrogen peroxide and ultraviolet radiation, *Ozone: Sci. Eng.* **9**(4), 335–352 (1987).

[8] M. Rea and K. Yan, Energization for pulse corona induced chemical processes. In B. M. Penetrante and S. E. Schultheis (eds.), *Non-thermal plasma techniques for pollution control*, vol. G34, *NATO ASI Series*, pp. 191–204. Springer-Verlag Berlin Heidelberg (1993).

[9] A. A. Joshi, B. R. Locke, P. Arce, and W. C. Finney, Formation of hydroxyl radicals, hydrogen peroxide and aqueous electrons by pulsed streamer corona discharge in aqueous solution, *J. Hazard. Mat.* **41**(1), 3–30 (1995).

[10] M. Sato, T. Ohgiyama, and S. Clements, Formation of chemical species and their effects on microorganisms using a pulsed high voltage discharge in water, *IEEE Trans. Ind. Appl.* **32**(1), 106–112 (1996).

[11] P. Sunka, V. Babický, M. Clupek, P. Lukes, M. Simek, J. Schmidt, and M. Cernak, Generation of chemically active species by electrical discharges in water, *Plasma Sources Sci. Technol.* **8**(2), 258 (1999).

[12] D. R. Grymonpré, A. K. Sharma, W. C. Finney, and B. R. Locke, The role of fenton's reaction in aqueous phase pulsed streamer corona reactors, *Chem. Eng. J.* **82**(1-3), 189–207 (2001).

[13] N. Sano, T. Fujimoto, T. Kawashima, D. Yamamoto, T. Kanki, and A. Toyoda, Influence of dissolved inorganic additives on decomposition of phenol and acetic acid in water by direct contact of gas corona discharge, *Sep. Purif. Technol.* **37**(2), 169–175 (2004).

[14] Z. Stará and F. Krcma. Influence of OH radicals on organic dyes in DC diaphragm discharge in water solutions. In *Proceedings of XXVIIth ICPIG*, p. 226. Veldhoven, The Netherlands (18–22 July, 2005).

[15] M. Sahni and B. R. Locke, Quantification of hydroxyl radicals produced in aqueous phase pulsed electrical discharge reactors, *Ind. Eng. Chem. Res.* **45**(17), 5819–5825 (2006).

[16] J.-L. Brisset, D. Moussa, A. Doubla, E. Hnatiuc, B. Hnatiuc, G. Kamgang Youbi, J.-M. Herry, M. Naitali, and M.-N. Bellon-Fontaine, Chemical reactivity of discharges and temporal post-discharges in plasma treatment of aqueous media: Examples of gliding discharge treated solutions, *Ind. Eng. Chem. Res.* **47**(16), 5761–5781 (2008).

[17] B. R. Locke and S. M. Thagard, Analysis and review of chemical reactions and transport processes in pulsed electrical discharge plasma formed directly in liquid water, *Plasma Chem. Plasma Process.* **32**(5), 875–917 (2012).

[18] P. Rumbach, M. Witzke, R. M. Sankaran, and D. B. Go. Plasma-liquid interactions: separating electrolytic reactions from plasma/gas phase reactions. In *Proceedings of the ESA Annual Meeting on Electrostatics*, pp. 1–8 (2013).

[19] C. Yang, Q. Guangzhou, L. Tengfei, N. Jiang, and W. Tiecheng, Review on reactive species in water treatment using electrical discharge plasma: Formation, measurement, mechanisms and mass transfer, *Plasma Sci. Technol.* **20**(10), 102001 (2018).

[20] P. J. Bruggeman, M. J. Kushner, B. R. Locke, J. G. E. Gardeniers, W. G. Graham, D. B. Graves, R. C. H. M. Hofman-Caris, D. Maric, J. P. Reid, E. Ceriani, *et al.*, Plasma–liquid interactions: A review and roadmap, *Plasma Sources Sci. Technol.* **25**(5), 053002 (2016).

[21] J. Pawlat. *Electrical discharges in humid environments: Generators, effects, application.* PhD thesis, Monografie Politechnika Lubelska, Lublin, Poland (2013, ISBN 978-83-63569-37-2).

[22] H. Akiyama, Streamer discharges in liquids and their applications, *IEEE Trans. Dielectr. Electr. Insul.* **7**(5), 646–653 (2000).

[23] P. Vanraes, A. Y. Nikiforov, and C. Leys, Electrical discharge in water treatment technology for micropollutant decomposition. In T. Mieno (ed.), *Progress in Physical States and Chemical Reactions*, pp. 428–478. Plasma Science and Technology, INTECH London (2016).

[24] J. E. Foster, Plasma-based water purification: Challenges and prospects for the future, *Phys. Plasmas.* **24**(5), 055501 (2017).

[25] Y. Yang, Y. I. Cho, and A. Fridman, *Plasma discharge in liquid: Water treatment and applications.* CRC press, Taylor & Francis, New York (2012).

[26] P. Vanraes and A. Bogaerts, Plasma physics of liquids — A focused review, *App. Phys. Rev.* **5**(3), 031103 (2018).

[27] Y. Akishev, F. Arefi-Khonsari, A. Demir, M. Grushin, V. Karalnik, A. Petryakov, and N. Trushkin, The interaction of positive streamers with bubbles floating on a liquid surface, *Plasma Sources Sci. Technol.* **24**(6), 065021 (2015).

[28] V. L. Goryachev, F. G. Rutberg, and V. N. Fedyukovich, Some properties of pulse–periodic discharge in water with energy per pulse of 1J for water purification, *High Temp.* **34**, 746–749 (1996).

[29] V. Bystritskii, T. Wood, Y. Yankelevich, S. Chauhan, D. Yee, and F. Wessel. Pulsed power for advanced waste water remediation. In *Digest of Technical Papers. 11th IEEE International Pulsed Power Conference (Cat. No. 97CH36127)*, pp. 79–84 (1997).

[30] B. Sun, M. Sato, A. Harano, and J. Clements, Non-uniform pulse discharge-induced radical production in distilled water, *J. Electrost.* **43**(2), 115–126 (1998).

[31] B. Sun, M. Sato, and J. Clements, Use of a pulsed high-voltage discharge for removal of organic compounds in aqueous solution, *J. Phys. D: App. Phys.* **32**(15), 1908 (1999).

[32] W. F. L. M. Hoeben, E. M. Van Veldhuizen, W. R. Rutgers, and G. M. W. Kroesen, Gas phase corona discharges for oxidation of phenol in an aqueous solution, *J. Phys. D: App. Phys.* **32**(24), L133 (1999).

[33] A. K. Sharma, G. B. Josephson, D. M. Camaioni, and S. C. Goheen, Destruction of pentachlorophenol using glow discharge plasma process, *Environ. Sci. Technol.* **34**(11), 2267–2272 (2000).

[34] E. M. van Veldhuizen, ed., *Electrical discharges for environmental purposes: Fundamentals and applications.* Nova Science, Huntington NY (2000).

[35] A. M. Anpilov, E. M. Barkhudarov, Y. B. Bark, Y. V. Zadiraka, M. Christofi, Y. N. Kozlov, I. A. Kossyi, V. A. Kop'ev, V. P. Silakov, M. I. Taktakishvili, *et al.*, Electric discharge in water as a source of UV radiation, ozone and hydrogen peroxide, *J. Phys. D: App. Phys.* **34**(6), 993 (2001).

[36] M. A. Malik, A. Ghaffar, and S. A. Malik, Water purification by electrical discharges, *Plasma Sources Sci. Technol.* **10**(1), 82 (2001).

[37] M. A. Malik, A. Ghaffar, K. Ahmed, *et al.*, Synergistic effect of pulsed corona discharges and ozonation on decolourization of methylene blue in water, *Plasma Sources Sci. Technol.* **11**(3), 236 (2002).

[38] M. A. Malik, Synergistic effect of plasma-catalyst and ozone in a pulsed corona discharge reactor on the decomposition of organic pollutants in water, *Plasma Sources Sci. Technol.* **12**(4), S26 (2003).

[39] S. Kunitomo, T. Ohbo, and B. Sun, The effects of using various types of pulsed discharge reactors for phenol removal in waste water, *J. Adv. Oxid. Technol.* **6**(1), 70–73 (2003).

[40] N. Sano, D. Yamamoto, T. Kanki, and A. Toyoda, Decomposition of phenol in water by a cylindrical wetted-wall reactor using direct contact of gas corona discharge, *Ind. Eng. Chem. Res.* **42**(22), 5423–5428 (2003).

[41] P. Lukes, M. Clupek, V. Babicky, V. Janda, and P. Sunka, Generation of ozone by pulsed corona discharge over water surface in hybrid gas–liquid electrical discharge reactor, *J. Phys. D: App. Phys.* **38**(3), 409 (2005).

[42] P. Bruggeman and D. C. Schram, On OH production in water containing atmospheric pressure plasmas, *Plasma Sources Sci. Technol.* **19**(4), 045025–045034 (2010).

[43] R. P. Joshi and S. M. Thagard, Streamer-like electrical discharges in water: Part I. Fundamental mechanisms, *Plasma Chem. Plasma Process.* **33**(1), 1–15 (2013).

[44] R. P. Joshi and S. M. Thagard, Streamer-like electrical discharges in water: Part II. Environmental applications, *Plasma Chem. Plasma Process.* **33**(1), 17–49 (2013).

[45] R. J. Wandell and B. R. Locke, Low-power pulsed plasma discharge in a water film reactor, *IEEE Trans. Plasma Sci.* **42**(10), 2634–2635 (2014).

[46] S. H. R. Ruma, H. Hosseini, K. Yoshihara, M. Akiyama, T. Sakugawa, P. Lukeš, and H. Akiyama, Properties of water surface discharge at different pulse repetition rates, *J. App. Phys.* **116**(12), 123304 (2014).

[47] Y. S. Akishev, V. Karalnik, M. Medvedev, A. Petryakov, N. Trushkin, and A. Shafikov, Streamers sliding on a water surface, *Eur. Phys. J. App. Phys.* **79**(1), 10803 (2017).

[48] Y. Akishev, M. Grushin, V. Karalnik, N. Trushkin, V. Kholodenko, V. Chugunov, E. Kobzev, N. Zhirkova, I. Irkhina, and G. Kireev,

Atmospheric-pressure, nonthermal plasma sterilization of microorganisms in liquids and on surfaces, *Pure Appl. Chem.* **80**(9), 1953–1969 (2008).

[49] Y. S. Akishev, G. I. Aponin, M. E. Grushin, V. B. Karalnik, A. V. Petryakov, and N. I. Trushkin, Determination of local ionic wind velocity in corona by laser interferometer taking into consideration the charging-discharging of probe particles, *IEEE Trans. Plasma Sci.* **37**(6), 1034–1046 (2009).

[50] Y. S. Akishev, M. E. Grushin, V. B. Karalnik, A. E. Monich, M. V. Pankin, N. I. Trushkin, V. P. Kholodenko, V. A. Chugunov, N. A. Zhirkova, I. A. Irkhina, *et al.*, Generation of a nonequlibrium plasma in heterophase atmospheric-pressure gas-liquid media and demonstration of its sterilization ability, *Plasma Phys. Rep.* **32**(12), 1052–1061 (2006).

[51] Y. S. Akishev, G. I. Aponin, M. E. Grushin, V. B. Karalnik, A. V. Petryakov, and N. I. Trushkin. Electrical discharges operating above and along a liquid surface. In *The 2nd Annual Meeting on COST Action TD1208 — Electrical Discharges with liquids for Future Applications*, Barcelona, Spain (23–26 February, 2015).

[52] Y. Shidong, C. Fengguo, and M. Jun. Preliminary study on control of water treatment foam by pulsed high voltage discharge. In *Proc. 2nd Int. Conf. on Bioinformatics and Biomedical Engineering*, pp. 3704–3706. Shanghai, China (2008).

[53] N. Y. Babaeva and M. J. Kushner, Effect of inhomogeneities on streamer propagation: I. Intersection with isolated bubbles and particles, *Plasma Sources Sci. Technol.* **18**(3), 035009 (2009).

[54] N. Y. Babaeva and M. J. Kushner, Effect of inhomogeneities on streamer propagation: II. Streamer dynamics in high pressure humid air with bubbles, *Plasma Sources Sci. Technol.* **18**(3), 035010 (2009).

[55] J. C. Bird, R. De Ruiter, L. Courbin, and H. A. Stone, Daughter bubble cascades produced by folding of ruptured thin films, *Nature.* **465**(7299), 759–762 (2010).

[56] A. M. Anpilov, E. M. Barkhudarov, V. A. Kopiev, and I. A. Kossyi. High-voltage pulsed discharge along the water surface: Electric and spectral characteristics. In *Proceedings of XXVIIIth ICPIG*, pp. 1030–1033. Prague, Czech Republic (15–20 July, 2007).

[57] Y. D. Korolev, O. B. Frants, N. V. Landl, V. G. Geyman, and I. B. Matveev, Glow-to-spark transitions in a plasma system for ignition and combustion control, *IEEE Trans. Plasma Sci.* **35**(6), 1651–1657 (2007).

[58] Y. S. Akishev, G. I. Aponin, M. E. Grushin, V. B. Karal'nik, M. V. Pan'kin, A. V. Petryakov, and N. I. Trushkin, Alternating nonsteady gas-discharge modes in an atmospheric-pressure air flow blown through a point-plane gap, *Plasma Phys. Rep.* **34**(4), 312–324 (2008).

[59] Y. D. Korolev, O. B. Frants, V. G. Geyman, V. S. Kasyanov, and N. V. Landl, Transient processes during formation of a steady-state glow discharge in air, *IEEE Trans. Plasma Sci.* **40**(11), 2951–2960 (2012).

[60] C. Yamabe, F. Takeshita, T. Miichi, N. Hayashi, and S. Ihara, Water treatment using discharge on the surface of a bubble in water, *Plasma Process. Polym.* **2**(3), 246–251 (2005).

[61] Y. S. Akishev, G. Aponin, M. Grushin, V. Karalnik, A. Petryakov, and N. Trushkin, Self-running low-frequency pulsed regime of DC electric discharge in gas bubble immersed in a liquid, *J. Optoel. Adv. Mat.* **10**(8), 1917–1921 (2008).

[62] P. Bruggeman, C. Leys, and J. Vierendeels, Experimental investigation of dc electrical breakdown of long vapour bubbles in capillaries, *J. Phys. D: App. Phys.* **40**(7), 1937 (2007).

[63] P. Lukes, M. Clupek, V. Babicky, E. Spetlikova, I. Sisrova, E. Marsalkova, and B. Marsalek, High power DC diaphragm discharge excited in a vapor bubble for the treatment of water, *Plasma Chem. Plasma Process.* **33**(1), 83–95 (2013).

[64] K. Y. Shih and B. R. Locke, Chemical and physical characteristics of pulsed electrical discharge within gas bubbles in aqueous solutions, *Plasma Chem. Plasma Process.* **30**(1), 1–20 (2010).

[65] H. Aoki, K. Kitano, and S. Hamaguchi, Plasma generation inside externally supplied ar bubbles in water, *Plasma Sources Sci. Technol.* **17**(2), 025006 (2008).

[66] R. P. Joshi, J. Qian, G. Zhao, J. Kolb, K. H. Schoenbach, E. Schamiloglu, and J. Gaudet, Are microbubbles necessary for the breakdown of liquid water subjected to a submicrosecond pulse?, *J. App. Phys.* **96**(9), 5129–5139 (2004).

[67] S. M. Korobeinikov, Breakdown initiation in water with the aid of bubbles, *High Temp.* **40**(5), 1670–1679 (2006).

[68] W. An, K. Baumung, and H. Bluhm, Underwater streamer propagation analyzed from detailed measurements of pressure release, *J. App. Phys.* **101**, 053302 (2007).

[69] P. H. Ceccato, O. Guaitella, M. R. Le Gloahec, and A. Rousseau, Time-resolved nanosecond imaging of the propagation of a corona-like plasma discharge in water at positive applied voltage polarity, *J. Phys. D: App. Phys.* **43**(17), 175202 (2010).

[70] H. Fujita, S. Kanazawa, K. Ohtani, A. Komiya, T. Kaneko, and T. Sato, Initiation process and propagation mechanism of positive streamer discharge in water, *J. App. Phys.* **116**(21), 213301 (2014).

[71] W. An, K. Baumung, and H. Bluhm, Underwater streamer propagation analyzed from detailed measurements of pressure release, *J. App. Phys.* **101**(5), 053302 (2007).

[72] H. Fujita, S. Kanazawa, K. Ohtani, A. Komiya, and T. Sato, Spatiotemporal analysis of propagation mechanism of positive primary streamer in water, *J. App. Phys.* **113**(11), 113304 (2013).

[73] M. Dors, E. Metel, and J. Mizeraczyk, Phenol degradation in water by pulsed streamer corona discharge and fenton reaction, *Int. J. Environ. Sci. Technol.* **1**(1), 76–81 (2000).

[74] T. Sato, R. Kumagai, T. Nakajima, and F. Yano. Continuous Visualization of Process of Positive Streamer Initiation, Propagation, Spark, and Bubble Expansion in Water. Water, SHIMADZU (www.shimadzu.com/an/),

Excellence in Science, Application Note, No. 46, Electric and Electronics (August, 2018).

[75] R. Kumagai, S. Kanazawa, K. Ohtani, A. Komiya, T. Kaneko, T. Nakajima, and T. Sato, Propagation and branching process of negative streamers in water, *J. App. Phys.* **124**(16), 163301 (2018).

[76] S. C. Troughton, A. Nominé, A. V. Nominé, G. Henrion, and T. Clyne, Synchronized electrical monitoring and high speed video of bubble growth associated with individual discharges during plasma electrolytic oxidation, *App. Surf. Sci.* **359**, 405–411 (2015).

[77] F. Jaspard-Mécuson, T. Czerwiec, G. Henrion, T. Belmonte, L. Dujardin, A. Viola, and J. Beauvir, Tailored aluminium oxide layers by bipolar current adjustment in the Plasma Electrolytic Oxidation (PEO) process, *Surf. Coatings Technol.* **201**(21), 8677–8682 (2007).

[78] P. Lukes, M. Clupek, V. Babicky, and P. Sunka, Pulsed electrical discharge in water generated using porous-ceramic-coated electrodes, *IEEE Trans. Plasma Sci.* **36**(4), 1146–1147 (2008).

[79] P. Lukes, V. Clupek, M. Babicky, E. Spetlikova, I. Sisrova, E. Marsalkova, and B. Marsalek, High power DC diaphragm discharge excited in a vapor bubble for the treatment of water, *Plasma Chem. Plasma Process.* **33**, 83–95 (2013).

[80] B. S. Sommers and J. E. Foster, Plasma formation in underwater gas bubbles, *Plasma Sources Sci. Technol.* **23**, 015020 (2014).

[81] N. Y. Babaeva and M. J. Kushner, Structure of positive streamers inside gaseous bubbles immersed in liquids, *J. Phys. D: App. Phys.* **42**, 132003 (2009).

[82] W. Tian, K. Tachibana, and M. J. Kushner, Plasmas sustained in bubbles in water: Optical emission and excitation mechanisms, *J. Phys. D: App. Phys.* **47**, 055202 (2014).

[83] B. S. Sommers, J. E. Foster, N. Y. Babaeva, and M. J. Kushner, Observations of electric discharge streamer propagation and capillary oscillations on the surface of air bubbles in water, *J. Phys. D: App. Phys.* **44**, 082001 (2011).

[84] K. Tachibana, Y. Takekata, Y. Mizumoto, H. Motomura, and J. M., Analysis of a pulsed discharge within single bubbles in water under synchronized conditions, *Plasma Sources Sci. Technol.* **20**, 034005 (2011).

[85] P. J. Bruggeman, C. A. Leys, and J. A. Vierendeels, Electrical breakdown of a bubble in a water-filled capillary, *J. App. Phys.* **99**, 116101 (2006).

[86] K. Y. Shih and B. R. R. Locke, Chemical and physical characteristics of pulsed electrical discharge within gas bubbles in aqueous solutions, *Plasma Chem. Plasma Process.* **30**, 1–20 (2010).

[87] H. Yui, Y. Someya, Y. Kusama, K. Kanno, and M. Banno, Atmospheric discharge plasma in aqueous solution: Importance of the generation of water vapor bubbles for plasma onset and physicochemical evolution, *J. App. Phys.* **124**, 103301 (2018).

[88] T. Aka-Ngnui and A. A Beroual, Bubble dynamics and transition into streamers in liquid dielectrics under a high divergent electric field, *J. Phys. D: App. Phys.* **34**, 1408–1412 (2001).

[89] Y. S. Akishev, G. Aponin, M. Grushin, V. Karalnik, F. Petryakov, and N. Trushkin. Dynamics of relay electric breakdown along gas bubble chain in a liquid. In *Proceedings of XXVIIIth ICPIG*, pp. 885–887. Prague, Czech Republic (15–20 July, 2007).

[90] A. M. Anpilov, E. M. Barkhudarov, V. A. Kopiev, I. A. Kossyi, and V. P. Silakov, Atmospheric electric discharge into water, *Plasma Phys. Rep.* **32**, 968–972 (2006).

[91] A. Aleksandrov, V. Bychkov, V. Chernikov, A. Ershov, S. Kamenshikov, and D. Vaulin. Pulsed discharge over a surface of a liquid. In *Proceedings of 47th AIAA Aerospace Sciences Meeting*, p. 1553. Orlando, Florida, USA (5–8 January, 2009).

[92] D. N. Vaulin, A. P. Yershov, S. A. Kamenschikov, and V. A. Chernikov, High-voltage pulse discharge propagating along a water surface, *High Temp.* **49**(3), 356–362 (2011).

[93] Y. S. Akishev, V. B. Karalnik, M. Medvedev, A. V. Petryakov, A. Shafikov, and N. I. Trushkin, Propagation of positive streamers on the surface of shallow as well as deep tap water in wide and narrow dielectric channels, *Plasma Sources Sci. Technol.* **26**, 025004 (2017).

[94] F. Pechereau, J. Jánský, and A. Bourdon, Simulation of the reignition of a discharge behind a dielectric layer in air at atmospheric pressure, *Plasma Sources Sci. Technol.* **21**, 055011 (2012).

[95] Y. V. Yurgelenas and H. E. Wagner, A computational model of a barrier discharge in air at atmospheric pressure: The role of residual surface charges in microdischarge formation, *J. Phys. D: App. Phys.* **39**(18), 4031 (2006).

[96] Y. S. Akishev, G. I. Aponin, A. Balakirev, M. E. Grushin, V. B. Karalnik, A. V. Petryakov, and N. I. Trushkin, Memory and sustention of microdischarges in a steady-state DBD: Volume plasma or surface charge?, *Plasma Sources Sci. Technol.* **20**, 024005 (2011).

Chapter 4

Plasmas as Sources of Nanostructured Materials: Plasma-Assisted Supersonic-Jet Deposition

S. Caldirola[*], H.E. Roman[†] and C. Riccardi[‡]

*Dipartimento di Fisica, Università di Milano-Bicocca,
Piazza della Scienza 3, 20126 Milano, Italy*
[*] *stefano.caldirola@unimib.it*
[†] *hector.roman@unimib.it*
[‡] *claudia.riccardi@unimib.it*

We review the application of plasma techniques to create new advanced materials displaying scale-invariance properties down to the nanoscale. The aim is to describe in sufficient detail to a general reader the principles and effective implementation of the plasma-assisted supersonic jet deposition (PA-SJD) technique. Indeed, its popularity has been increasing during the last decades as a result of its precision and easy controllability in the fabrication of thin films, yielding both a broad range of widths and a variety of morphologies. The presentation puts emphasis on the different processes taking place in the plasma chamber, where the nanoparticles constituting the building blocks of the film are generated, as well as on the behavior in both space and time of the supersonic jet inside the deposition chamber. In the latter, the supersonic jet carries the nanoparticles which impact onto a substrate, on top of which a growing nanostructured thin film is deposited. In addition to the description of the experimental set-up and results, Monte Carlo (MC) simulations are discussed illustrating the dynamics and expansion of Ar^+ ions in a pure argon plasma, that is without the presence of a precursor for generating nanoparticles. The chapter closes with a description of the different morphologies effectively produced in the laboratory using TiO_2-based nanoparticles as the building blocks of the deposited thin films.

Contents

1. Introduction

Newly developed nanostructured films have changed our understanding of material's surface properties, opening up a variety of contemporary applications, which in most cases are tightly related to their morphology [1–10].

Porous materials, for instance, are particularly suitable when the efficiency of a physical/chemical process increases with the increase of their effective surface area. Examples include their use as photocatalysts, to facilitate, for instance, the extinction of pollutants, as photovoltaics and in sensor technologies [11–13], as well as specific electrodes in devices devoted to both the conversion or accumulation of energy, such as solar cells, supercapacitors, etc. [14–20]. Alternatively, more dense and compact nanostructured films find useful applications as protective coatings of different types of surfaces, by exploiting their particular mechanical properties. Recently, a great advance in the field of biomedicine has been possible due to the experience accumulated in the manipulation of sophisticated nanomaterials [21–29].

The large-scale morphology and the small-scale organization of thin films can effectively be tailored by using different deposition techniques, aimed at tuning film thickness, porosity, as well as its electrical conductivity and optical properties. Plasma-based deposition techniques, as discussed also in Chapter 2, have gained popularity among all available deposition processes due to their low temperatures involved, low energy consumption and the multiple chemical scenarios associated with their intrinsic high reactivity [11–13]. This is in keeping with both the energetic radicals and charged particles responsible for the etching and activation of surface reactions determining the film formation [30, 31].

Plasma-assisted supersonic-jet deposition (PA-SJD) is an innovative technique which combines a reactive plasma chamber with a deposition one [32]. The former constitutes a parameter-controlled chemical-reactions environment for the generation of seed aggregates, while in the latter the seeds are transported by a supersonic jet and deposited on a substrate, yielding a nanostructured thin film [33–45].

The set-up of the reactive chamber depends on the type of discharge required to ionize the gas therein present. Here, we consider an inductively coupled plasma (ICP) to an RF field [46–49], and the gas used is a mixture of Ar and O_2. The associated discharge is produced using external electrodes, so that the plasma is generally much cleaner than in the case of capacitively coupled discharges. In fact, the latter typically introduce contaminants into the discharge, which deteriorate the whole process. The ICP source generates a high-density argon–oxygen plasma inside the reactive chamber, constituted by a cylindrical vacuum vessel. By adding a metalorganic precursor into the chamber, such as titanium tetra-isopropoxide (TTIP) used here, it is possible to induce dissociation and oxidation reactions within the chamber, leading to the formation of TiO_2-based aggregates. The latter are present in the form of nanoclusters, denoted also as nanoparticles, having a specific size according to the plasma parameters employed. The reactive chamber is connected through a converging nozzle to the deposition chamber.

The deposition chamber is kept at a much lower pressure than at the reactive one. As the gas enters the deposition chamber, it expands, forming a supersonic plasma jet when the pressure ratio between both chambers is sufficiently large [50]. The nanoparticles are carried by the expanding jet, where the lighter Ar and O_2 mixture accelerates them up to the mean fluid speed. The nanoparticles impact on a substrate, typically placed perpendicularly to the jet and located at a precise distance from the nozzle, on top of which a growing nanostructured thin film is deposited. Variations of the plasma parameters in the first chamber, and/or the parameters controlling the supersonic jet, such as kinetic energy and impinging direction of the NPs onto the substrate, can crucially modify both the stoichiometry and structure of the deposited thin film [37]. Examples are illustrated in Fig. 1, where two different types of morphologies are shown: a columnar-like one (top panel) and a more structured one in which the elements have grown in a self-similar tree-like (cauliflower) shape (bottom panel).

A better control of thin film morphology can be achieved if, in addition, information about the collisional processes among jet components, supplemented by the ions energy distribution functions (IEDFs), can be used [47], since charged particles are known to play a prominent role in the process [51]. To elucidate the effects of ion–ion and ion–neutral

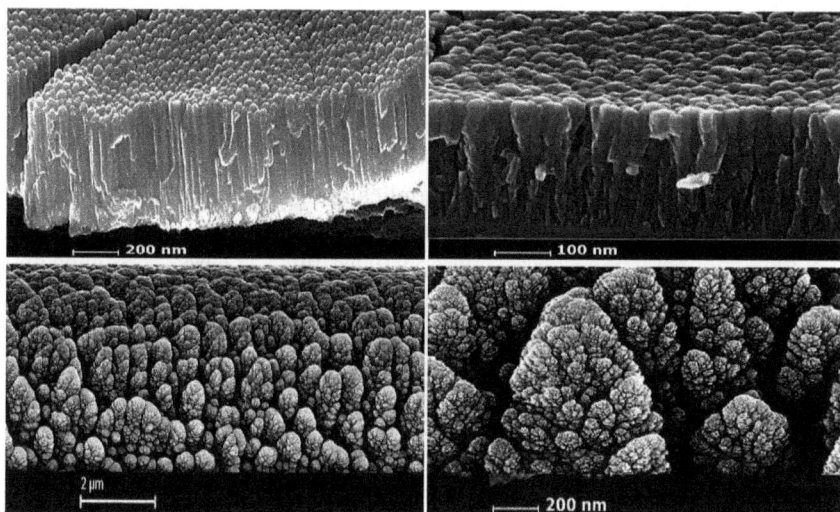

Fig. 1. Illustrative examples of nanostructured thin films created by deposition of TiO₂-nanoparticles onto a silicon-based substrate. Images to the right side of the figure are higher resolutions of the ones shown to the left side, from which one can appreciate the building grain shapes at the nanoscale.

components collisions, one can study a simpler case, without a precursor, that is a plasma containing only Ar and Ar^+ inside the supersonic jet. Here, we review experiments dealing with a purely argon plasma in which a quadrupole mass spectrometer (QMS) is used to sample Ar^+ ions, allowing us to determine their total energy. To understand the experimental results, Monte Carlo (MC) simulations of the dynamics of ions along the jet are discussed.

2. Synthesis of Hierarchically Mesoporous Materials: State of the Art

TiO_2 materials play an emerging role in many areas of physics, chemistry and materials science, due primarily to their natural abundance leading to low production costs and their proved environmental benignity. This is in addition to their recognized multiple polymorphs, chemical stability and useful optical properties. Indeed, TiO_2 nanomaterials are amenable to a variety of applications, specially in their porous forms, showing extraordinary potential as a result of their huge specific surface areas, accompanied by a large porous volume and easily tunable morphologies down to nanoscales (Section 2.1). Applications to environmental issues

(Section 2.2), and recent advances in biomedicine (Section 2.3) are also discussed.

2.1. *Nanomaterials and architectures*

Zhang *et al.* [1] review synthesis and applications of TiO_2-based hierarchical mesoporous materials. The different titanium-oxide architectures discussed include nanofibers, nanosheets, microparticles, films, spheres, core–shell and multi-level structures, with a particular emphasis on their synthesis and associated controlled parameters. Among the several applications discussed, they consider energy storage and environmental protection, photocatalytic fuel generation and pollutants degradation, water splitting based on photo-electrochemistry, catalyst support and the hot issue of lithium- and sodium-ion batteries.

Graves *et al.* [2] are concerned with the synthesis of materials using continuous plasma processes. By starting from bulk industrial powders they are able to produce hierarchical structures for energy storage applications. The scalability of the process enables the production of low-cost precursors due to the high plasma densities involved. The synthesized material is based on aggregated particles of iron and aluminum oxides, and carbon nanotubes. Based on TEM observations, and the use of a novel aerosol approach, they could characterize the resulting morphologies of the the primary particles and the final aggregated material. They showed the suitability of the new materials as battery (anode) components showing a high rate charge cycling. In addition, the synthesized hierarchical materials were optimized with plasma for energy storage materials with the aim to develop a global supply chain.

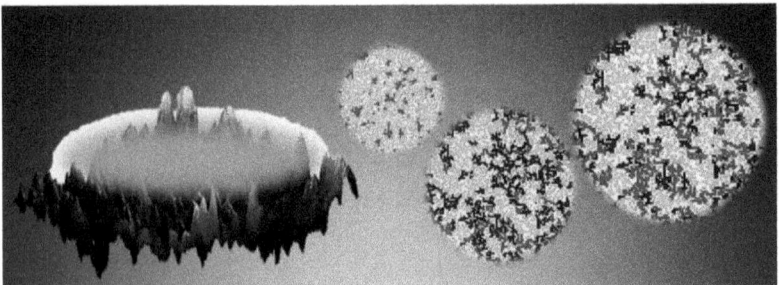

Fig. 2. Nanomaterials: Fractal films (left) and composite surfaces (right).

Maduraiveeran *et al.* [3] are engaged with the ambitious development of electrochemical biosensor platforms at the nanoscale, which are expected to have an impact in early-stage detection and diagnosis of diseases. They review recent developments in designing sensing and biosensing platforms based on functional nanomaterials, having many good attributes such as low-cost, miniaturization, easy fabrication and energy efficiency. Possible applications include neuroscience issues, clinical and medical diagnostics, health monitoring and the extent to which medical industries can play a fundamental role in future developments.

Welch *et al.* [4] discuss various types of 0D, 1D, 2D and 3D nanostructures to improve biosensor sensitivity, selectivity and limits of detection. Recent advances in the use of nanostructured materials, integrated into electrochemical, optical and other types of biosensors, are reported. The structures considered cover a very broad range of possibilities at the nanoscale such as, particles, rods, fibers, pillars, wires, sheets, indented patterns (holes and slits), gaps, channels, pores, functionalized surfaces and complex hierarchical structures. Clinical applications in diagnostics are discussed.

Jeong *et al.* [5] are concerned with the status and future developments of 2D nanomaterials for energy and environmental applications, based on novel assembly mechanisms consisting of multiple-dimensional-type of structures such as, colloidal 1D fibers obtained within a liquid crystalline phase, thin (2D) films driven by interfacial tension (the so-called Marangoni effect), and finally, 3D nanoarchitectures emerging in electrochemical processes. The authors stress the advantages of 2D materials over their one- and three-dimensional counterparts, illustrating several applications related to secondary batteries, supercapacitors, catalysts, desalination, water decontamination and gas sensors.

Li *et al.* [6] have produced hybrid nanomaterials by combining three different techniques, i.e., core-shell synthesis, 3D printing and plasma grafting. Metal oxide nanoparticles, such as ZnO and TiO$_2$, and Fe-based metal organic framework (Fe-MOF) nanounits, were succesfully grafted on 3D-printed fractal substrates using cold plasma discharges (CPD), with the aim of producing photocatalysts for the degradation of organic pollutants in water. The methods discussed open the possibility for the functionalization of any type of polymeric surface using inorganic nanocompounds. The hybrid material's activity has been monitored by photodegradation of Rhodamine B dye by metal oxides, and removal of Ciprofloxacin antibiotic by Fe-MOF, and their reusability has been assessed. Moreover, by using

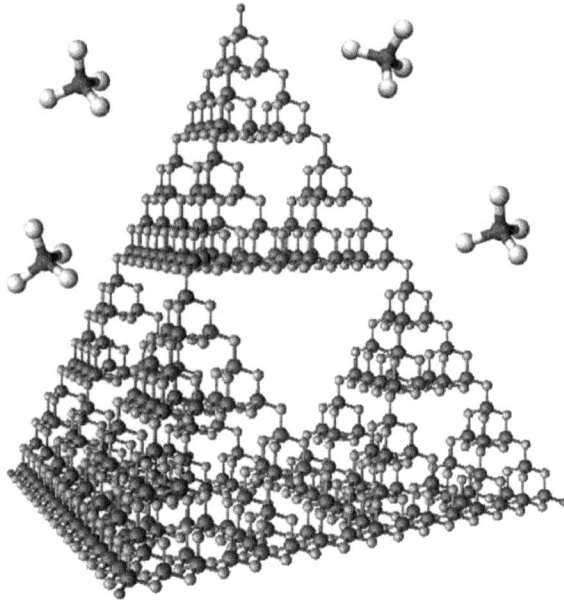

Fig. 3. Nano-architectures displaying a fractal structure.

a liquid crystal diode-based stereolithography, a resin fractal support has been printed, and subsequently surface-functionalized by means of Fe-MOF nanoparticles. The results point to the possibility of using well-known laser-based 3D printing technologies to create substrates displaying high specific surface areas aimed at providing the immobilization of catalysts by means of traditional plasma grafting.

Carrola *et al.* [7] explore a rather new possibility in nanomaterials research by combining additive manufacturing (AM) with nanocomposites (NCs). The idea is to get cooperative, evolving properties of NCs in AM, which can be actually realized along the following three main lines: processing, morphological and architectural levels. The end material is studied in detail, suggesting that indeed the expected relationship between nanomaterials and AM can lead to enhanced material performances, yielding innovative and advanced structures amenable to the bioengineering, defense and transportation sectors.

Xu *et al.* [8] consider the emerging field of Nature-inspired assemblies with potential applications in the development of high-performance wearable electronic devices. Nature-inspired structures do exhibit impressive properties, including ultrahigh sensitivity, excellent energy density, and ultralong cycling stability and durability. More importantly, these properties can be tuned, and even enhanced, by finely controlling the emerging synergy between the individual components arising from interfacial interactions. The review addresses recent breakthroughs in the newly developed Nature-inspired sensing and energy storage materials.

Finally, the book edited by Pogrebnjak *et al.* [9] contains selected reviews on innovative technologies for the development, modeling, chemical/physical and biomedical (in vitro and in vivo) investigations of materials and composites at the nanoscale. The book highlights promising new frontiers in the use of metal/metal oxide nanoparticles, hierarchical nanostructures and organic coatings, aimed to be used as sensors for the detection of gases, inorganic and organic materials, including, in addition, biosensors for bacteria and cancer-related environments. Within the realm of medical applications, questions such as tissue engineering, tissue replacement, regeneration, etc., are described in detail, in addition to the associated in vitro and preclinical investigations. Furthermore, recent findings on orthopedic and dental implant coatings using nanoparticles are also reviewed, together with their biological efficacy and safety.

2.2. *Environment*

Biorenewable and sustainable resources are the only way out to prevent irremediable degradation of our planet. In this section, we list some of the recent works dealing with these basic and other related environmental issues. In line with the whole chapter, the focus will be on applications of materials built primarily at the nanoscale.

Ates *et al.* [14] focus on the advancement of new materials extensively used as a matrix and/or reinforcement in several applications. In particular, nanocomposites often show a great advantage over their peers due to the ease and low-costs of fabrication, high mechanical and thermally stable properties, to name a few. These new materials have found applications in food delivery, biomedical devices, energy storage, wastewater treatment and transportation. They are reviewed regarding the chemistry and structural behavior, with a special emphasis on advanced applications developed recently using biorenewable resources.

Thangadurai *et al.* [15] consider an important class of mesoporous nanoparticles, as building blocks of stable nanomaterials displaying well-defined chemical and thermal behavior, as well as exhibiting controllable morphology and porosity. Mesoporous nanoparticles have started to play a

Fig. 4. Environment displaying a complex organization of different building blocks.

prominent role as low-cost adsorbents in system decontaminations from a large number of pollutants. Inspired by natural silicon- and carbon-based nanoparticles, as in plants, man-made mesoporous nanoparticles emerge, for instance, as good candidates in agricolture as suitable nanocarriers of different molecules. The authors review the unique chemical and physical properties of mesoporous nanomaterials to be used in large-scale environmental applications, with a critical discussion of the associated toxicological effects on various biological systems.

Zhang and Liu [16] discuss the important and currently hot issue of radioactive wastes produced in nuclear power generation, together with several applications of radioactive materials used in the commercial sector. Clearly, radioactive wastes must be isolated from the environment, but despite the large efforts made by the scientific and engineering communities, a fully satisfactory progress toward a sustainable waste management has not been obtained yet. Due to their unique physical and chemical properties, such as large specific surface areas, and their intrinsic high reactivity and selectivity, nanomaterials have gained a great deal of attention for effectively dealing with radioactive wastewater decontamination. The list of nanomaterials considered is extensive and are based on nanoparticles constituted of carbon, metal sulfides and oxides, several types of hydroxides compounds, metal–organic frameworks, cellulose, and biogenic nanocomposite particles. The challenges associated with large-scale applications of nanomaterials for radioactive wastewater decontamination are discussed in detail pointing to future research directions.

Rajashekharaiah *et al.* [17] discuss the synthesis of Dy^{3+}, activated by a $Bi_2Zr_2O_7$ nanophosphor, indicating that it can be successfully used as an efficient anti-oxidant and anti-bacterial, in addition to their cytotoxicity potentials.

Chen *et al.* [18] develop an strategy for the fabrication of controlled hierarchical structures based on the self-assembling of folded synthetic polymers. Linear poly(2-hydroxyethyl methacrylate) of different lengths are folded into cyclic polymers, and the process of formation of hierarachical structures is elucidated by both experimental techniques and molecular dynamics simulations. The work stresses out the essential role played by polymer folding in dealing with macromolecular self-assembly, thus establishing a novel and amenable approach for building biomimetic hierarchical structures.

Wang *et al.* [19] address the issue of antibiotic contaminants, of which biochar adsorbents are considered to be highly suitable materials due to their intrinsic safety for human health, in addition to their well-documented adsorption performances. The authors discuss in detail the preparation of loofah activated carbon (LAC), consisting of a highly ordered hierarchical laminae-trestle-laminae (L-T-L) microstructure, fully covered by nanoscale protrusions. They show that LAC can contribute significantly to the adsorption of antibiotics, such as tetracycline (TC), ofloxacin (OFO) and norfloxacin (NFO), which are significantly larger than the currently available adsorbents.

Huang *et al.* [20] discuss biomass-derived nanomaterials, which are becoming increasingly attractive both in academy and industry, as they comply with green chemistry and renewable energy requirements. The authors review major eco-friendly and green synthetic approaches based on biomass resources, showing that the abundant presence of active functional groups, and their intricate biostructures, endow these new nanomaterials with unique properties and functionalities. A summary of the different specific material properties, such as size, shape, morphology and structure, is presented.

2.3. Biomedicine: Hydrogels and proteins

Hydrogels are widely recognized as essential materials in many biomedical applications due to their softness, flexibility, hydrophilicity, in addition to their almost-solid nature. A great deal of interest has been recently drawn to the possibility of creating new hydrogel materials aimed at displaying superior performance over that of their bare counterparts. As a result, hydrogels combined with nanomaterials have been proposed as a new route toward the functional transformation of standard hydrogels. The incorporation of suitable nanomaterials into the hydrogel matrix converts the traditional hydrogel into nanocomposite ones, exhibiting multi-functionality, in addition to the biocompatibility features typical of the original hydrogels.

Ajdary *et al.* [21] discuss the novel area of plant-based hydrogels derived from wood as a hierarchical source of nanomaterials. The authors discuss in detail water interactions, hydration and swelling, as a basis for understanding the formation of hydrogels, and the associated phenomena of fluid transport, diffusion, capillarity and ionic effects. These features are thoroughly discussed since they are intimately involved in the formation of porous structures produced after removal of water, such as foams, sponges, cryogels, xerogels and of course aerogels themselves, and in the way in which hydrogel properties develop.

Cha *et al.* [22] study the influence of adding nanoparticles, nanowires and nanosheets to hydrogel matrices, in the search for new applications in catalysis, environmental purification, bio-imaging, sensing and controlled drug delivery. Novel technologies based on material engineering and processing are developed for obtaining hydrogel nanocomposites having a predetermined shape. As a result, it is expected that the newly developed nano-hydrogels can play the role of advanced soft components in future electronic, electrochemical and biomedical devices.

Fu *et al.* [23] study the formation of self-assembling peptide-based hydrogels (SAPHs), in which nanofibers are entangled to form a complex nano-fibrous network. Due to their extremely high biocompatibility, SAPHs are expected to find a variety of applications in the field of biotechnology. The authors review the fabrication, properties and biological applications of SAPHs, specially regarding the main factors that determine the synthesis process as well as their properties.

Zeng *et al.* [24] review recent developments in the active field of protein-based nanomaterials, discussing promising applications in a large number of fields related to the problems of encapsulation, bioimaging, biocatalysis, biosensors, electron transport, photosynthetic apparatus, magnetogenetic applications, vaccine development and antibodie's design. The authors

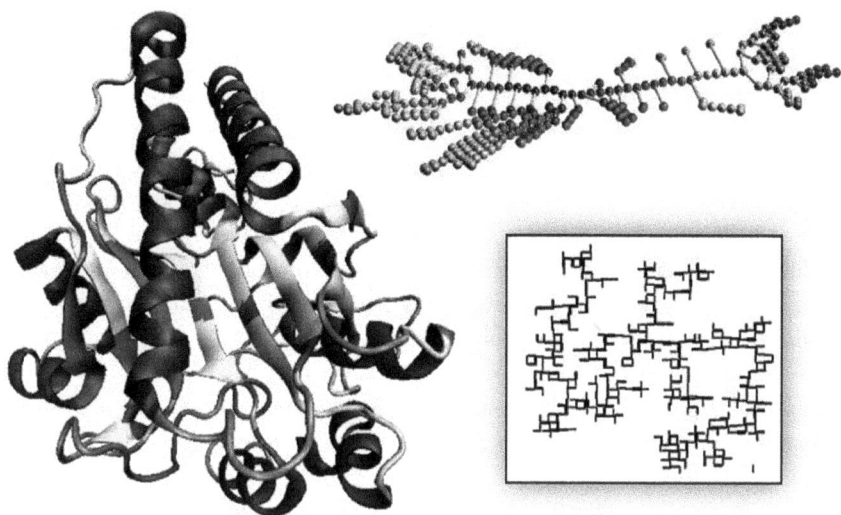

Fig. 5. Proteins (left) and Self-assembly of biological networks (right).

present detailed discussions about the design and building of 0D to 3D protein assembled nanomaterials, a key feature of which is the fabrication of highly sophisticated nanostructures based on specific rules from supramolecular chemistry.

Mohan *et al.* [25] are concerned with the combination of bio- and nanotechnology, in which nanoscale molecular biology is actually converged with microbiology at larger scales. They discuss how rapid-response biosensors can be obtained by combining biological-sensing elements such as DNA, RNA, enzymes and antibodies together with a transducer. Examples include cardiac immunosensors for early detection of a heart attack, ultrasensitive nanosensor to be used in agriculture and future nanosensors for diagnostics of highly lethal diseases such as HIV, Alzheimer's disease, Nipah and Zika virus.

Esmaeili *et al.* [26] discuss the emerging field of nanotheranostics as a way to detect, among others, MUC-1-positive tumor cells, and be used as a therapy of breast and colon tumors based on non-invasive fluorescence imaging. Fluorescence microscopy and flow cytometry techniques are discussed to assess the capability and usefulness of the designed platforms in imaging applications, drug delivery evaluation and in vitro therapies. In particular, regarding cell toxicity applications, the platform is shown to display a very accurate selective performance, pointing to its use as a multifunctional cancer nanotheranostics system, able to trace specific biomarkers.

Kanioura *et al.* [27] discuss the important field of cancer diagnosis based on the use of poly(methyl-methacrylate) films characterized by a wide range of length scales, from about 20 nm to 2 μm, obtained by applying oxigen plasma techniques. They consider normal skin and lung fibroblasts, and four different cancer cell lines, A431 (skin cancer), HT1080 (fibrosarcoma), A549 (lung cancer) and PC3 (prostate cancer). Remarkably, the proliferation of cancer cells was found to increase on hierarchical nano/microstructured surfaces with respect to untreated ones. The authors stress the fact that such high proliferation of cancer cells is achieved without the use of specific binding molecules for cancer cell development.

Ganguly *et al.* [28] consider cellulose nanocrystals (CNCs), used as nanomaterials displaying excellent physiochemical properties and self-assembly of randomly oriented species, the latter responsible for the formation of a hierarchical cholesteric structure in CNCs. The self-assembly process depends on several factors, including chemical properties of the nanoparticles, intermolecular forces, and thermodynamics. Remarkably, good self-assembling properties have been observed in CNCs-grafted polymer nanocomposites, opening the door to an immense number of applications in modern biotechnology and medicine. The authors discuss effects of different external factors such as pH, temperature and electric/magnetic fields, on the actual mechanism of formation of self-assembled CNCs.

Liu *et al.* [29] review key phenomena related to cell signaling as a promising way to monitor disease development, the signaling normally taking place at the interface between organisms/cells or between organisms/cells and abiotic materials. The idea is therefore to build more efficient biomedical interfaces to regulate the information transmission from the cell as an opportunity to improve the therapeutic results. The state-of-the-art of plasma-activated interfaces is reviewed, with the aim of providing useful guidelines for selecting the most appropriate plasma processing conditions for the design of interfaces with specific biological functions.

3. The Plasma-Assisted Supersonic-Jet Deposition Technique

3.1. *Experimental set-up*

The reactive plasma and deposition chambers are located inside a main vacuum vessel, a stainless-steel cylinder of 200 mm length and 160 mm inner radius (see Fig. 6). The plasma chamber is also of cylindrical shape, of

Fig. 6. Photograph of the PA-SJD device taken in the laboratory. The main vacuum chamber can be seen at the left side of the device. The reactive chamber containing the plasma is positioned to the left side, and the remaining space is occupied by the deposition chamber. The quartz window is clearly visible at the center of the main vacuum vessel, from which one can inspect the evolution of the plasma processes, and also operate some of the diagnostics employed in the experiments. The quadrupole mass spectrometer, Hiden-EQP 1000, is located at the right side of the device, and it can be moved inside the deposition chamber up to a desired distance from the reactive chamber.

95 mm length and 63 mm inner radius. The remaining 105 mm vessel length is devoted to the deposition chamber.

The chambers are connected through a converging nozzle located at the central axis of the cylinder (see Fig. 7). The nozzle has a rounded shape, closed by a 1 mm thick foil film having a circular orifice of 6.9 mm diameter. The nozzle favors the formation of a supersonic jet in the deposition chamber by gradually accelerating the gas particles to a sonic speed.

A vacuum pumping system, consisting of turbo-molecular and rotary pumps, is connected at the bottom of the deposition chamber providing about 130 L/s of pumping speed. A gate valve can be operated if variations in the pumping speed are required. The small conductance of the nozzle orifice, measured to be about 4.8 L/s at working conditions, allows one to change the pressure inside the deposition chamber, P_d, without altering the pressure in the plasma chamber, P_p. The limiting pressure reached inside the deposition chamber when no gas is injected was about 10^{-5} Pa.

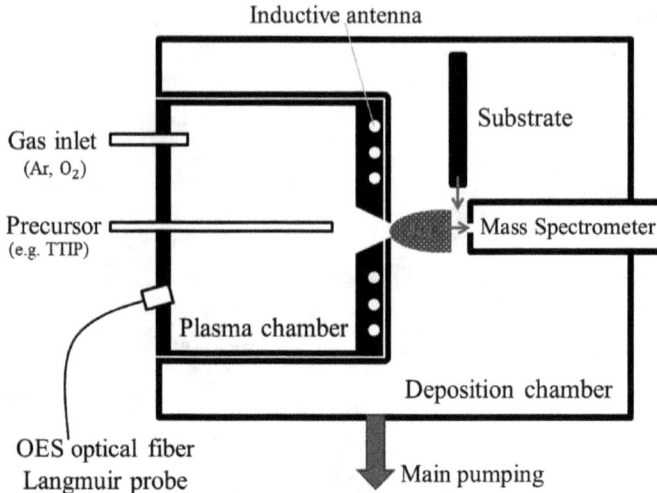

Fig. 7. Schematic representation of the PA-SJD set-up. The plasma and deposition chambers are connected by a converging nozzle, having a circular orifice of 6.9 mm diameter, through which the plasma jet is expelled. The deposition chamber is connected to a pumping system which keeps its internal pressure at a lower value than at the plasma chamber. Ar and O_2 gases, and the metal–organic precursor, can be injected into the plasma chamber as indicated in the figure. The configuration of the inductive antenna used inside the latter is shown. The positions of the diagnostics used: OES optical fiber, Langmuir probe and mass spectrometer, are indicated. The deposition substrate is removed to allow the mass spectrometer to operate. The latter can then be moved horizontally inside the deposition chamber for the measurements. To be noted is that the precursor is injected close to the nozzle, to improve the performance of the plasma reactions. To the right of the nozzle, the emerging supersonic jet is depicted, pointing in the direction of the QMS.

A relevant role in our problem is played by the parameter R, the ratio between the pressures in the two chambers,

$$R = \frac{P_{\mathrm{p}}}{P_{\mathrm{d}}} > 1. \tag{1}$$

Typically, operative ranges are: $P_{\mathrm{p}} = (1 - 10)\,\mathrm{Pa}$, in the plasma chamber and $P_{\mathrm{d}} = (0.01 - 10)\,\mathrm{Pa}$, in the deposition chamber. The experiments here reported correspond to the range $1 \leq R \leq 40$. During the measurements, the pressures can be monitored using two capacitance pressure gauges.

On the right side of the deposition chamber (see Fig. 7), a second vacuum system allows the insertion and the movement of a quadrupole mass spectrometer (Hiden EQP-1000 Analyzer) [41, 52, 53], which keeps operating to maintain the pressure at above $10^{-4}\,\mathrm{Pa}$. This additional system is constituted by a turbo-molecular pump and a rotary pump providing $60\,\ell/s$ of pumping speed. The QMS can collect the gas particles from a small circular orifice, of $0.1\,mm$ diameter, placed at the center of a flat circular cap having $50\,mm$ diameter.

3.2. *Plasma chamber: ICP and precursor*

After the success achieved using a chemical vapor deposition technique to deposit thin films from a gas state to a solid state on a substrate, such as plasma-enhanced chemical vapor deposition (PECVD) (see also Chapter 2), it was concluded that a plasma source was the most appropriate solution for producing molecules to be aggregated in clusters and nanoparticles.

A plasma, in fact, is considerably more reactive than a gas, because the presence of free ions and electrons can lower the activation energy of many chemical reactions, making possible processes that in a non-ionized gas would require much higher temperatures. Indeed, in a plasma, the reactions take place in a well-controlled manner, in which the plasma parameters can be varied accurately.

In our experiments, we have chosen TiO_2 as the building unit for the deposited thin films. Titanium oxide can be released during the dissociation of a suitable precursor injected into the source, i.e., titanium tetra-isopropoxide (TTIP), as we discuss in what follows. TTIP decomposition in a plasma is the result of either the interactions between the precursor and free electrons and radicals, or thermal dissociation. The final quest is to decide which type of plasma source, that is, which type of plasma discharge, is going to be used for ionizing the gas and precursor present in the chamber. Here, we have used an inductively coupled plasma as described next.

3.2.1. *Inductively coupled plasmas*

ICPs are plasma sources working at low pressures and at relatively high discharge densities charaterized by two operation regimes, the inductive and the capacitive ones, denoted as H-mode and E-mode, respectively [46–49]. In the former, the discharges are larger, by at least one order of magnitude than those present in the E-mode, thus making them very attractive when large rates of dissociated precursors are needed, as in the case of large-scale industrial applications.

Regarding the inductively coupled source antenna, there are two types of configurations: (1) the cylindrical one, in which the antenna is winded helically around the chamber; and (2) the planar one, in which the antenna consists of a multi-turn winding facing the reactor through a window of a dielectric material. In our experimental set-up, we have used this second type of antenna (cf. Fig. 7).

Argon and oxygen gases can be injected directly into the plasma chamber using mass-flow controllers, and the plasma is generated using an inductive source [54]. A two- and three-quarters loop planar copper wire antenna is placed inside the right side of the first chamber (as shown in Fig. 7), shielded by a Teflon scaffold and covered by an alumina disk to reduce sputtering. The antenna is fed with an RF power generator (Huttinger PFG 1600 RF) working at 13.56 MHz and 450 W, connected through an L-type matching box whose parameters can be manually modified in order to keep the proper tuning (no reflected power). In this way, it is possible to obtain a uniform and stable plasma discharge inside the reactive plasma chamber. See Section 3.3.3 for further details.

3.2.2. *Metalorganic precursor (TTIP)*

A metalorganic precursor is injected in the first chamber once the plasma discharge becomes stable and is sustained at a sufficiently high power level.

TiO_2 films, considered here, can be obtained from different organic precursors, titanium alkoxides are preferable, such as $Ti(OiPr)_4$, where Pr stands for a prochiral sulfide [55], which already displays the TiO_4 tetrahedral motif of the titanium dioxide lattice in its chemical structure [56]. In our experiments, we have considered titanium tetra-isopropoxide (TTIP), $Ti(OCH(CH_3)_2)_4$ (see Fig. 8), which is a liquid at 20°C. It can be heated with a power transformer to reach the vapour phase. The temperature can be monitored with a thermocouple over the precursor tank and varied to obtain different precursor flows (from 0.25 g/h to 0.75 g/h). At operating temperatures between 40° and 47°C, the vapour pressure of TTIP is sufficiently high to create a stable flow inside the plasma chamber.

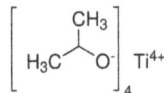

$$\left[\begin{array}{c} CH_3 \\ H_3C \diagup O^- \end{array} \right]_4 Ti^{4+}$$

Fig. 8. Chemical formula of titanium tetraisopropoxide. This molecule is the metalorganic precursor used to deposit TiO_2-based thin films.

3.3. Diagnostics: Measurement devices

3.3.1. Hiden analytical quadrupole mass spectrometer

The QMS (see Fig. 9) is employed for the acquisition of ion fluxes and their energy distribution functions along the supersonic jet [57]. The measurements can be performed at different positions, with an accuracy of 1 mm, along the main chamber axis. Ions produced outside the instrument enter the 0.1 mm sampling orifice with an initial energy E_i, and are rapidly focused by a lens into the drift space.

The spectrometer is grounded at the deposition chamber, which acts as a reference potential, and an extractor lens can be set to promote ion accelerations within the device. Inside the drift region, charged particles are accelerated to a higher energy, $E = E_i + qV_{axis}$, where V_{axis} is set to 40 V. Then, a 45° energy filter allows the transit of only those ions having

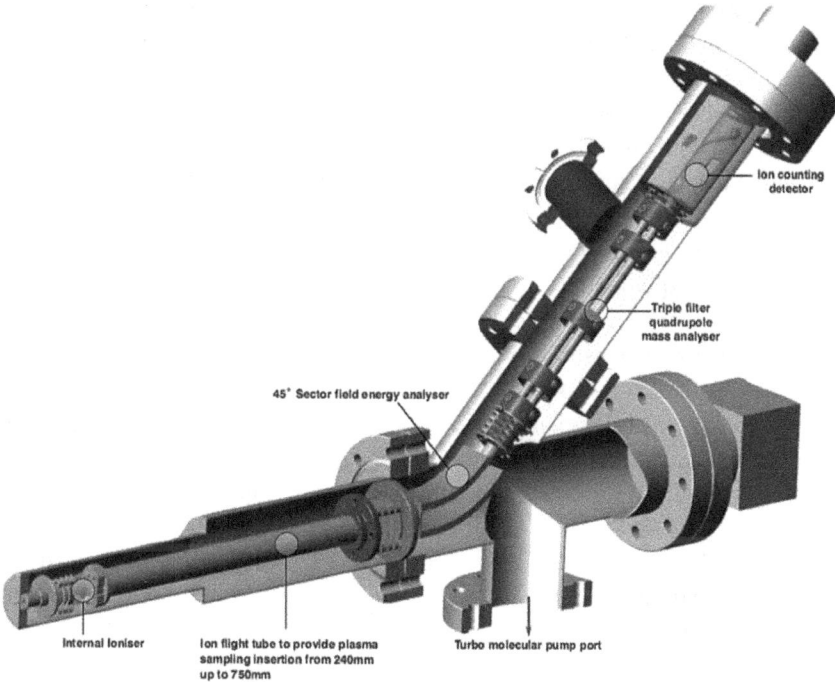

Fig. 9. The QMS: Design and structure of the Hiden EQP-1000 Analyzer.

a predetermined energy E. Their energy is then slowed down to the transit energy, selected by a mass filter, to scan the particle masses inside the quadrupole.

A secondary electron multiplier pulse-counting detector provides mass and energy spectra in the range $1–10^7$ counts/s. Mass spectra can be acquired for both neutral and ion species with a resolution of 0.02 amu. For the latter, the mass/charge ratio falls in the range 0.4–1000 amu/q, where q is the particle charge. Trends for relevant masses up to 100 amu may also be measured with a time resolution of 100 ms for each mass, providing a truly real-time and quick diagnostic tool. Since the particle's detection is very dependent on the ion trajectories, the instrument requires an internal low pressure (less than 10^{-4} Pa) to ensure almost no collisions inside.

Ion energy spectra can be acquired, for a fixed mass, at the high energy resolution, 0.05 eV, in the range $(-150$ to $+150)$ eV. The gas ionization is obtained by electron-impact. An oxide-coated iridium filament emits thermal electrons which are accelerated toward a grid at a determined energy. For the ion source, an electron current of 10 μA and 70 eV electron energy have been used. These values are commonly employed, corresponding to a high first-ionization cross-section value for both argon and molecular oxygen. Moreover, at this electron energy there is a small production (few percent) of doubly charged ions and fragments. Successively, a focusing system drives the ions into the quadrupole mass filter, where an RF-voltage applied to the quadrupole unit allows only ions of a certain mass-to-charge ratio to reach the detector.

For charged species, the QMS signal can be affected by the transmission (energy dependent) function and acceptance angle [58, 59]. A careful tuning needs to be implemented by focusing the system to reduce the energy-dependent phenomena, such as chromatic aberration, and/or high acceptance angles [60, 61]. The reported experiments here have been obtained by using a very small value for the extractor voltage $(-2$ V$)$ to reduce the acceptance angle. Since the sampling orifice is drilled on a grounded cap kept at the same reference potential as the vacuum vessel, the ions coming from the plasma region are accelerated, due to the electric potential difference, before being collected.

The mass spectrometer scans were performed on the ion fluxes by using an energy filter set to transmit ions at 8 eV. The latter was found optimal under the experimental plasma conditions since most of the masses were found at this energy. For each single mass, the measurements of the ion distribution functions were performed by setting the quadrupole mass filter at the weighted average mass of the measured mass spectra.

For neutral species, the measured signal is directly proportional to the sampled gas density, the electron impact ionization cross-section and the instrument mass-dependent transmission function of secondary ions. A value of the absolute density of neutral particles of a given mass can be obtained if the instrument transmission function is known for that specific mass. Since the measured signal is linearly dependent on the gas density, the latter can be calculated also using a simple conversion factor obtained from a calibration performed in a region where the density of the gas species is known.

3.3.2. *Optical emission spectrometer*

Optical emission spectroscopy (OES) is a commonly used non-intrusive diagnostic (see Fig. 10), which can be easily adapted to study plasma discharges in quite general circumstances [53, 62]. The free electrons produced in a discharge can excite atoms, molecules and ions within the plasma. As a result, they emit electromagnetic radiation, typically at and near the visible range. The analysis of the emitted light can yield useful

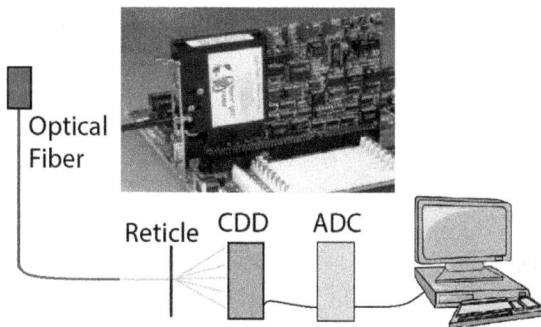

Fig. 10. An optical emission spectrometer: Plasma radiation is transmitted through an optical fiber (red line) onto a diffraction reticle. The transmitted photons (yellow lines) are detected by a charged coupled device (CCD) and converted into a digital signal by an analog-to-digital converter (ADC). The digital signal can then be stored in a computer for the analysis.

information to understand the main reaction processes taking place in the plasma.

The intensity of the emission lines can identify the chemical species present in the ionized gas mixture, allowing to estimate their abundances. As described schematically in Fig. 7, there is a quartz window (left side of the plasma chamber), through which the emission spectra generated within the chamber can be collected by an optical fiber, connected to the OES.

The data have been acquired using two different low resolution spectrometers (see [36, 53] for details). The first one is a wide band, low resolution spectrometer (AvaSpec-2048 by Avantes) equipped with a 10 μm slit, a holographic grating (300 lines/mm, blazed at 300 nm), a coated quartz lens to increase sensitivity in the UV and a 2048 pixels CCD. The spectrometer has a resolution of about 0.8 nm and a spectral band extending from 180 nm to 1150 nm, which was used in connection with an IR-enhanced optical fibre.

The second instrument is a lower resolution spectrometer (PS2000, Ocean Optics), which acquires spectra through a UV-enhanced optical fiber. This device was equipped with a 10 μm slit and a holographic grating (600 lines/mm, blazed at 400 nm) having a resolution of 1.02 nm as well as a spectral band extending from 200 nm to 850 nm.

3.3.3. *Electric diagnostics: HV and Langmuir probes*

The electrical parameters regulating the plasma can be determined by measuring the associated times series of voltage and current at the two different sides of the planar inductive antenna (see Fig. 11). The resulting voltage at the exit of the matching box is measured using a high-voltage (HV) probe (Tektronix P6015A), while the current flowing into the antenna is acquired by means of a Rogowski probe, positioned at the grounded side of the coil.

Both signals are acquired simultaneously using a digital oscilloscope, working at 4 GHz sampling frequency, allowing to obtain a detailed electrical characterization of the plasma. The RMS values for voltage and current can be obtained from the time series, as well as the associated phase shifts. Once these quantities are available, one can determine the load resistance and reactance, and, more importantly, decide whether the discharge is operating in a capacitive or in an inductive regime. The Langmuir probe can also be used to measure plasma parameters inside the plasma chamber alternatively to OES measurements (Section 3.3.2). The probe can

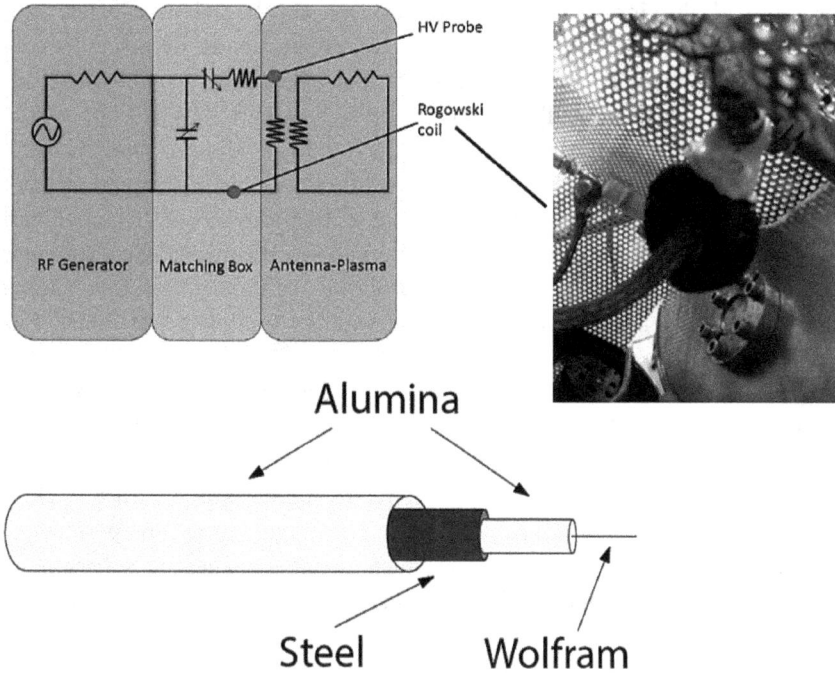

Fig. 11. Scheme of the measurement set-up for the plasma-regulated electrical parameters. (Topper panel left) The circuit showing the position of the matching box. The red dots show the locations of the electrical probes, the HV and the Rogowsky probes (Topper panel right). The latter, positoned at the grounded side of the antenna, measures the current flowing inside the wire. (Lower panel) A Langmuir probe basically consists of a conductive wire (commonly Wolfram, due to its high thermal resistance), shielded by dielectrics (typically alumina) and steel (to reduce signal interferences). Only the final tip of the wire is left exposed to the plasma, which should be longer than the Larmor radius of the charged particles inside the plasma.

be moved inside the plasma chamber from the quartz window, in a similar fashion as done for the optical spectroscopy measurements.

For experiments operating in an RF-plasma regime, the analysis of the Langmuir current-voltage characteristic curve can be affected by the plasma potential fluctuations. Several methods are known for reducing RF perturbation effects [63–67]. To obtain the actual current-voltage characteristics in a high RF environment, one can use a Langmuir probe with a compensating circuit. In our experiments, the probe had an exposed tungsten tip, 5 mm length and 1 mm diameter. The probe tip connection was insulated and shielded in a probe holder (15 cm long) made out of two coaxial alumina tubes. RF compensation can be achieved using well-known methods [68, 69]. A miniature RF choke, made of a ferrite core having $Q \simeq 200$ and an impedance of about $85\,k\Omega$, was embedded near the probe tip. Then, a floating electrode constituted by a copper foil, $10\,mm$ wide and $3\,mm$ diameter, glued on the probe alumina tube at $3\,mm$ from the tip, is exposed to the plasma, while it is coupled through a large capacitor to the probe tip $(100\,pF)$ to ensure an almost constant impedance load during the probe voltage scan, aimed at recording the Langmuir characteristics [69]. The Langmuir probe is calibrated by fine tuning the choke capacitance such that the frequency matches the $13.56\,MHz$ RF supply.

This method has been employed in other plasma discharges and confronted with traditional Langmuir probes [53]. Typically, a fixed voltage ramp is applied in a $100\,ms$ time scan [70], the probe current is measured via an equivalent resistance of $100\,\Omega$ using an Agilent scope with a $100\,kHz$ sampling rate, and an average of over 20 voltage ramps are performed. Analysis of the curve can be performed using a dedicated software, in order to extract the plasma density and potential, as well as the electron temperature [70].

Unfortunately, RF disturbance from the plasma is quite strong at high power inputs. As a result, the analysis of the whole probe parameters within the H-mode discharge region is greatly affected for RF power inputs larger than $200\,W$, thus making it difficult to study the electron capture region. It is still possible, however, to evaluate the charge density for an H-mode discharge (see Section 3.2.1) by using accurate methods [64, 66].

3.3.4. *Thin film analysis: FTIR spectrometer*

Infrared spectroscopy is a widely used technique for investigating both chemical and structural properties of a material or fluid. The Fourier transform infrared–attenuated total reflectance (FTIR-ATR) spectroscopy exploits the property of total internal reflection within an ATR crystal (see Fig. 12).

The IR beam interacts with the surface sample in the form of an evanescent wave. The chemical composition of a sample surface, such as a deposited thin film as in our case, is investigated by means of a Fourier transform infrared (FTIR) spectrometer (Nicolet iS10, Thermo Scientific), equipped with an ATR sampling accessory (Smart iTR). For each IR spectrum, 64 scans were recorded at a spectral resolution of $2\,cm^{-1}$.

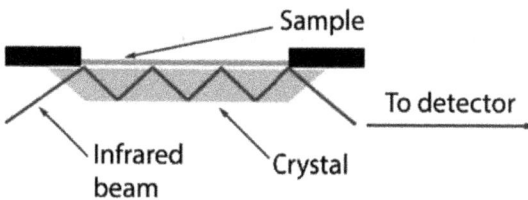

Fig. 12. Illustration of the Fourier transform infrared–attenuated total reflectance (FTIR-ATR) technique. An infrared laser ray impinges onto the sample to be probed and keeps reflecting inside a thin ATR crystal, as shown in the scheme. This can only happen when the refractive index of the sample is smaller than the ATR crystal one. In some cases, just one reflection occurs. Usually, a mirror is employed to select the angle of incidence, changing the distribution of light inside the ATR crystal. As the laser beam interacts with the sample, it produces an evanescent wave which enters the sample for a small distance (typically from $0.5\,\mu$m to $2\,\mu$m), thus producing a variation in the IR light intensity of the exiting beam. The latter is then collected by a detector. In this way, structural and compositional information of the sample can be obtained.

A basic IR spectrum is essentially a graph of infrared light absorbance or transmittance as a function of the incident light wavenumber. The infrared spectrum of a sample is obtained by letting a beam of IR light go through the target. If the frequency of the IR coincides with the vibrational frequency of a bond, then absorption will occur. Examination of the transmitted light reveals how much energy is absorbed at each frequency. Infrared light is guided through an interferometer and then through the sample (or vice versa). A moving mirror inside the apparatus alters the distribution of IR light that passes through the interferometer (Fig. 12). The recorded signal, called an interferogram, actually represents the intensity of the output light as a function of mirror position. A data-processing technique, based on a Fourier transform of the raw signal, turns the raw data into the associated sample spectrum.

3.3.5. *Thin film imaging: Atomic force microscopy*

Atomic force microscopy (AFM) is performed with a device consisting of a cantilever with a sharp tip at its end, which is used to scan the sample surface. Typically, the tips are made of silicon or silicon nitride, with a radius of curvature of the order of nanometers (Fig. 13).

As the tip is brought sufficiently close to the sample surface, forces between them start acting leading to a deflection of the cantilever according to Hooke's law. The deflection is typically measured using a

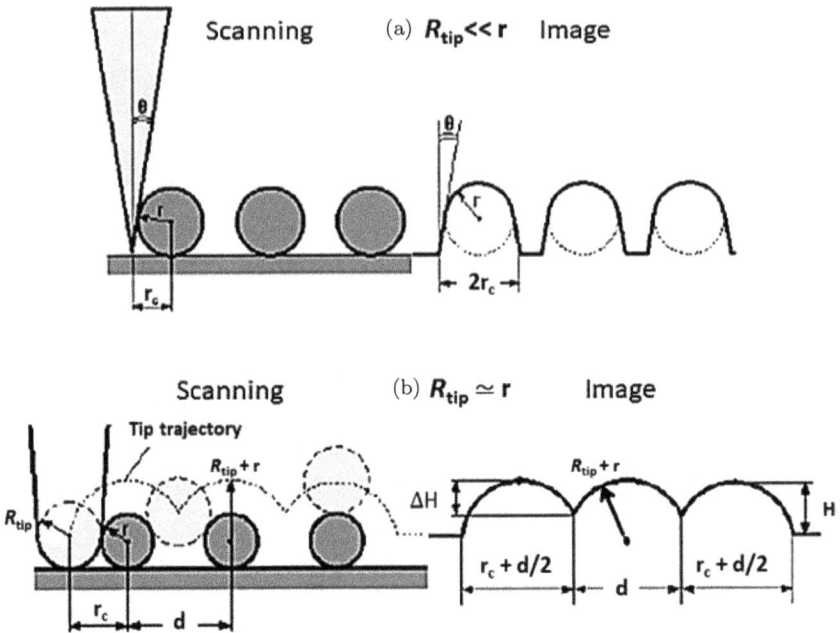

Fig. 13. AFM in the case of a tip radius R_{tip} and a local surface radius of curvature r. (a) $R_{tip} \ll r$: The angle of contact θ and the critical distance r_c, from the sharp tip to the center of the surface sphere of radius r, are shown. The tip trajectory, defining the measured surface profile, is indicated by the thick line. (b) $R_{tip} \simeq r$: Here the tip has a finite radius $R_{tip} \simeq r$, where r_c has the same meaning as in (a), while d is the distance between 'local' (spherical) protrusions on the surface. The tip trajectory in this case looks smoother than in (a).

laser spot which is reflected from the top surface of the cantilever into an array of photodiodes. The problem is that the tip can damage the surface if it is kept scanning at a constant height hitting the sample protrusions.

To overcome this issue, a feedback mechanism is employed for adjusting the tip-to-sample distance by keeping a constant force between them. Typically, the tip (or the sample) is mounted on a piezoelectric tube, which allows scanning in the three spatial directions. A topographic image of the sample is obtained by plotting the deflection of the cantilever versus its position on the sample.

Care must be exercised when analyzing the produced AFM image to disentangle measured artifacts from true surface topography. During scanning, two major AFM artifacts can occur: (1) a profile broadening effect due to the tip–sample convolution, and (2) a height lowering effect due to the elastic deformation of the surface. In the case of a sharp tip (Fig. 13(a)), the object lateral width r_c is a function of the contact angle θ, yielding an apparent object broadening proportional to $2(r_c - r)$, while the measured sample height yields the correct value, $2r$. In the second case, i.e., $R_{tip} \simeq r$ (Fig. 13(b)), the motion of the tip on the sample surface is given approximately by the position of the center of a moving sphere of radius R_{tip}, describing an arc of radius $(R_{tip} + r)$ as shown in the figure. Now, the object lateral dimension becomes $r_c = 2\sqrt{R_{tip}r}$.

Our AFM experiments have been implemented using a Solver P47-PRO, in semicontact (tapping) mode on dry samples, using a high-accuracy non-contact silicon tip having a spring constant of $3.5\,\text{N/m}$. We have used a circular tip of radius $R_{tip} = 10\,\text{nm}$, corresponding to the second tip type (Fig. 13(b)), which is well suited for evaluating the size of surface grains present in the nanostructured depositions. From the 2D images it is also possible to estimate the surface roughness quite accurately from the measured height distributions of the thin films.

4. Plasma Chamber: Precursor–TiO$_2$ Nanoparticles

The basic aim of a PASJD technique is to achieve an effective control of the deposited film morphology required for the expected applications. To this end, one needs to characterize both the seed particles chemistry and the physics of the expanding plasma jet related to the film nanostructure and stoichiometry properties. The chemistry of the process involves the study of nanoparticles produced in the reaction chamber, where the plasma and the precursor products can be monitored using optical emission spectroscopy. The resulting nanoparticles, which will become the building blocks of the thin films, rapidly cross the nozzle to the deposition chamber, where they are carried by the supersonic jet.

In this section, we first review the dissociation of the precursor in the plasma chamber as measured by OES. The newly created nanoparticles are then analyzed by a quadrupole mass spectrometer, whose ion flight tube (see Fig. 9) can be moved along the central axis of the deposition chamber, determining both the neutral and ion species masses, as well as their kinetic energies.

4.1. *Optical emission spectra from plasmas*

Optical emission spectra were measured inside the plasma chamber (see Section 3.3.2) using previously developed systematics [53], in order to quantify the chemical reactions taking place inside it. The measured intensity ratios of nearby emission lines, resulting from the main reactive radicals and inert argon, complemented by an actinometry analysis [71], yield useful information about radical concentrations. Of special interest are the modifications in plasma reactions in the presence of the precursor TTIP, providing evidence for its dissociation and emerging combustion reactions.

OES results are shown in Fig. 14. Different emission lines of excited atoms and molecules could be identified and their intensity ratios were determined for each plasma deposition experiment.

In the absence of the TTIP precursor, the optical spectra are dominated by the emission lines of neutral argon and oxygen atoms (red lines in the figure). In the presence of TTIP precursor, several new lines emerge in the spectra, corresponding to hydrogen, carbon monoxide and hydroxyl radicals. As we mentioned in Section 3.2.2, the precursor injection flow into

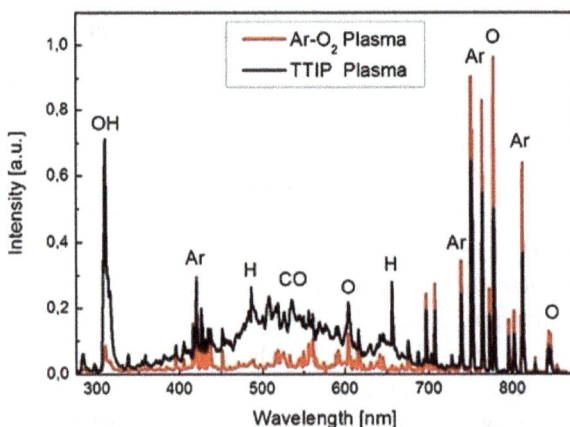

Fig. 14. Optical emission spectra measured within the plasma chamber for: A simple argon–oxygen plasma (red line) and, a resulting plasma in the presence of TTIP precursor (black line). The peaks represent the fraction of different atomic species identified by their specific emission lines.

the plasma chamber (Fig. 7) can be modified by varying the temperature at which the TTIP is vaporized from the container.

The intensity of the 796 nm wavelength line of argon (Fig. 14) was used as a reference to estimate the radical concentration of carbon monoxide and oxygen at the different precursor fluxes injected into the plasma chamber. The precursor flux increases with the TTIP container temperature, i.e. the higher the temperature, the larger the amount of precursor molecules entering the reactor per unit of time. The amount of precursor can be monitored by means of OES, since it is proportional to the concentration of CO, and other dissociation products. In contrast, radical oxygen concentration decreases during the precursor combustion process. In this way, we can easily monitor precursor abundance inside the plasma chamber and, in addition, control the deposition rate, typically in the range from few nm/min to about 200 nm/min, in the second chamber.

4.2. *Precursor dissociation and nanoparticle mass spectra*

The QMS yielded the species summarized in Table 1. For illustration, they are also reported in Fig. 15 as a function of the m/q ratios.

The rather large peak observed at $m/q = 269$ amu/q in Fig. 15 suggests that the precursor is not fully dissociated. A rough estimation of monomer dissociation, based on the assumption that the mass transmition function of the QMS scales as $\sim m^{-1}$, yields a value of about 80%, suggesting that about 4/5 of titanium isopropoxide monomers are well dissociated. This value can be used as a relative parameter to compare different experimental conditions.

Table 1. Main products of TTIP dissociation.

m/q	Ion	m/q	Ion
15	CH_3	139	$TiO_2(OCH(CH_3)_2)$
43	$CH(CH_3)_2$	167	$Ti(OCH(CH_3)_2)_2H$
59	$OCH(CH_3)_2$	181	$TiO(OCH(CH_3)_2)_2-H$
64	TiO	211	$Ti(OCH(CH_3)_2)_3H-CH_3$
81	TiO_2H	225	$Ti(OCH(CH_3)_2)_3$
99	TiO_3H_3	243	$Ti(OCH(CH_3)_2)_4-CH(CH_3)_2$
125	$TiO(OCH)(CH_3)_2)$	269	$Ti(OCH(CH_3)_2)_4-CH_3$

Note: The precursor has $m/q = 284$ amu/q.

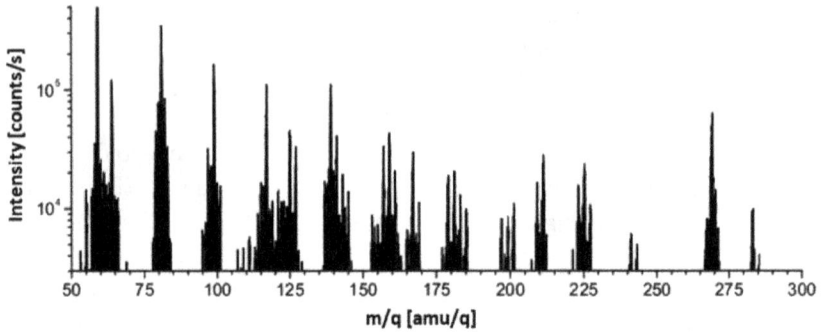

Fig. 15. QMS spectrum [counts/s] vs nanoparticle mass-charge ratio m/q, of a super-sonic jet carrying TiO_2 nanoparticles, measured at a distance $z = 5$ mm from the nozzle. An electron impact ionization of 70 eV was used. No events were found above 300 amu/q.

5. Deposition Chamber: Theory of Supersonic Jets

To understand the effects of a supersonic jet carrying NPs to be deposited onto a substrate in the deposition chamber, we need to know few essential features of gas transport inside the plasma chamber, and the way the gas transits toward the low-pressure chamber through the nozzle. In general, one needs to characterize the type of transport regime the gas particles undergo in a system. To do this, one considers the Knudsen number K_n, defined as the ratio between the gas molecules/particle's mean free path, ℓ, and the system linear size, L, in which they are contained, $K_n = \ell/L$. Inside the plasma chamber, the gas flow is said to be found in a continuum flow regime, meaning that $\ell \ll L$, i.e., $K_n \ll 1$. In contrast, high Knudsen numbers, $K_n \gg 1$, denote a collision-free molecular regime. The intermediate case, $0.01 \lesssim K_n < 1$, is referred to as a transitional regime.

Within the plasma chamber, the mean free paths are $\ell \lesssim 2$ mm, while the vessel length is about 100 mm in linear size, yielding $K_n \lesssim 0.02$. As the gas flows into the deposition chamber through the sonic nozzle, its density decreases and the flow regime becomes transitional, with K_n roughly in the range $0.02 \lesssim K_n < 1$. Indeed, as the plasma gas enters the deposition chamber, it undergoes a free expansion process as a result of the lower pressure environment [72]. If the pressure ratio R is sufficiently high, say $R \gtrsim 2$, the expansion process can be considered isentropic (i.e. adiabatic and reversible) and a supersonic jet is formed [50]. A photograph of a jet is reported in Fig. 16.

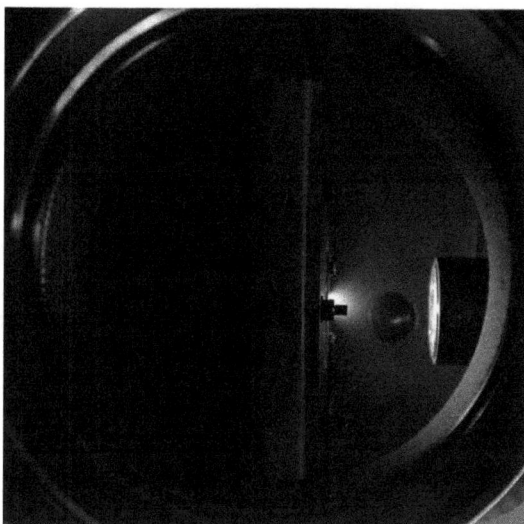

Fig. 16. A photograph of the supersonic jet exiting the plasma chamber (left side) through the nozzle, taken from the main quartz window located at the center of the vessel. The QMS head is visible within the deposition chamber, on the right side of the figure, sampling the plasma jet.

In the process, the gas particles get accelerated forming a low collisional and nearly mono-energetic jet, leading to a rapid cooling of the gas in which its thermal energy is quickly converted into traslational fluid flow. In this regime, both the density and pressure of the particle jet decrease significantly as the distance from the nozzle increases. At the same time, the Mach number M, defined as the ratio between the jet speed and the local speed of sound, $M = v_{\text{jet}}/v_{\text{s}}$, increases. The jet geometry is sketched in Fig. 17.

The expansion ends with a sharp shock, called Mach disk, located at a distance z_{M} from the nozzle, where the jet temperature, pressure and density suddenly shrink to the local background values, and the gas flow becomes subsonic (Fig. 17). For distances $z < z_{\text{M}}$, the particle's mean free paths are of the order of few centimeters. The properties of the supersonic beam are mainly determined by the size and shape of the nozzle, and depend to a large extent on the pressure ratio parameter R [73, 74].

The distance z_{M} is a function of the nozzle orifice diameter, d_n, and the parameter R, and it can be estimated using the following empirical

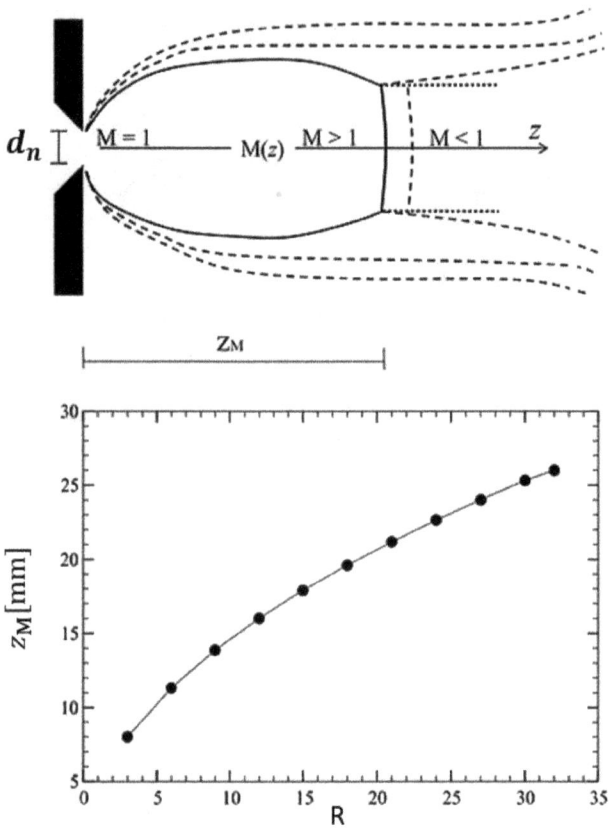

Fig. 17. (Top panel) Representation of the supersonic expansion geometry formed in the deposition chamber right away from the converging nozzle. The supersonic jet extends up to a distance z_M, denoted as the Mach disk, and it is delimited by the continuous lines shown. The dashed lines represent non-supersonic streamlines of the gas. The Mach number $M(z)$ depends on the distance z from the nozzle, and becomes $M(z) < 1$ outside the supersonic zone. (Lower panel) Values of z_M vs pressure ratio R according to Eq. (2).

relation [74],

$$z_M \simeq \frac{2}{3} \cdot d_n \sqrt{R}. \tag{2}$$

In our experiments, we have $d_n = 6.9\,\text{mm}$. Using this relation, we can estimate the smallest R value for which we can expect the presence of a supersonic jet. This value is obtained by taking the Mach disk distance

$z_M \simeq d_n$ in Eq. (2), yielding $R \simeq 9/4 = 2.25$, consistent with the lowest R value mentioned above (see discussion previous to Fig. 16).

5.1. *Supersonic jet for a single gas component*

In PASJD, the supersonic jet consists of an expanding argon–oxygen gas mixture carrying the TiO_2 nanoparticles produced in the plasma chamber, which are the seeds of the thin film to be deposited on a substrate. It turns out that the properties of the expanding Ar-O_2 gas are not appreciably affected by the much heavier TiO_2 species added to the jet, while all the particles in the jet get accelerated in more or less the same fashion [75]. Therefore, in order to understand the behavior of the composite jet, it is sufficient to study a simpler system made of an Ar-O_2 gas mixture.

A further simplification can be achieved if we consider a single expanding gas component, which we chose to be argon. The reason for this simplification is motivated by the need to perform supersonic jet experiments with the noble gas, which are used to test our theoretical results on an expanding argon gas, to be discussed later in Section 7.3.

For a single-component jet, the particle speeds admit an upper limit, here denoted as v_{Lim}, i.e., $v_{jet} \leq v_{Lim}$, which is given by [50],

$$v_{Lim} = \sqrt{2\gamma R_g T_0/[(\gamma - 1)m_g]}, \tag{3}$$

where T_0 is the temperature inside the plasma chamber, R_g, the perfect gas constant, γ, the gas specific heat ratio, and m_g, the gas particle mass. For instance, in the case of argon at room temperature, we find $v_{Lim} = 550\,\text{m/s}$, while during a plasma discharge we obtain $v_{Lim} = 730\,\text{m/s}$.

The argon gas density, $\rho(z)$, at a distance z from the nozzle with orifice diameter d_n, can be estimated from numerical calculations leading to the following parametrization, valid for $z > d_n/2$ [76],

$$\rho(z) = \rho_0 \left[1.44\, x^2 - 0.65\, x + 0.87 \right]^{-1/\gamma}, \tag{4}$$

where $x = z/d_n$, ρ_0 is the argon mass density in the plasma chamber, and $\gamma = 5/3$. In our case, Eq. (4) holds for $x > 1/2$, i.e. $z > 3.5\,\text{mm}$, yielding very good agreement with the experimental results [38, 41].

Outside the supersonic zone, i.e. for distances beyond the location of the Mach disk, $z > z_M$, one can expect that pressure and density reach their background (local) values after few collisions, provided the gas is not fully expanded. In this case, the gas density along the z axis can be well fitted by the approximate expression,

$$\rho(z) = \rho_d + \frac{A}{z^2}, \tag{5}$$

where ρ_d is the background density inside the deposition chamber and A is a fitting parameter proportional to the mean free path length.

5.2. *Supersonic jet for a gas mixture*

The case of a well-mixed gas mixture, having components of similar molar fractions, can be described as a simple fluid in terms of mean mixture mass and average specific heat ratio. For instance, for an argon–oxygen gas mixture of molar fractions $c_{Ar} = 0.4$ and $c_O = 0.6$, respectively, one finds $\langle \gamma \rangle = 3/2$. In real cases, the situation is a bit more complex and additional phenomena need to be taken into account. For instance, a strong radial pressure gradient is present along the jet expansion, such that the gas components are drifted perpendicularly to the jet axis due to pressure diffusion, depending on their mass m [77–79]. The diffusion 'speed', v_D, is inversely proportional to the mass according to the relation, $v_D \simeq m^{-1/2}$, so that heavier species are more likely to remain along the jet axis. This effect is, however, negligible for high-pressure jets [50], while for low-pressure expansions it may lead to the separation of different gas species. For a binary mixture, the lateral diffusion can be described by introducing a separation factor, α, defined as [80],

$$\alpha = \frac{(\rho_H/\rho_L)}{(\rho_H/\rho_L)_0}, \tag{6}$$

where the subscripts H and L refer to the heavier and the lighter species in the gas, respectively, and 0 denotes the reference values of the corresponding densities inside the plasma chamber.

In a more general context, a supersonic expansion is widely used as an effective tool for the production of intense molecular beams for a variety of applications [33, 81].

6. Hierarchical Deposition of TiO_2

The actual functionality of a thin film is determined by its morphology. Compact thin films are mainly used as protective coatings, whose main properties are high mechanical resistance, low friction coefficient and, in some cases, good resistance to high temperatures. If, instead, the deposit is not compact, i.e., it displays a conspicuous porous or foamy (fractal) structure, then it is more suitable for applications requiring a maximum exposed surface with a minimum occupied volume. This is required in the case of heterogeneous catalysis.

The PASJD is able to produce complex morphologies characterized by a hierarchical porous structure. This means that different types of porosities are present at different length scales, often developing self-similar (fractal) features. In our experiments, we have obtained structures displaying a high degree of ramification, associated to varying types of building blocks. The latter are the constituting elements or nanoparticles, which are created in the reactive plasma chamber. The intrinsic porosity of these films favors electrical conduction across the structure, and at the same time it is sufficiently dense to facilitate catalytic processes.

6.1. *Nanoparticle flux and pressure ratio effects on morphology*

We start our discussions on TiO_2 thin film morphologies deposited using PASJD by discussing the effects of incident particle flux carried by the jet and the consequencies of varying the parameter R, i.e. the ratio between pressures at the two chambers.

The effect of particle flux is illustrated in Fig. 18, showing that two types of morphologies emerge depending on the intensity of the nanoparticles flux. That is, a compact and homogenous structure at higher fluxes (Fig. 18(a)), and a more interesting one displaying a column-like configuration at lower NP fluxes (Fig. 18(b)).

The effect of the parameter $R = P_p/P_d$ can be studied by keeping the nanoparticles flux fixed, and performing different PASJD experiments at each of the chosen ratios R. Specifically, we keep the pressure P_p constant, which is controlled by the amount of precursor gas injected into the plasma chamber. This is important since the film stoichiometry depends sensitively on P_p. To vary R, one can adjust the pressure at the deposition chamber, P_d, by changing the pumping speed in the latter until the desired value is reached. Once the pressure P_d has been obtained, one can still consider different nozzle–substrate distances. Indeed, film morphology depends on this distance yielding different types of film structures. Examples for the

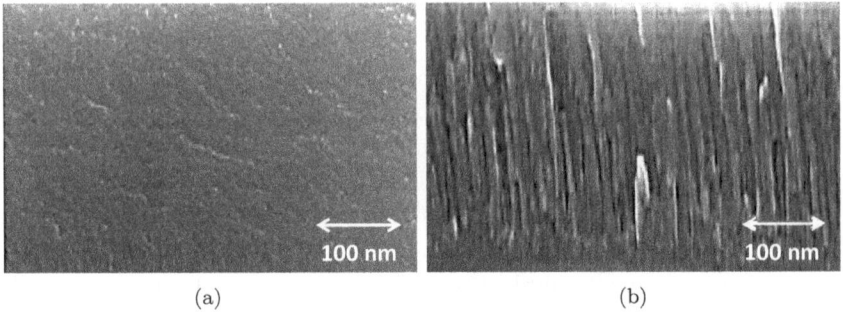

(a)   (b)

Fig. 18. Effects of incident nanoparticles flux. We observe two types of morphologies: (a) A high particle flux yields very compact structureless-type of depositions. (b) A low particle flux produces depositions characterized by column-like structures.

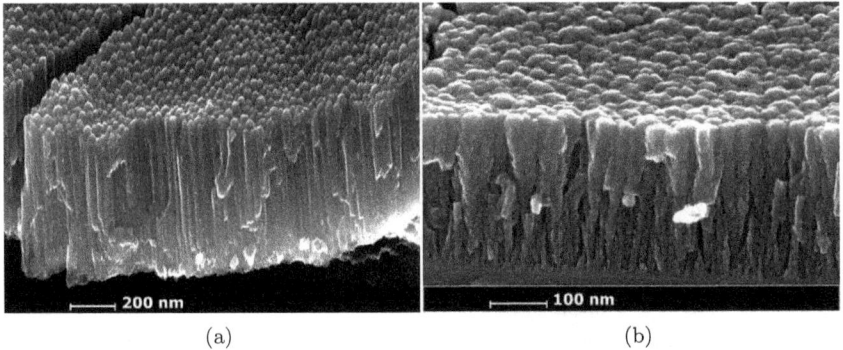

(a)   (b)

Fig. 19. Effects of varying the parameter R on morphology: (a) $R = 16$, and (b) $R = 4$.

cases, $R = 16$ and $R = 4$, for fixed nozzle–substrate distance, are reported in Fig. 19.

For $R = 16$, the film is dense and compact, displaying a well-defined ordering of quite straight columnar elements, in close contact to each other (Fig. 19(a)). The case obtained for $R = 4$ looks different, in the sense that the elements still appear to be of columnar shape, but now more complex structures are formed, displaying a conspicuous degree of ramification similar to tree-like shapes (Fig. 19(b)).

6.2. *Film width vs precursor temperature and rate of deposition*

The phenomenology associated to the formation of thin films using the PASJD is quite complex as there are several controlling parameters to be considered. For instance, one question of interest is at which precursor temperatures a columnar growth is facilitated. In fact, the precursor temperature, T, plays a fundamental role in determining the morphology of the thin film, in addition to its width. The latter depends also on the total deposition time, t, as one can intuitively expect. Illustrative results for these two dependencies, $W(T)$ and $W(t)$, are reported in Fig. 20.

In the first case (Fig. 20(a)), we have considered four different precursor temperatures, by keeping fixed the deposition time and the nozzle-substrate distance, z. The results suggest that the width is quite sensitive to temperature, displaying an exponential dependence, $W(T) \sim \exp(T/t_0)$. The question whether there is a single exponential law, as displayed in the figure, or a transition to a faster growth at higher temperatures, here around $52-55°$C, remains open.

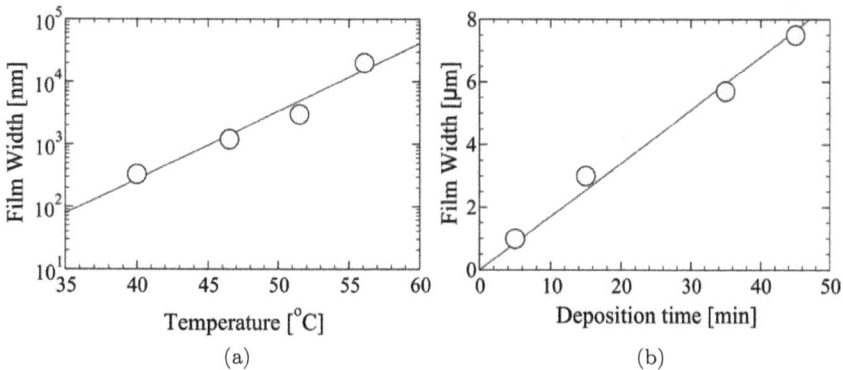

Fig. 20. Film width's characterisitics. (a) Film width [nm] vs precursor temperature [°C], in semi-logarithmic scale. These results were obtained for 15 min deposition time and a nozzle–substrate distance $z \approx 9$ mm. The straight line is a fit with the exponential form, $W(T) = W_T \exp(T/T_0)$, with $W_T = 0.013$ nm and $T_0 = 4°$C. (b) Film width [μm] vs deposition time [min], in linear scale. The straight line is a fit, $W(t) = W_D\, t$, with $W_D = 0.17\,\mu$m/min. These results were obtained for a precursor temperature of $50°$C and nozzle-substrate distances: $z \approx 8$ mm (for 5 min and 15 min), and $z \approx 6$ mm (for 35 min and 45 min). In all cases, the size of the circles is a measure of the experimental error bars.

To be noted is that, in order to perform the structural analysis, the sample needs to be annealed at room temperature after the deposition is completed. We have found that the most favorable working temperature is about 50°C, for which the film growth is most homogeneous and does not show any cracks or fractures in its structure. The second case (Fig. 20(b)), therefore, has been performed at this 'optimum' temperature, by keeping also the distance z fixed. The results are consistent with a linear dependence, $W(t) \sim W_D t$. The corresponding coefficient, W_D, represents the film deposition rate, which appears to be constant, an important feature in view of the possible applications of this growth technique.

6.3. *Fractal growth of TiO$_2$ films near a staircase-like substrate*

To get some insight about the possible types of morphologies expected for our thin films, we have performed simple Monte Carlo simulations of deposition processes, consisting of vertically falling identical particles onto a given substrate. The particles are cubes of size d (i.e., radius $r = d/2$, cf. Section 3.3.5), and the simulations are performed on a cubic lattice with lattice constant $a_L = d$. It is assumed that all particles fall one after the other, and a new particle is 'released' once the previous one has been attached to the film.

The MC rules are very simple. Once a particle touches a substrate site, or a previously attached one at its top, it gets stuck there forever with a vertical sticking probability P_V, here taken as $P_V = 1$. There is a second parameter, $0 \leq \beta \leq 1$, yielding the probability, $P_L = \beta$, that a falling particle gets stuck at a previously deposited particle by a lateral sticking interaction (see Fig. 21).

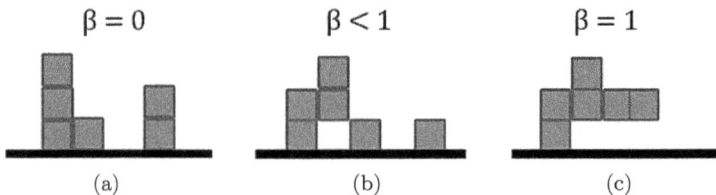

Fig. 21. Monte Carlo deposition rules. The probability for vertical sticking is unity, while it is $0 \leq \beta \leq 1$ for lateral sticking. Three typical cases are illustrated in the cases: (a) $\beta = 0$, (b) $0 < \beta < 1$, (c) $\beta = 1$.

More complex situations arise if one considers a flux of falling particles and takes into account suitable dynamics for them. In our present zeroth-order approximation, we do not allow for transversal diffusive motion of particles, neither on the substrate nor on previously stuck particles forming the film. However, we may still induce additional disorderly deposition behavior if we assume that the substrate is 'corrugated' presenting an intrinsic roughness of significant magnitude. To test these ideas, we have implemented MC deposition simulations for vertically falling particles on a flat substrate, and on a 'rough' surface presenting a random surface modulation. In both cases, we have used the same lateral sticking parameter $\beta = 0.1$ (see Fig. 22).

The idea of studying particle deposition on a spatially 'modulated' substrate, discussed in Fig. 22, can be realized experimentally. The simplest perturbation is a single rectangular step separating a two-plateau substrate, such as the one illustrated in Fig. 23.

In this case, deposition would remain vertically directed away from the step, while around it perturbations of the process are expected. Intuitively, the step would enhance a lateral diffusion, that is perpendicularly to the step direction, of the (slowly) falling particles. As a result, they would interact with each other randomizing their trajectories. This sort of chaotic dynamics will certainly modify the shape of the growing film near the step.

It is well-known that deposition processes governed by diffusion of falling particles can give rise to fractal growth, examples of which are diffusion limited aggregates (DLAs) [82]. In our case, we have a directional growth, but the essential features of diffusion limited aggregation remain valid to a large extent.

If our depositions behave according to this picture, they should display self-similar scaling over some range of length scales. We expect that the fractal film would possess a well-defined degree of ramification, as in the case of DLA. The experimental result is quite surprising and beautiful at the same time (Fig. 24). Close-up views of the obtained fractal film are shown in Fig. 25.

6.4. *Deposition of TiO₂ films using PASJD technique*

We summarize experimental results obtained with PAJSD, for different nozzle-substrate distances, precursor temperatures, deposition times, parameter R (see Fig. 26, and Table 2 for details).

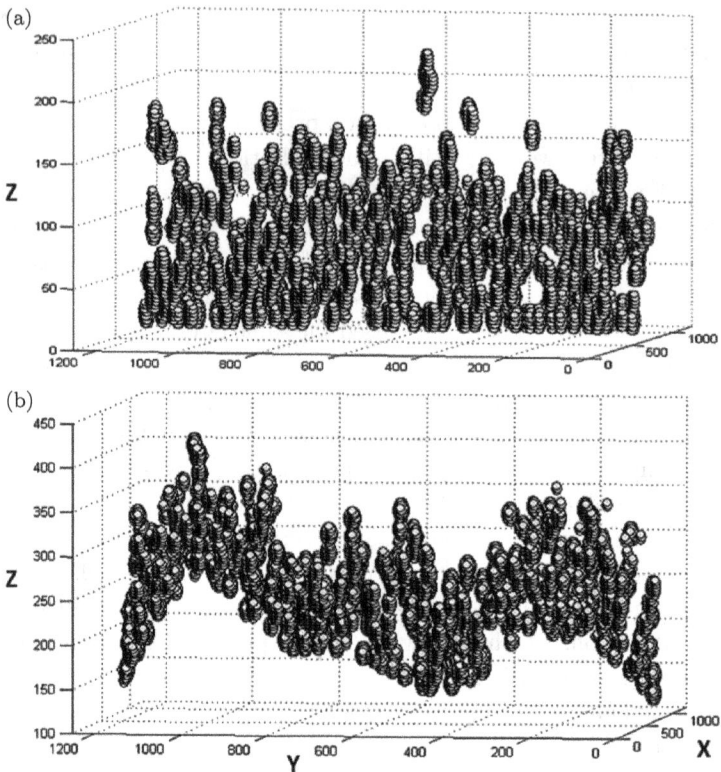

Fig. 22. Monte Carlo simulation of particle deposition on a: (a) Flat substrate. (b) Vertically modulated substrate. The results corrrespond to vertical sticking probability $P_V = 1$, and lateral one $P_L = \beta = 0.1$. The lattice used had size $(1200 \times 1000 \times 500)$ lattice units. The lattice unit is equal to the particle size d. In (b), the amplitude of the assumed roughness is about 250 lattice units. Only vertical motion is allowed, and a new particle starts falling, from a random (x, y) location at height $H = 500$, once the previous one has been attached to the forming film. The growth is stopped when the film reaches the height H for the first time. In this model, the deposition rate corresponds to 1 particle/MC time step. For simplicity of the presentation, we have shown only a thin slab of the deposited material.

As one can see from Fig. 26, the TiO_2 films display a typical well-ordered columnar structure, at least for the cases (a,b,c,e). The depositions reported in (d) and (f) appear to be anomalously grown. Case (d) corresponds to a higher NPs flux resulting from the higher precursor temperature used. In fact, the wide columns appear to not be well connected to each other, yielding both a higher porous structure and at the same time a quite fragile

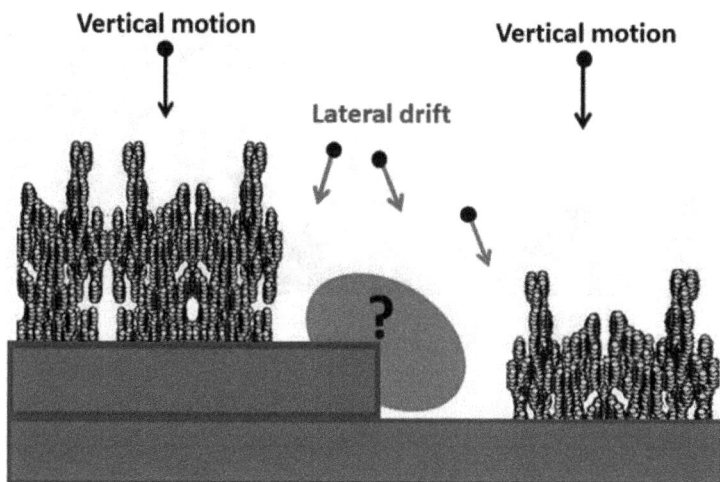

Fig. 23. Lateral enhanced diffusion for falling particle deposition near a single step on the substrate.

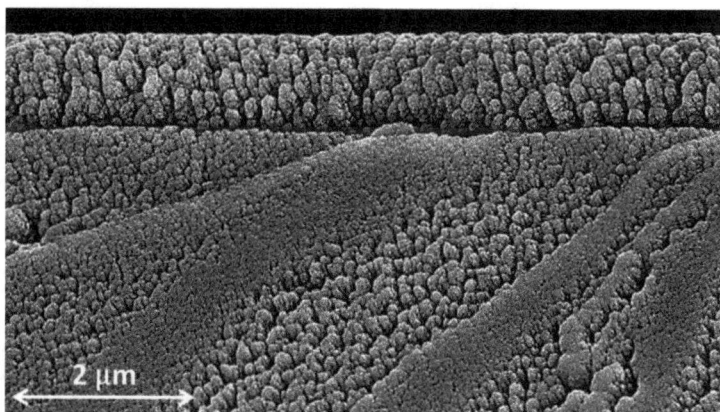

Fig. 24. A TiO_2 deposition on a flat two-plateau silicon substrate separated by a single staircase step.

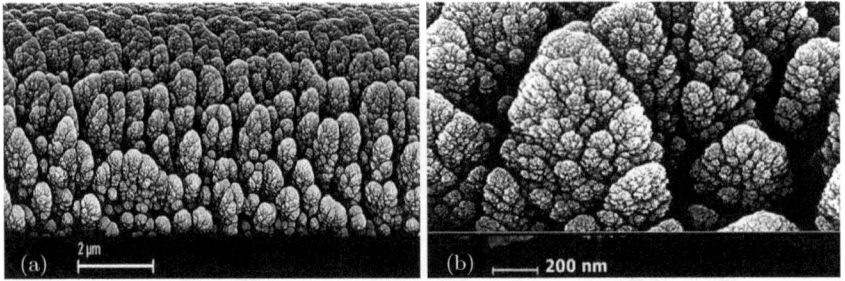

Fig. 25. Close-up views of the TiO$_2$ deposition in Fig. 24: (a) 2 μm scale, and (b) 200 nm scale.

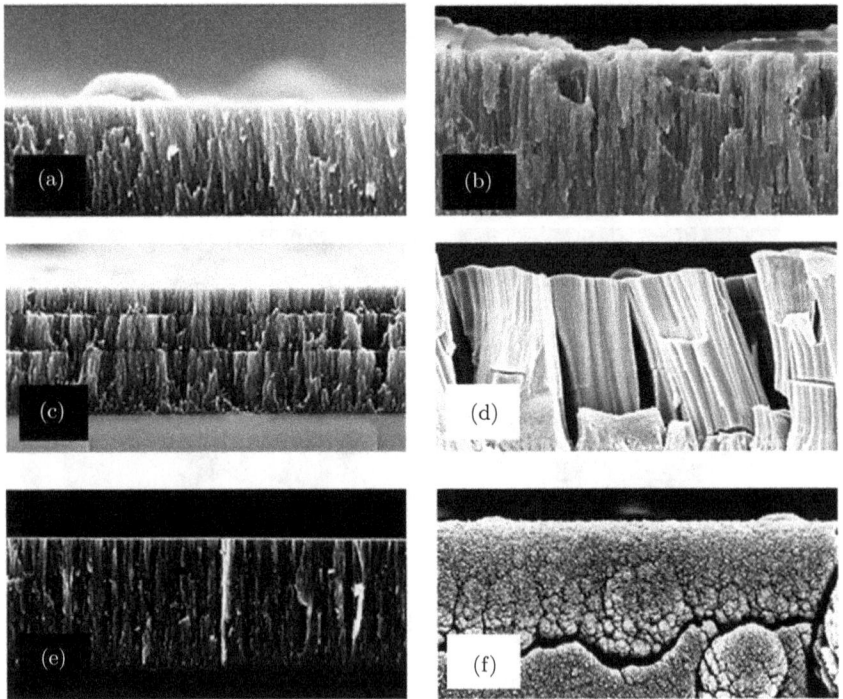

Fig. 26. Depositon of TiO$_2$ films on a substrate in the cases labeled from (a) to (f). For details, see Table 2.

Table 2. Summary of PASJD parameters for the samples in Fig. 26.

Sample	N-S distance (mm)	Temperature (°C)	Deposition time (min)	P_p (mbar)	P_d (mbar)	R
(a)	8.4	40.5	15	0.136	0.020	6.8
(b)	8.4	51.5	15	0.144	0.020	7.2
(c)	9.2	46.5	15	0.136	0.020	6.8
(d)	9.2	56.1	15	0.155	0.020	7.8
(e)	8.4	51.0	30	0.151	0.025	6.1
(f)	6.3	50.0	35	0.155	0.020	7.8

Notes: (1) sample index, (2) nozzle-substrate distance [mm], (3) precursor temperature [°C], (4) deposition time [min], (5) pressure at the plasma chamber P_p [mbar], (6) pressure at deposition chamber P_d [mbar], (7) parameter $R = P_p/P_d$.

one. Case (f) was obtained near a staircase-type of perturbed substrate which helps the formation of fractal shapes, as we discussed in Section 6.3, and Fig. 24.

6.5. *Effects of flat and rough substrates on TiO$_2$ films*

The roughness of a PASJD thin film depends on the type of substrate chosen for depositing the TiO$_2$ nanoparticles. Here, we discuss two different scenarios, one in which the substrate is 'flat', such as a crystalline silicon waffer, and a second one displaying 'large' surface variations, exemplified by fluorine-doped tin oxide (FTO) glass. The former is associated to roughness variations which are much smaller than the typical size of the TiO$_2$ NPs. The second case corresponds to cases in which the RMS are comparable or much larger than the NP's size.

The study of the dependence of film morphology on substrate roughness is relevant for applications [83]. It is well known, for instance, that TiO$_2$ materials develop remarkable photo-induced response to ultraviolet (UV) light, leading to the formation of photo-generated charge carriers [84, 85]. Substrate roughness generally determines TiO$_2$ film's nanostructure and, as a result, many of the thin film electro-chemical characterisitics such as photo-catalytic performance, adsorption, reflectance, adhesion and charge transport properties [36]. In particular, columnar structures, like nanorods and nanowires, favor charge carriers transfer rates and electron–hole recombinations, effects which are useful for increasing photo-catalytic reactions and solar cells efficiency.

6.5.1. *Deposition of TiO₂ NPs on different substrates*

The two substrates considered were: a flat crystalline silicon wafer, and a FTO glass, which will be referred to simply as silicon wafer and FTO glass. The deposited films were analyzed using AFM imaging technique, examples of which are reported and discussed in Fig. 27, for silicon wafer on the top panel, and FTO glass in the lower panel. Both substrates have been studied prior to TiO_2 deposition in order to characterize their original substrate properties.

At the left side of Fig. 27, we show the original surfaces, for images of 2000 nm linear size, before deposition. To be noted, the surface for silicon is quite flat, having a (RMS) roughnes of about 0.2 nm, while for FTO glass, it is much larger, of about 9 nm. Upon TiO_2 deposition, these values increase to about 2 nm and 12 nm, respectively. Images are shown at the 2000 nm and 500 nm scales. The precursor was heated at 44°C,

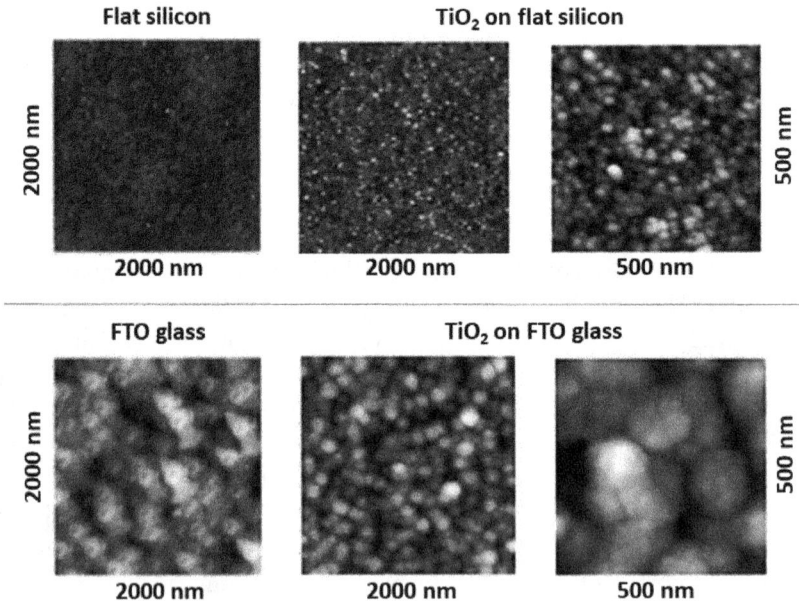

Fig. 27. AFM images of our TiO_2 deposition experiments: (Upper panel) Flat silicon substrate. (Lower panel) FTO glass substrate. The leftmost images correspond to the untreated samples, the central ones to the treated ones at 2000 nm scale, and the rightmost ones to the close-ups of treated samples at 500 nm scale.

the parameter $R = 30$, the distance nozzle-substrate was 9 mm (other values around this one yielded similar results), and the deposition rate was about 130 nm/min.

Using the theoretical results from Section 3.3.5, we can evaluate a typical size, r, for the elementary units building up the film. The latter is given by $r = (r_c/2)^2/R_{tip}$, where the tip size is $R_{tip} = 10$ nm and $r_c = (15 \pm 2)$ nm, yielding $r = (5.6 \pm 1.5)$ nm.

A careful examination of the AFM images suggests that the roughness increases dramatically on silicon wafer upon deposition of TiO_2, relative to its originally very small value, i.e., from 0.2 nm to (2–3) nm. While on FTO the increment is fractional, with respect to the already large glass roughness, from about 9 nm to about (10–12) nm. To be noted is that NPs radii are around 6 nm, and they do not contribute much to FTO glass roughness. Eventually, they aggregate more easily around the glass protrusions, yielding apparently larger TiO_2 grains, as the close-up at 500 nm scale in the lower panel of Fig. 27 suggests.

6.5.2. *Monte Carlo simulations of deposition*

We have implemented MC simulations to understand how the films get structured, and whether simple models can explain the main observed morphological features (see also [86]). In particular, our approach is based on the modeling discussed in Section 6.3, while here we discuss for illustration only the flat substrate case (Fig. 28).

We use our convention of vertical sticking probability, $P_V = 1$, and lateral one, $P_L = 0.1$. Some modifications to the choice of the lattice and particle size are introduced for convenience. The MC simulations where performed on a simple cubic lattice of $166 \times 166 \times 250$ sites. The lattice constant was chosen to be of 3 nm, yielding a substrate effective area of 500×500 nm^2, while the NPs were represented by a cube of 4^3 sites, hence they correspond to a linear size (diameter) of 12 nm. The mismatch between particle size and lattice constant produces a more realistic particle distribution within the film, also yielding a more porous structure. The height of the lattice, i.e., 750 nm, was enough to get films with stable RMS roughness. We found that 10^6 NPs were sufficient to reach our goals, in keeping with the requirement that lateral correlations of the film remain smaller than the substrate linear size chosen. In this sense, wider films (in the vertical direction) can be produced with the algorithm, but the lattice size may need to be increased accordingly.

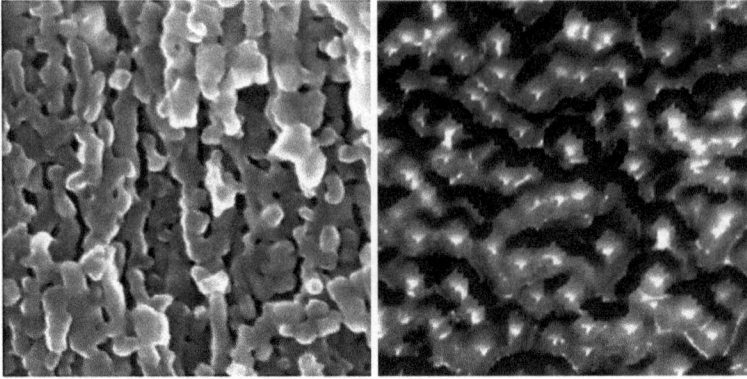

Fig. 28. (Left panel) High resolution SEM images of TiO_2 deposition on the flat silicon substrate area of about $500 \times 500 \, nm^2$. From this figure, one can estimate an average particle size, yielding a diameter of about $(12 \pm 2) \, nm$, in good agreement with the previously reported values from AFM measurements in Fig. 27. (Right panel) Simulation of TiO_2 deposition on a flat substrate. The substrate lattice is aimed at describing an area of $500 \times 500 \, nm^2$, and assuming a lattice constant of $3 \, nm$, it consists of 166×166 sites. Each TiO_2 nanoparticle is represented by a cube of linear size $12 \, nm$, consistent with TEM and AFM measurements. This model differs from the simpler one in which the linear size of the particle coincides with the lattice constant (see Section 6.3 and Fig. 22).

7. Experimental Results and Modeling of Ar Plasma Jets

In this section, we discuss experimental results and theoretical calculations in the case of a pure argon plasma. The modeling, based on MC simulations of a supersonic jet, is described in detail and a comparison of the measured ion energy distribution functions with the theoretical results is presented. Before entering this problem, we review the experimental setup for neutral gas species, in order to check the reliability of the measuring system.

7.1. *Neutral gas species in the deposition chamber*

The mass spectrum of neutral species effectively present in the deposition chamber was determined at the pressure conditions, $P_d \simeq 10^{-6} \, mbar$ and $P_{QMS} \simeq 10^{-7} \, mbar$. The spectrum is used to test the reliability of the vacuum system, allowing the search for possible leakages and to find out whether impurities are already present in the deposition chamber, before injecting a gas and/or turning on the plasma source.

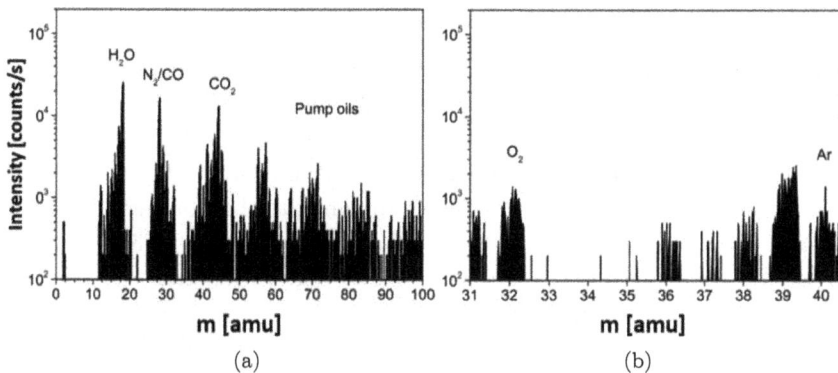

Fig. 29. (a) QMS spectrum [counts/s] vs mass [amu] of neutral species present inside the deposition chamber without injection of gas from the plasma chamber. The main peaks correspond to the substances indicated in the plot. The residual molecules detected are attributed to the pump oil generated by the rotative pumps, to those molecules retained inside the vacuum vessel from previous operations, and to air due to small leakages. (b) close-up of the spectrum from (a) showing the range of masses associated to molecular oxygen (32 amu) and argon (40 amu), possibly present due to trace impurities. Note their diminished peak intensities as compared to the main peaks in (a).

Figure 29(a) shows a QMS scan of neutral species at a distance of 99 mm from the nozzle. Many peaks are clearly seen, of which the most significant ones correspond to water molecules, nitrogen atoms, different carbon oxides and pump oil hydrocarbons.

Figure 29(b) displays the spectrum in the range 31 amu $< m <$ 40.5 amu, where one can identify the signals associated to molecular oxygen and atomic argon. The intensities of the latter ($\lesssim 10^3$ counts/s) are much lower than those corresponding to the main peaks in Fig. 29(a), which are about $(10^4$–$10^5)$ counts/s.

As a second test, we considered an argon-oxygen mixture (see Section 5.2), of respectively (0.4–0.6) molar fractions and 8 Pa total gas pressure, injected into the plasma chamber. Several QMS measurements were performed in the deposition chamber, by varying its pressure P_d, to evaluate the degree of stability of the whole system for different R values. In all cases, the expected argon and oxygen peaks were clearly visible, and the measurements showed an excellent stability and reproducibility. The statistical errors obtained by averaging the measured quantities over different scans were smaller than 1%. We consider next charged Ar^+ gas expansion phenomena.

7.2. *Charged Ar^+ gas expansion*

We consider an argon plasma generated with an ICP set up (see Section 3.2.1). The applied pressures were, $P_p = 3.2$ Pa and $P_d = 0.11$ Pa, in the plasma and deposition chambers, respectively, yielding $R = 32$ (see Fig. 17) and the Mach disk distance $z_M = 26$ mm from the nozzle. The RF input power was set at 450 W, corresponding to the working regime for thin film depositions, typically used in plasma applications [36,37]. We then used the QMS to obtain the ion energy distribution functions (IEDFs) at different positions along the jet.

In the ICP reactor, a uniform and low plasma potential value (<20 eV) produces ions described by a mono-energetic energy distribution function, exhibiting a peak centered around their mean thermal energy. The latter is much smaller than the typical electron temperature $T_e \simeq 1$ eV in the plasma [36]. Upon crossing the collisionless sheath, the ions gain an energy equal to the potential drop ΔV. Once the ion energy spectra has been measured by a grounded instrument, such as a QMS or a retarding potential analyzer, it can be analyzed to obtain the plasma potential [87].

A typical IEDF for Ar^+, measured inside the deposition chamber, is shown for illustration in Fig. 30. While Ar^+ represents the main ion

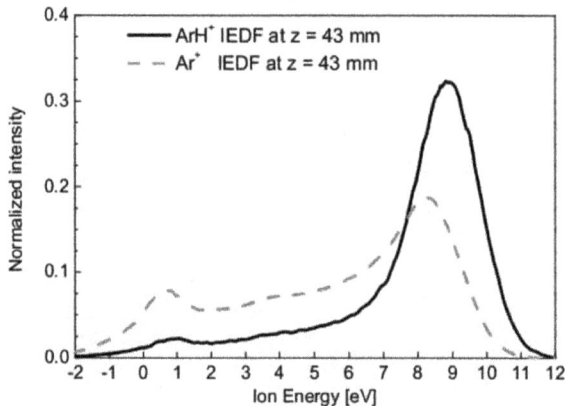

Fig. 30. Normalized ion energy distribution functions vs energy [eV], for Ar^+, obtained by positioning the QMS at a distance $z = 43$ mm from the nozzle, in the case of $R = 32$, corresponding to a Mach disk position $z_M = 26$ mm. A second ion component, ArH^+, is also present (see text). The RF plasma discharge was inductively coupled to the plasma and maintained at 450 W. Adapted with permission from [89]. © 2016 American Phyical Society.

component in the spectrum, a second species, here ArH^+, is clearly visible. This additional ion constitutes a minor species formed by the reaction of Ar^+ with residual water molecules present in the plasma chamber [88].

In this example, we considered the position $z = 43\,mm$, far away from both the Mach disk and the nozzle, for illustration purposes only. Later in what follows, we consider other more interesting positions along the z-axis, both inside and outside the supersonic plasma jet.

As is apparent from Fig. 30, both energy spectra exhibit a main peak around 8.3 eV, corresponding to the fast ion population entering the mass spectrometer without undergoing dissipative collisional processes. Such an energy value is close to the plasma potential since the mass spectrometer is grounded [87].

In both spectra, the presence of a second, less prominent peak is also apparent at low energies. To be noted is that, in the case of Ar^+, this second peak occurs at an energy smaller than 1 eV. The latter can be related to ions formed by charge transfer processes in which a thermal ion is generated. The difference between both spectra can be understood by noticing that ArH^+ ions do not undergo symmetric resonant charge exchange collisions since their equivalent neutral species do not exist, making electron tunneling processes very unlikely [90]. In fact, other minor ion species, having much larger masses than Ar, do not generally show a low-energy peak, but only a high-energy peak around 8.3 eV. This result highlights the importance of charge transfer phenomena that argon ions undergo outside the plasma chamber.

According to this picture, we can expect a rather high plasma potential inside the plasma chamber, here around 8.3 eV, and two different scenarios governing Ar^+ collisions: (i) elastic collisions, characterized by small momentum transfers describing intermediate energy processes ending at the main 'elastic' peak (here at 8.3 eV), and (ii) charge transfer collisions, in which ions are slowed down to thermal energies yielding a secondary, low energy peak (here around and below 1 eV).

In the following, we discuss the results on ion energy distributions in the deposition chamber, both inside and outside the supersonic jet. The measured IEDFs for Ar^+ are shown in Fig. 31. The spectra display a well-separated two-peak structure. The high-energy peaks, around 8.3 eV, are due to ions carrying maximum available energy, while the low-energy ones, from thermalized ions, are found around 2.2 eV inside the jet, and localized around 0.2 eV outside the Mach disk. Indeed, when the ions enter the deposition chamber, they undergo a low-collisional supersonic

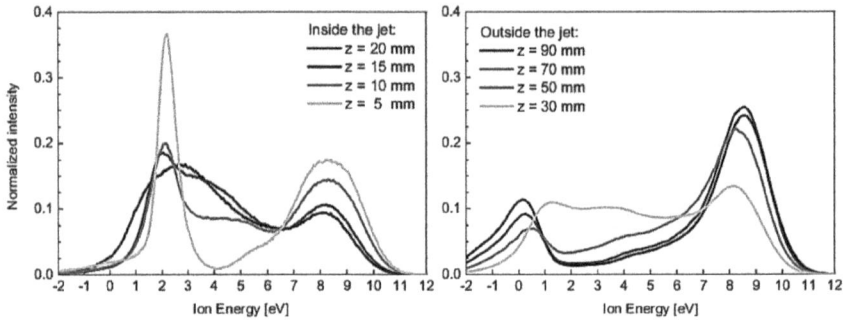

Fig. 31. (Left panel) Normalized IEDFs vs ion energy [eV] obtained with the QMS for Ar^+ inside the supersonic jet expansion, $z < z_M$. (Right panel) Same as for the left panel but outside the supersonic jet expansion, $z > z_M$. Here, $R = 32$ and $z_M = 26$ mm. The RF discharge was inductively coupled to the plasma and maintained at 450 W, and averages over different measurements have been performed in all cases. The distances from the QMS to the nozzle z are shown in the insets for each case. Adapted with permission from [89]. © 2016 American Phyical Society.

expansion regime, getting accelerated by the potential drop between the plasma potential, V_P, and the potential along the jet, V_{jet} [91], but they can still lose part of their energy during collisions before reaching the QMS.

Since in our experimental set-up the QMS is grounded, its sheath can be considered almost collisionless, and the lowest ion energy attainable corresponds to the thermal energy of neutral species, plus the potential energy $V_{jet} \cong 2$ eV (see left panel in Fig. 31). Outside the Mach disk (right panel in Fig. 31), the low-energy peak drops to a lower value, about 0.2 eV, thus indicating a rapid decrease of the jet potential in this region, $V_{jet} \cong 0$.

In general, ion collisions with neutral species within the jet yield a complex redistribution of strength of the IEDFs as a function of energy and position z. As is apparent from the figures, negative energy events are measured which are known to be due to an instrument artifact [92].

The suggested scenario regarding the role of the potential energy V_{jet} was an attempt to interpret the IEDFs shown in Fig. 31, and it is supported by similar measurements performed using different experimental conditions based on the choice of different gas mixtures in the plasma chamber.

A more accurate measurement of the plasma potential was attempted by using a Langmuir probe, designed to compensate the radiofrequency signal [68, 69]). Despite the RF shielding, it was very difficult, however, to obtain a reliable value of the plasma potential, or of the electron temperature, in this way due to the presence of the 13.56 MHz noise, affecting the Langmuir current–voltage characteristics both inside and outside the plasma chamber [68]. In what follows, we discuss an *ab initio* theoretical approach to interpret the IEDFs measured experimentally.

7.3. *Modeling of* Ar^+ *plasma jets*

The inductively coupled plasma generated in the plasma chamber reaches a steady temperature by keeping the RF input power constant [36]. Inside the plasma chamber, ions are produced mainly by electron impact ionization and their initial energy distribution can be assumed, to a very good approximation, to be of Maxwellian shape, with its peak centered at the corresponding mean thermal energy of the ions [47].

The latter governs the properties of the resulting supersonic jet in the deposition chamber. In fact, ions trajectories in the jet are determined by both the jet gas expansion and the electrostatic interactions produced by free electrons and local electric fields [34].

In particular, a potential drop is expected in the region between the plasma chamber and the Mach disk, behaving similarly to a pre-sheath [39]. The latter produces a shift in the ion energy distribution functions proportional to the potential energy drop [87]. The ions undergoing collisions within the sheath will be partially slowed down, contributing to the low energy tail observed in the IEDF [93]. In the following, we discuss the theoretical model developed by us to describe the experimentally measured IEDFs.

7.3.1. *Type of interactions: Elastic and charge transfer events*

In our theoretical approach, we assume an adiabatic jet expansion process and consider an isentropic flow model to describe both neutral and charged gas expansions [34,41]. To a first approximation, the neutral gas component of the jet can be treated as an ideal one, due to the low collision rates and the low charge density in the rarefied plasma [38, 41]. Since the charge density in the jet is about two orders of magnitude lower than the neutral mass density, we restrict our simulations to solely ion–neutral collisions, the most relevant within the supersonic jet.

The interaction potential between an Ar^+ ion and a neutral Ar atom can be described by means of a modified type of Lennard-Jones potential, frequently used in the literature [94,95], in which, in addition to the short distance repulsive term, $+1/r^{12}$, and the standard long-range van der Waals attractive term, $-1/r^6$, there occurs an additional charge-induced dipole attractive interaction, $-1/r^4$. The extra term arises from the electric dipole induced by Ar^+ on a neutral Ar atom (see, e.g., [96]).

The effective interaction potential, $U_{eff}(r)$, between Ar^+-Ar, located at a relative distance r, can thus be parametrized in the form

$$U_{eff}(r) = \frac{A}{r^{12}} - \frac{B}{r^6} - \frac{C}{r^4}, \qquad (7)$$

where A, B and C are constants describing atomic Ar. The first two, A and B, represent the Lennard-Jones potential constants [97,98], while the third one, C, is proportional to the atomic polarizability, α_{Ar}, where $C = \alpha_{Ar}/2$ in atomic units. It turns out that the second term, B/r^6, plays a lower role than the last one, C/r^4. This is because of the rarefied nature of the jet, where the relevant events altering the trajectory of Ar^+ occur more frequently at relatively large distances between the species.

In our case, charge transfer reactions describe proximity events, taking place during a collision, in which an electron tunnels from the slow moving neutral atom, Ar_{slow}, to the fast moving charged species, Ar_{fast}^+ (see [99] for a general discussion). If such an electron exchange takes place, one can describe the process by the following reaction:

$$Ar_{fast}^+ + Ar_{slow} \rightarrow Ar_{fast} + Ar_{slow}^+. \qquad (8)$$

In the case of Eq. (8), the process is generally elastic and it is referred to as symmetric resonant charge transfer [100]. The associated cross-sections are larger at lower collision energies, becoming common collisional processes for low-temperature plasmas [101,102].

7.3.2. Geometric aspects of the model

We consider a bottom-up approach, by discussing first the geometry of the system and the scattering problem at the nanoscale. To do this, we restrict the simulations inside an effective volume within which the interactions take place.

This volume has a cylindrical shape, whose cross-section is sufficiently large as compared to the relevant Ar^+–Ar interaction range, but sufficiently small in order to save computing time. The cylinder diameter was fixed at 10 nm and its length at 500 μm, corresponding to the extreme aspect ratio of 50,000. The choice could be made possible by exploiting the fact that the ion trajectories remain largely confined around the cylinder axis over long distances as compared to its width.

To describe the Ar^+ dynamics, we chose a frame of reference moving with the mean speed of neutral Ar species, given by Eq. (3), implying that they are initially at rest and placed at random inside the cylinder. In the latter, the number of neutral Ar is chosen to be consistent with their density in the jet, described by Eq. (4), which turns out to be rather small, typically about 10 atoms/cylinder. Using the Ar^+-Ar interaction potential, Eq. (7), we solve the equations of motion of the moving Ar^+ to a high degree of precision (see Section 7.3.3).

The main axis of the cylinder coincides with the initial direction of the Ar^+ velocity (Fig. 32(a)). The ion moves along the cylinder and interacts with the neutral species (Fig. 32(b)). We calculate the total force acting on the ion by summing over the contributions of all neutral Ar present within the cylinder. Contributions from nearby cylinders are not considered.

Whenever the ion exits the cylinder, we stop the calculation, attach a new cylinder whose axis now coincides with the direction of the exiting ion velocity, and continue solving the equations of motion with these new initial conditions (Fig. 32(c)). Each time a new cylinder is considered, new neutral atoms are placed at fixed random positions inside it.

When the ion kinetic energy becomes comparable to that of the neutral species, the simulation is stopped since the method would become too time-consuming. This is consistent with our central assumption that we solve the ions dynamics in the presence of a 'frozen' distribution of neutral atoms in space. The ions contribution to the low energy IEDF is treated

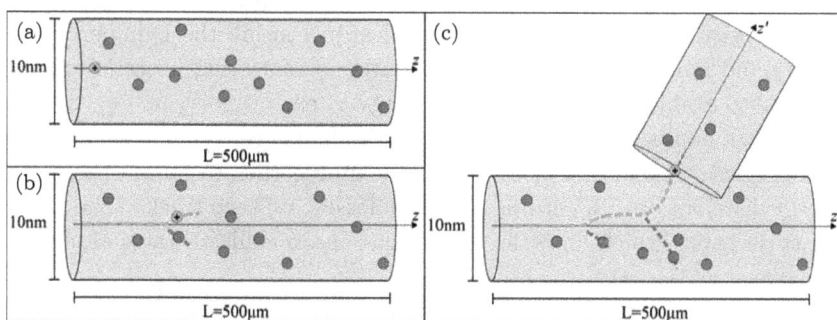

Fig. 32. Schematic representation of an Ar^+ trajectory inside the interaction space (cylinder). The dots represent the neutral Ar atoms, located at random inside the cylinder. (a) Starting cylinder: The ion velocity is assumed to coincide with the z-axis (the cylinder main axis). (b) Example of an Ar^+ trajectory inside the cylinder. (c) Case of the Ar^+ laterally exiting the cylinder, in which the new orientation axis, z', of the second cylinder coincides with the direction of the exiting ion velocity. Adapted with permission from [89]. © 2016 American Phyical Society.

separately, assuming a Gaussian distribution centered at the mean neutral kinetic energy.

7.3.3. *Simulations of Ar^+ energy distribution functions*

In this section, we discuss the simulation results obtained for the IEDFs Ar^+ using the model described in Section 7.3.2, and compare the theoretical predictions with the experimental observations (Fig. 31). This study sheds some light about the phenomenology of Ar^+-Ar collision rates, which are otherwise difficult to extract experimentally.

We perform first principles simulations of the ions dynamics by numerically integrating the equations of motion of charged Ar^+ and neutral Ar species, using the geometrical model of Section 7.3.2. The particular geometry of the model allows us to treat the many particle dynamics from the nanoscale to centimeters. We use an integration step of about 10 ps, which turns out to be sufficiently small to get accurate results.

In practical calculations, we consider only one Ar^+ at a time, which can be justified due to the rather small density of Ar^+ ions in the jet. The single ion interacts with the neutral Ar atoms, via the interaction potential given in Eq. (7), which are assumed to be at rest inside the cylinder. For simplicity, we neglect the B/r^6 interaction term, which is found to yield a small contribution compared to the dominant C/r^4 induced dipole term. Upon an interaction with the ion, a neutral atom attains a finite speed, which is obtained by solving its equation of motion assuming it does not interact with the remaining few neutral atoms inside the cylinder. Both energy and momentum are conserved during a scattering process, and we are able to evaluate the Ar^+ kinetic energy at each time-step.

A large number of single ion trajectories have been simulated, and the ion energies at position z from the nozzle are recorded to obtain the mean energy distributions. Regarding the second issue, we keep track of the single ion trajectories which allow us to estimate mean collision rates along the supersonic jet.

Considering the Ar^+ initial conditions, it should be possible in principle to take the value,

$$E_{initial} = E_{peak} - qV_{jet}, \quad \text{with} \quad E_{peak} \simeq 8.3\,\text{eV}, \quad (9)$$

as the kinetic energy of Ar^+ at the nozzle, at $z = 0$ and $t = 0$, to solve the equations of motion for the ion within the deposition chamber [91]. The problem is that the ions undergo scattering processes inside the nozzle, which modify their initial energy, $E_{initial}$. In addition to this complication, we need to know the density of neutral species rather accurately in order to use it in our simulations, which can be estimated using Eq. (4), but it is valid only for $z \gtrsim 3.5$ mm.

To overcome these problems, we consider only kinetic energies, E_k, in our simulations. From the kinetic energy we extract the speed associated to Ar$^+$ as the inital condition to start integrating the equations of motion. We determine the initial value of E_k from the actual energy distribution, measured by the QMS for positive energies at $z = 5$ mm, shown in the left panel of Fig. 31 (and by the continuous line in Fig. 33(a)). This distribution displays essentially two peaks, one centered around 8.3 eV, representing high energy ions which have undergone some momentum transfer collisions, and a low-energy peak located around 2.2 eV (see also [34]). Note the low number, or even absence of ions, at intermediate energies.

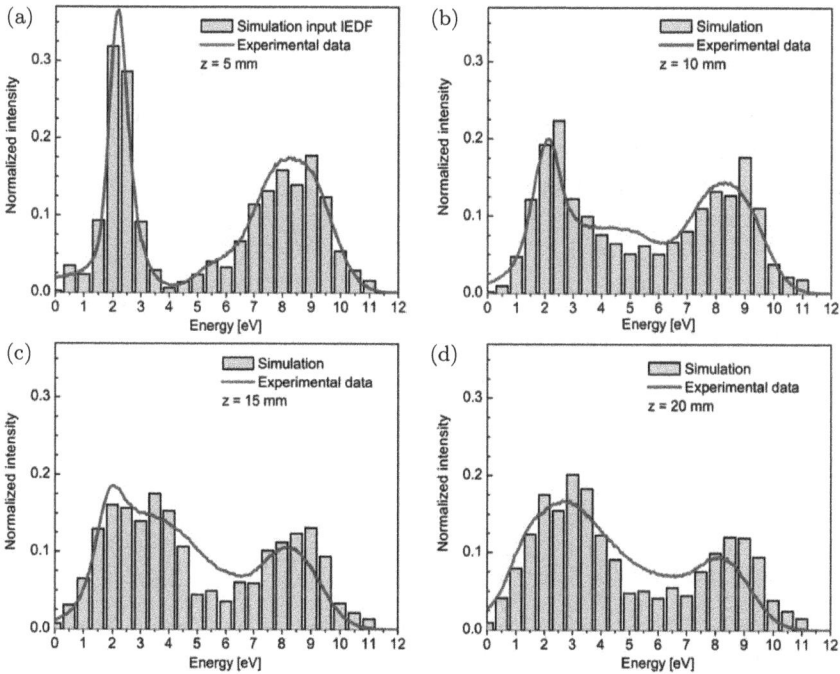

Fig. 33. Ar$^+$ normalized energy distribution functions vs energy [eV] for different z positions: (a) 5 mm, (b) 10 mm, (c) 15 mm and (d) 20 mm. The model predictions (bars) are compared with the experimental results (continuous lines). The initial distribution function in (a) was assumed for Ar$^+$ to start off the simulations. About 1000 Ar$^+$ single trajectories for each value of z have been simulated. Adapted with permission from [89]. © 2016 American Phyical Society.

Figure 33 shows the simulation results (green bars) compared to the experimentally measured IEDFs (red lines). For illustration, the numerically generated initial energy distribution for Ar^+ is reported in Fig. 33(a).

The case $z = 10$ mm is shown in Fig. 33(b), where one can see that the simulated IEDF displays an increased population of ions at intermediate energies, between 2 eV and 8 eV, in very good agreement with the measured distribution function. Also for larger z values, 15 mm and 20 mm shown in Figs. 33(c) and 33(d), respectively, the model predicts a rather conspicuous change in the shape of the distribution displaying a further increase of ion population between 2 eV and 4 eV, in good agreement with the experimental results. For illustration, we show few typical ion trajectories in Fig. 34.

In what follows, we discuss the collision events predicted by the model.

7.3.4. *Charge transfer processes*

In general, the effects of charge transfer phenomena on IEDFs are important in spatial regions where ions are more energetic than neutrals, as in the case of the sheath or pre-sheath regions of a plasma. During a collision between a fast moving ion and a slow neutral species, Eq. (8), a charge transfer can

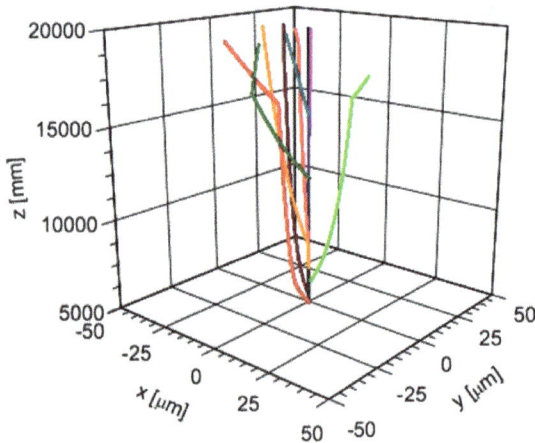

Fig. 34. Full trajectories over macroscopic lengths, $z \leq 15$ mm, simulated for Ar^+ in a supersonic jet. Note the much smaller scale used along the (x, y) directions, here in the range $-50\,\mu m \leq (x, y) \leq 50\,\mu m$, relevant for the scattering processes, and roughly corresponding to the sampling orifice diameter of the QMS. Adapted with permission from [89]. © 2016 American Phyical Society.

take place in which only part of the kinetic energy of the impinging ion is transferred to the neutral target. As a result, this process generally leads to the formation of low-energy peaks in the IEDFs [103].

In our jet experiments, this process is quite significant since the cross-sections for elastic scattering and charge transfer are of the same order of magnitude in the relevant energy range, $0.1\,\mathrm{eV} < E < 10\,\mathrm{eV}$ [101]. Therefore, it is necessary to consider charge transfer collisions between Ar^+ and Ar in our model, too. We do this by means of an approximation based on quantum considerations.

The probability of charge transfer, $P_{\mathrm{CT}}(r)$, depends on the distance r between the colliding species. Typically, one assumes the existence of a sharp cut-off, R_c, such that $P_{\mathrm{CT}}(r) = 0$ for $r > R_c$. For $r < R_c$, the probability P_{CT} displays a strong oscillating behavior, varying between zero and one, and described by a sine-squared function (see, e.g., [99]). The probability $P_{\mathrm{CT}}(r)$ can be estimated from the overlap of the wave-functions, between the relevant electronic states of the two colliding partners, involved in the tunneling process [100]. In a simplified picture, we disregard the oscillations for distances $r < R_c$, and assume P_{CT} to be constant [96].

From the known cross-sections reported in the literature (see, e.g., [101]), we determine the cut-off distance to be $R_c \simeq 0.4\,\mathrm{nm}$. The effective picture implies that, whenever the ion and the neutral argon get closer than R_c, an electron can be shared between Ar^+ and Ar, forming a sort of quasi-molecule. This is possible since the ion speeds in play are much smaller than those of the electron participating in the transfer process. When they get farther apart, the electron can be found with the same probability, $1/2$, in either one of the pair members. If the electron remains on Ar^+, the collision can be considered as a simple elastic scattering event, otherwise it is denoted as a charge transfer event. As a result, the charge transfer probability is taken simply as $P_{\mathrm{CT}} = 1/2$ for $r < R_c$, and $P_{\mathrm{CT}} = 0$ for $r > R_c$.

In order to get further insight into the charge transfer process, we consider the interaction energy involved when the two partners collide. Their interaction energy, $U_{\mathrm{eff}}(r)$, actually used in our simulations reads (cf. Eq. (7)),

$$U_{\mathrm{eff}}(r) = \frac{A}{r^{12}} - \frac{C}{r^4}, \tag{10}$$

which is shown in Fig. 35(a), together with the location of the cut-off distance, R_c. As one can see from the figure, the potential displays a sharp

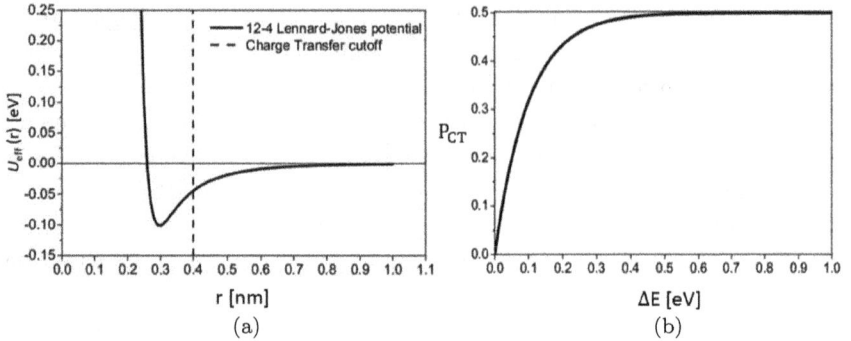

Fig. 35. (a) The interaction potential $U_{eff}(r)$ [eV], vs distance r [nm], for Ar^+-Ar system used in the simulations. The vertical dashed line indicates the cut-off distance $R_c = 0.4$ nm. (b) Probability P_{CT} for charge-transfer processes as a function of transfer energy ΔE [eV] for Ar^+–Ar, as described by Eq. (11). Adapted with permission from [89]. © 2016 American Phyical Society.

repulsive barrier below $r \simeq 0.25$ nm, suggesting that for such distances the interaction energy between the ion and the atom can be large, thus favoring the electron tunneling process. In this regime, the charge transfer is expected to take place as described by the simple scheme discussed above. The situation for distances around R_c is different.

For distances close to R_c, where the interaction potential is much softer, the interaction energy for electron transfer can be very small, therefore the assumption that $P_{CT} = 1/2$ does not seem accurate. Indeed, in our simulations we have found many scattering events taking place for distances close to R_c, with almost no appreciable interaction energy contributions.

To soften the sharp condition at $r = R_c$, we have incorporated an energy-dependent transfer factor for P_{CT}, which is now assumed to be given by the empirical relation

$$P_{CT} = \frac{1}{2}\left[1 - \exp\left(-\frac{\Delta E}{E_0}\right)\right], \qquad (11)$$

where $\Delta E \geq 0$ is the total transfer energy from the ion to the neutral atom in the single scattering process, and $E_0 = 0.1$ eV, corresponding to the minimum Ar^+-Ar interaction energy (see Fig. 35(a)), representing the characteristic energy for the process.

Notice that $P_{CT} \to 0$ when $\Delta E \to 0$ in Eq. (11), yielding the expected result when the particles are far apart, also softening the discontinuity introduced above, i.e., $P_{CT} = 0$ for $r > R_c$. The new form for P_{CT}

is displayed in Fig. 35(b). Notice that this energy-correction factor is important for energies $\Delta E < 0.2\,\text{eV}$. However, as a result of Eq. (11), the number of unrealistic charge transfer events gets reduced, in better agreement with the experimental results, as we discuss in the following section.

7.3.5. *Analysis of the scattering processes*

The IEDFs shown in Fig. 33 stress the importance of collisional phenomena inside the plasma jet. In our analysis, a collisional event takes place when there is an energy transfer between an Ar^+ and a neutral Ar. A collisional event can be either a simple scattering event or a scattering with a charge transfer, the latter occurring with a probability described by Eq. (11). Both types of elastic scatterings redistribute the strength determining the shape of the two-peak IEDFs. We find that for all the ion trajectories simulated, almost every ion experiences a collision, resulting in a charge transfer with a probability very close to 50%. Specifically, about 40% of the ions undergo a charge transfer reaction.

In order to estimate the global effect of charge transfer events, we have performed a simulation of the dynamics of about 1000 ions, for $z = 10\,\text{mm}$, setting the probability $P_{\text{CT}} = 0$, thus considering only simple collisions. The results are shown in Fig. 36, where it is possible to compare them with the case in the presence of charge exchange phenomena.

The Ar^+ energy distribution obtained without charge transfer becomes slightly larger than its charge-transfer counterpart for energies $E > 6\,\text{eV}$, while it becomes slightly smaller for energies $E < 5\,\text{eV}$. As it is apparent from Fig. 36, charge transfer events seem to provide a better agreement with the experimental data, suggesting that indeed they are playing an important role.

It is also possible to consider the ion energy loss after each collision, defined as the difference between the final ion (or newly produced ion) energy and the initial ion energy. Ion–neutral collisions clearly result in the formation of a high-energy neutral particles population [104], as testified by Eq. (8). The latter can exhibit a wide energy spectra, depending on the ion collisional processes. Considering the simulation results shown in Fig. 33, we distinguish between charge transfer and simple elastic scattering events to evaluate the energy loss along the ion's trajectories. This energy loss also corresponds to the energy which is acquired by a single neutral particle, and it is shown as a normalized histogram in Fig. 37 for both types of elastic scatterings, without and with charge exchange reactions.

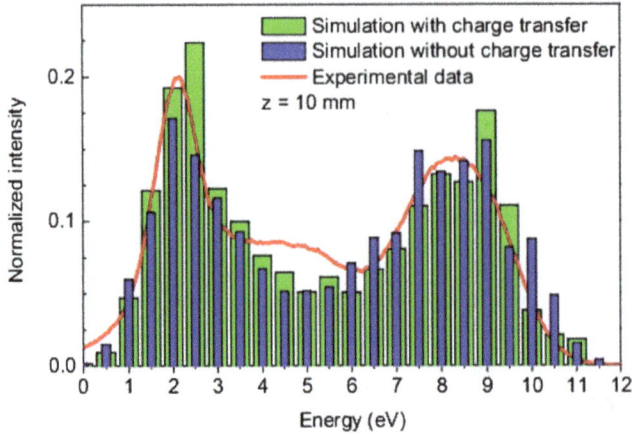

Fig. 36. Probability distribution function (IEDF) for Ar^+ vs energy [eV] at $z = 10\,mm$ obtained including a charge transfer probability in the model (green vertical bars) and without charge transfer (blue vertical bars). The experimental results (see also Fig. 33(b)) are represented by the continuous lines. Adapted with permission from [89]. © 2016 American Phyical Society.

In general, the energy loss is mostly limited to the range $0 < E < 2\,eV$, however, the two types of collisions yield very different results. Let us first consider scattering events without charge transfer. As one can see from Fig. 37, there occurs a prominent peak at energies $E < 0.5\,eV$, suggesting that in the majority of the scattering events the impinging ion does not lose much kinetic energy. In the case of charge transfer events, however, this low-energy peak decreases substantially, while increasing the probability of higher energy loss events. Interestingly, these low-energy events correspond to high-energy ions which have lost a large fraction of their kinetic energy, and simultaneously have captured an electron from the neutral target. The latter thus becomes the newly created ion carrying a large amount of the first ion initial kinetic energy (cf. Eq. (8)).

Another quantity of interest predicted by the model is the mean free path λ of Ar^+ in the supersonic jet, which can eventually be measured experimentally. We determine λ numerically by evaluating the distance between two collisions along a given ion trajectory. For convenience, this calculation was performed dividing the space into three adjacent sectors: $(5, 10)\,mm$, $(10, 15)\,mm$, and $(15, 20)\,mm$. The mean free path was then evaluated in each sector in the case of simple scatterings only, and in the presence of charge transfer events (see Fig. 38).

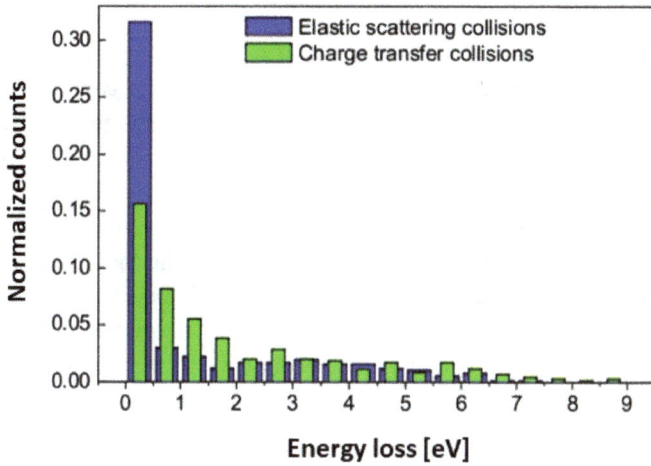

Fig. 37. Energy loss due to elastic scattering without charge transfer (blue columns), and scattering events when charge transfer occurs (light green columns), expected in a supersonic jet. These energy distributions can be associated with the energy acquired from the neutral atoms involved in collisions with Ar^+. Adapted with permission from [89]. © 2016 American Phyical Society.

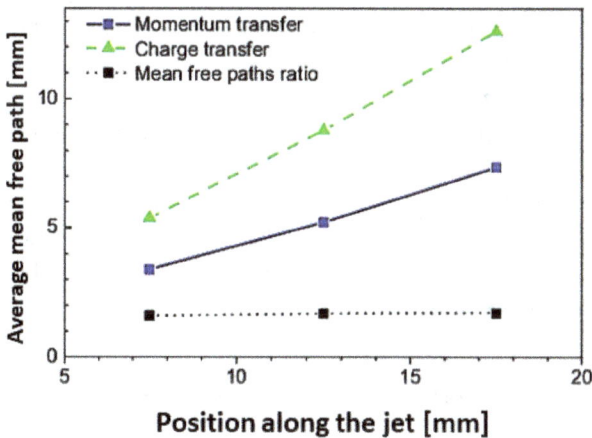

Fig. 38. Ar^+ mean free path vs sector distance z, for simple scattering events, λ_s (full squares), and including charge transfer processes, λ_c (full triangles). Their ratios, $\lambda_c/\lambda_s \simeq 1.66$, are also shown (black dotted line). Adapted with permission from [89]. © 2016 American Phyical Society.

As we mentioned in the introduction to Section 7.2, the experiments for pure Ar^+ jets have been performed for $R = 32$, corresponding to a Mach disk $z_M = 26$ mm. Such a 26 mm long supersonic jet can be considered nearly collisionless for the low density of neutral atoms present (Eq. (4)), since their mean free paths $\lambda_N \in 20$–50 mm. In contrast, we find that Ar^+ ions undergo collisions in the jet due to their relatively long-range interactions with the neutral ones (Eq. (10)). The results suggest that λ increases with z (see Fig. 38), in agreement with the fact that Ar density decreases with z, as one can see from Eq. (4). We find that for scattering including the charge exchange mechanism, the mean free path is about 5/3 times larger than for simple scattering events, that is without charge exchange.

We can now estimate the cross-sections associated to simple scattering events, σ_s, and for scattering events including charge transfer, σ_c. To do this, we resort to the general definition of the mean free path (see, e.g., [50]), $\lambda(z) = (\sigma_{s,c}(z)n(z))^{-1}$, where $\sigma_{s,c}(z)$ represents the scattering cross-section $[m^2]$ and $n(z)$, the number density $[m^{-3}]$ of the neutral Ar target at position z. Using the expression for the argon gas density in Eq. (4), we obtain the scattering cross-sections reported in Table 3, for the three sectors z (see Fig. 38). The latter can be averaged yielding the mean total cross-sections inside the jet shown in the last row of the table.

These cross-sections are in very good agreement with the values reported in the literature from beam crossing and drift tube experiments (see, e.g., [101], [105]). The values thus obtained for the energy loss (Fig. 37), and the cross-sections for Ar^+-neutral species collisions in a supersonic jet, could be employed for estimating Ar^+ energy distributions using Monte Carlo simulations on larger systems. The present results can therefore be very useful as additional information to be used in a number of PASJD technological applications.

Table 3. Cross-sections σ_s and σ_c.

z (mm)	σ_s $(10^{-19} m^2)$	σ_c $(10^{-19} m^2)$
5–10	8.3	5.3
10–15	9.1	5.4
15–20	9.5	5.6
5–20	9.0	5.4

7.4. *Summary and conclusions*

We have studied a simple Ar plasma using a PASJD technique, aimed at obtaining accurate ion energy distribution functions at different positions along the jet, by directly measuring the ion fluxes as a function of their energies using a QMS. These experimental results are complemented by extensive Monte Carlo simulations, in which the equations of motion of Ar^+ and Ar are solved numerically by using a (12-4) Lennard-Jones interaction potential between the two species.

The Ar^+ measured energies are distributed up to about 10 eV and their associated IEDFs exhibit a shape that depends on the distance z from the nozzle. In particular, as the argon gas expands supersonically, two peaks emerge, one around 2 eV and the second about 8 eV, while outside of the supersonic region, i.e., for $z \gtrsim 26$ mm, the low-energy peak is shifted down to thermal energies, $E \simeq 0.1$ eV. These results suggest that, in a supersonic jet, the ions energy behaves as in a long pre-sheath region.

The ions are produced in the plasma chamber getting an energy of about 8 eV, which broadens down to few eV as they cross the nozzle orifice. As they enter the deposition chamber, they undergo additional collisions which are assumed to be elastic in our calculations. The resulting ion energies are then measured by the QMS device. Along the expanding jet there is a plasma region at a nearly constant potential $V_{jet} = 2$ eV. Among the different collisional phenomena, we consider simple scattering events and scatterings in which a charge transfer process between the Ar^+ and the neutral Ar target takes place. These are found to be the dominant scattering processes within the jet.

The numerical calculations are performed by considering one Ar^+ at a time, interacting with randomly located neutral Ar atoms inside a suitably chosen finite system displaying the appropriate geometry for this problem. The equations of motion are integrated using a small time step of 10 ps. The interaction potential includes a charge-induced dipole interaction of the form $1/r^4$, and neglects the small contribution due to the van der Waals interaction. In the model, the additional process of a charge transfer is described by an empirically derived probability distribution, in which the associated tunneling for electron transfer depends on the transferred energy between the colliding Ar^+ and Ar. The dynamics of about 1000 ions has been simulated from $z = 10$ mm up to $z = 20$ mm, by using the measured IEDF at $z = 5$ mm as the input information for the initial kinetic energies of Ar^+ ions in the jet.

The agreement between the simulations and the experimental results is very good, thus validating the model assumptions and approximations used. In addition to these features, the scattering phenomenology was further studied by analyzing the Ar^+ energy loss in each collision, and calculating the ionic mean free paths also in the presence of charge transfer events. We find that charge transfer collisions occur in nearly 50% of the total number of scattering events. The global effect of the charge transfer mechanism on the final distribution functions is found to be quite significant. Indeed, the simulations performed without charge exchange reproduce the observed IEDFs only partially, while the inclusion of charge transfer favors medium-low ionic energies in better agreement with experiments. In addition to this, the mean free paths relative to general collision events and those displaying charge transfers can be calculated separately, thus allowing to estimate their mean cross-sections, $9.0\,10^{-19}\,m^2$ and $5.4\,10^{-19}\,m^2$, respectively, in very good agreement with estimates known in the literature.

References

[1] W. Zhang, Y. Tian, H. He, L. Xu, W. Li, and D. Zhao, Recent advances in the synthesis of hierarchically mesoporous TiO_2 materials for energy and environmental applications, *Natl. Sci. Rev.* **7**(11), 1702–1725 (2020).

[2] B. Graves, S. Engelke, C. Jo, H. G. Baldovi, J. De la Verpilliere, M. De Volder, and A. Boies, Plasma production of nanomaterials for energy storage: Continuous gas-phase synthesis of metal oxide CNT materials via a microwave plasma, *Nanoscale.* **12**(8), 5196–5208 (2020).

[3] G. Maduraiveeran, M. Sasidharan, and V. Ganesan, Electrochemical sensor and biosensor platforms based on advanced nanomaterials for biological and biomedical applications, *Biosens. Bioelectron.* **103**, 113–129 (2018).

[4] E. C. Welch, J. M. Powell, T. B. Clevinger, A. E. Fairman, and A. Shukla, Advances in biosensors and diagnostic technologies using nanostructures and nanomaterials, *Adv. Funct. Mater.* **31**(44), 2104126 (2021).

[5] G. H. Jeong, S. P. Sasikala, T. Yun, G. Y. Lee, W. J. Lee, and S. O. Kim, Nanoscale assembly of 2D materials for energy and environmental applications, *Adv. Mat.* **32**(35), 1907006 (2020).

[6] K. Li, Y. de Rancourt de Mimérand, X. Jin, J. Yi, and J. Guo, Metal oxide (ZnO and TiO_2) and Fe-based metal-organic-framework nanoparticles on 3D-printed fractal polymer surfaces for photocatalytic degradation of organic pollutants, *ACS Appl. Nano Mater.* **3**(3), 2830–2845 (2020).

[7] M. Carrola, A. Asadi, H. Zhang, D. G. Papageorgiou, E. Bilotti, and H. Koerner, Best of both worlds: Synergistically derived material properties via additive manufacturing of nanocomposites, *Adv. Funct. Mater.* **31**(46), 2103334 (2021).

[8] C. Xu, A. R. Puente-Santiago, D. Rodríguez-Padrón, M. Muñoz-Batista, M. A. Ahsan, J. C. Noveron, and R. Luque, Nature-inspired hierarchical materials for sensing and energy storage applications, *Chem. Soc. Rev.* **50**, 4856–4871 (2021).

[9] A. D. Pogrebnjak, M. Pogorielov, and R. Viter, eds., *Nanomaterials in Biomedical Application and Biosensors (NAP-2019)*. Springer Nature, Singapore (2020).

[10] I. Levchenko, O. Baranov, C. Riccardi, H. E. Roman, U. Cvelbar, E. Ivanova, M. Mandhakini, P. Ščajev, T. Malinauskas, S. Xu, and K. Bazaka, Nanoengineered carbon-based interfaces for advanced energy and photonics applications: A recent progress and innovations, *Adv. Mater. Interfaces.* **10**(1), 2201739 (2023).

[11] Y. Lin, G. Yaun, R. Liu, S. Zhou, S. W. Sheehan, and D. Wang, Semiconductor nanostructure-based photoelectrochemical water splitting: A brief review, *Chem. Phys. Lett.* **507**(4–6), 209–215 (2011).

[12] T. Zhai, X. Fang, M. Liao, X. Xu, H. Zeng, B. Yoshio, and D. Golberg, A comprehensive review of one-dimensional metal-oxide nanostructure photodetectors, *Sensors.* **9**(8), 6504–6529 (2009).

[13] K. Ostrikov, U. Cvelbar, and A. B. Murphy, Plasma nanoscience: Setting directions, tackling grand challenges, *J. Phys. D: Appl. Phys.* **44**(17), 174001 (2011).

[14] B. Ates, S. Koytepe, A. Ulu, C. Gurses, and V. K. Thakur, Chemistry, structures, and advanced applications of nanocomposites from biorenewable resources, *Chem. Rev.* **120**(17), 9304–9362 (2020).

[15] D. Thangadurai, V. Ahuja, J. Sangeetha, J. Naik, R. Hospet, M. David, A. K. Shettar, A. Torvi, S. C. Thimmappa, and N. Pujari, *Mesoporous Nanomaterials: Properties and Applications in Environmental Sector*, In eds. O. V. Kharissova, L. M. Torres-Martínez, and B. I. Kharis, *Handbook of Nanomaterials and Nanocomposites for Energy and Environmental Applications*, pp. 403–420. Springer, Cham. (2021).

[16] X. Zhang and Y. Liu, Nanomaterials for radioactive wastewater decontamination, *Environ. Sci.: Nano.* **7**(4), 1008–1040 (2020).

[17] A. S. Rajashekharaiah, G. P. Darshan, H. B. Premkumar, P. Lalitha, S. C. Sharma, and H. Nagabhushana, Hierarchical $Bi_2Zr_2O_7:Dy^{3+}$ architectures fabricated by bio-surfactant assisted hydrothermal route for anti-oxidant, anti-bacterial and anti-cancer activities, *Mat. Chem. Phys.* **242**, 122468 (2020).

[18] C. Chen, M. K. Singh, K. Wunderlich, S. Harvey, C. J. Whitfield, Z. Zhou, M. Wagner, K. Landfester, I. Lieberwirth, G. Fytas, and K. Kremer, Polymer cyclization for the emergence of hierarchical nanostructures, *Nature Commun.* **12**(1), 1–7 (2021).

[19] Z. Wang, G. Wang, W. Li, Z. Cui, J. Wu, I. Akpinar, L. Yu, G. He, and J. Hu, Loofah activated carbon with hierarchical structures for high-efficiency adsorption of multi-level antibiotic pollutants, *Appl. Surf. Sci.* **550**, 149313 (2021).

[20] J. Huang, J. Liu, and J. Wang, Optical properties of biomass-derived nanomaterials for sensing, catalytic, biomedical and environmental applications, *TrAC Trends Anal. Chem.* **124**, 115800 (2020).

[21] R. Ajdary, B. L. Tardy, B. D. Mattos, L. Bai, and O. J. Rojas, Plant nanomaterials and inspiration from Nature: Water interactions and hierarchically structured hydrogels, *Adv. Mater.* **33**(28), 2001085 (2021).

[22] G. D. Cha, W. H. Lee, C. Lim, M. K. Choi, and D. H. Kim, Materials engineering, processing, and device application of hydrogel nanocomposites, *Nanoscale.* **12**(19), 10456–10473 (2020).

[23] K. Fu, H. Wu, and Z. Su, Self-assembling peptide-based hydrogels: Fabrication, properties, and applications, *Biotechnol. Adv.* **49**, 107752 (2021).

[24] R. Zeng, C. Lv, C. Wang, and G. Zhao, Bionanomaterials based on protein self-assembly: Design and applications in biotechnology, *Biotechnol. Adv.* **52**, 107835 (2021).

[25] H. Mohan, A. Dalal, M. Prasad, and J. S. Rana, *Chapter 3 — Biosensor fabrication with nanomaterials*, In eds. C. M. Hussain and S. K. Kailasa, *Micro and Nano Technologies*, pp. 31–55. Handbook of Nanomaterials for Sensing Applications. Elsevier (2021).

[26] Y. Esmaeili, A. Zarrabi, S. Z. Mirahmadi-Zare, and E. Bidram, Hierarchical multifunctional graphene oxide cancer nanotheranostics agent for synchronous switchable fluorescence imaging and chemical therapy, *Microchim. Acta.* **187**(10), 1–15 (2020).

[27] A. Kanioura, P. Petrou, D. Kletsas, A. Tserepi, M. Chatzichristidi, E. Gogolides, and S. Kakabakos, Three-dimensional (3D) hierarchical oxygen plasma micro/nanostructured polymeric substrates for selective enrichment of cancer cells from mixtures with normal ones, *Colloids Surf. B Biointerfaces.* **187**, 110675 (2020).

[28] K. Ganguly, D. K. Patel, S. D. Dutta, W. C. Shin, and K. T. Lim, Stimuli-responsive self-assembly of cellulose nanocrystals (CNCs): Structures, functions, and biomedical applications, *Int. J. Biol. Macromol.* **155**, 456–469 (2020).

[29] P. Liu, G. Wang, Q. Ruan, K. Tang, and P. K. Chu, Plasma-activated interfaces for biomedical engineering, *Bioact. Mat.* **6**(7), 2134–2143 (2021).

[30] H. B. Profijt and W. M. M. Kessels, Ion bombardment during plasma-assisted atomic layer deposition, *ECS Trans.* **50**(13), 23–34 (2013).

[31] D. M. Mattox, Particle bombardment effects on thin-film deposition: A review, *J. Vac. Sci. Technol. A: Vacuum, Surfaces, Films.* **7**(3), 1105 (1989).

[32] C. Riccardi, M. Piselli, F. S. Fumagalli, F. Di Fonzo, and C. E. Bottani, Method and apparatus for depositing nanostructured thin layers with controlled morphology and nanostructures: U.S. Patent 8,795,791 (2010–2014).

[33] O. F. Hagena, Cluster formation in expanding supersonic jets: Effect of pressure, temperature, nozzle size, and test gas, *J. Chem. Phys.* **56**(5), 1793 (1972).

[34] M. C. M. van de Sanden, J. M. de Regt, and D. C. Schram, Recombination of argon in an expanding plasma jet, *Phys. Rev. E.* **47**(4), 2792–2797 (1993).

[35] S. E. Selezneva and M. I. Boulos, Supersonic induction plasma jet modeling, *Nucl. Instr. Meth. Phys. Res. B.* **180**(1–4), 306–311 (2001).

[36] I. Biganzoli, F. Fumagalli, F. Di Fonzo, R. Barni, and C. Riccardi, A supersonic plasma jet source for controlled and efficient thin film deposition, *J. Mod. Phys.* **03**(10), 1626–1638 (2012).

[37] V. Trifiletti, R. Ruffo, C. Turrini, D. Tassetti, R. Brescia, F. Di Fonzo, C. Riccardi, and A. Abbotto, Dye-sensitized solar cells containing plasma jet deposited hierarchically nanostructured TiO_2 thin photoanodes, *J. Mat. Chem.* **1**(38), 11665 (2013).

[38] S. Caldirola, R. Barni, and C. Riccardi, Characterization of a low pressure supersonic plasma jet, *J. Phys.: Conf. Series.* **550**, 012042 (2014).

[39] S. Caldirola, H. E. Roman, and C. Riccardi, Ion energy distribution functions in a supersonic plasma jet, *J. Phys.: Conf. Series.* **550**, 012043 (2014).

[40] E. C. Dell'Orto, S. Caldirola, H. E. Roman, and C. Riccardi. Nanostructured TiO_2 film deposition by supersonic plasma jet source for energetic application. In *Oral, A. Y. and Bashi Oral, Z. B. and Ozer, M. (eds.) Proceedings of the 2nd International Congress on Energy Efficiency and Energy Related Materials (ENEFM2014)*, pp. 349–355, Oludeniz, Fethiye/Mugla, Turkey (16–19 October, 2014).

[41] S. Caldirola, R. Barni, H. E. Roman, and C. Riccardi, Mass spectrometry measurements of a low pressure expanding plasma jet, *Plasma Sources Sci. Tech.* **33**(6), 061306 (2015).

[42] E. C. Dell'Orto, S. Caldirola, A. Sassella, V. Morandi, and C. Riccardi, Growth and properties of nanostructured titanium dioxide deposited by supersonic plasma jet deposition, *Appl. Surf. Sci.* **425**, 407–415 (2017).

[43] G. Nava, F. Fumagalli, S. Neutzner, and F. Di Fonzo, Large area porous 1D photonic crystals comprising silicon hierarchical nanostructures grown by plasma-assisted, nanoparticle jet deposition, *Nanotechnology.* **29**(46), 465603 (2018).

[44] C. Carra, E. C. Dell'Orto, V. Morandi, and C. Riccardi, ZnO nanostructured thin films via supersonic plasma jet deposition, *Coatings.* **10**(8), 788 (2020).

[45] C. Piferi, C. Carra, K. Bazaka, H. E. Roman, E. C. Dell'Orto, V. Morandi, I. Levchenko, and C. Riccardi, Controlled deposition of nanostructured hierarchical TiO_2 thin films by low pressure supersonic plasma jets, *Nanomaterials.* **12**(3), 533 (2022).

[46] R. B. Piejak, V. A. Godyak, and B. M. Alexandrovich, A simple analysis of an inductive RF discharge, *Plasma Sources Sci. Tech.* **1**(3), 179–186 (1992).

[47] U. R. Kortshagen and M. Zethoff, Ion energy distribution functions in a planar inductively coupled RF discharge, *Plasma Sources Sci. Tech.* **4**(4), 541–550 (1995).

[48] U. Kortshagen, N. D. Gibson, and J. E. Lawler, On the E-H mode transition in RF inductive discharges, *J. Phys. D: Appl. Phys.* **29**(5), 1224 (1996).

[49] U. R. Kortshagen, U. V. Bhandarkar, M. T. Swihart, and S. L. Girshick, Generation and growth of nanoparticles in low-pressure plasmas, *Plasma Sources Sci. Tech.* **71**(10), 1871–1877 (1999).

[50] G. Sanna and G. Tomassetti, *Introduction to Molecular Beams Gas Dynamics*. Imperial College Press, London (2005).

[51] I. Petrov, L. Hultman, U. Helmersson, J. E. Sundgren, and J. Greene, Microstructure modification of TiN by ion bombardment during reactive sputter deposition, *Thin Solid Films.* **169**(2), 299–314 (1989).

[52] R. Barni, S. Zanini, and C. Riccardi, Characterization of the chemical kinetics in an O_2/HMDSO RF plasma for material processing, *Adv. Phys. Chem.* **2012**, 1–6 (2012).

[53] R. Barni, S. Zanini, and C. Riccardi, Diagnostics of reactive RF plasmas, *Vacuum.* **82**(2), 217–219 (2007).

[54] M. A. Lieberman and A. J. Lichtenberg, *Principles of Plasma Discharges and Materials Processing (2nd Ed.)*. John Wiley & Sons, Inc., Hoboken, NJ, USA (2005).

[55] S. H. Zhao, O. Samuel, and H. B. Kagan, Asymmetric oxidation of sulfides mediated by chiral titanium complexes: Mechanistic and synthetic aspects, *Tetrahedron.* **43**(21), 5135–5144 (1987).

[56] H. Nizard, M. L. Kosinova, N. I. Fainer, Y. M. Rumyantsev, B. M. Ayupov, and Y. V. Shubin, Deposition of titanium dioxide from TTIP by plasma enhanced and remote plasma enhanced chemical vapor deposition, *Surf. Coat. Technol.* **202**(17), 4076–4085 (2008).

[57] P. A. Delattre, T. Lafleur, E. Johnson, and J. P. Booth, Radio-frequency capacitively coupled plasmas excited by tailored voltage waveforms: Comparison of experiment and particle-in-cell simulations, *J. Phys. D: Appl. Phys.* **46**(23), 235201 (2013).

[58] J. Benedikt, A. Hecimovic, D. Ellerweg, and A. von Keudell, Quadrupole mass spectrometry of reactive plasmas, *J. Phys. D: Appl. Phys.* **45**(40), 403001 (2012).

[59] M. D. Logue, H. Shin, W. Zhu, L. Xu, V. M. Donnelly, D. J. Economou, and M. J. Kushner, Ion energy distributions in inductively coupled plasmas having a biased boundary electrode, *Plasma Sources Sci. Tech.* **21**(6), 065009 (2012).

[60] E. A. G. Hamers, W. G. J. H. M. van Sark, J. Bezemer, W. J. Goedheer, and W. F. van der Weg, On the transmission function of an ion-energy and mass spectrometer, *Int. J. Mass Spectrometry and Ion Proc.* **173**(1–2), 91–98 (1998).

[61] D. O'Connell, R. Zorat, A. R. Ellingboe, and M. M. Turner, Comparison of measurements and particle-in-cell simulations of ion energy distribution functions in a capacitively coupled radio-frequency discharge, *Phys. Plasmas.* **14**(10), 103510 (2007).

[62] R. Barni, P. Esena, and C. Riccardi, Chemical kinetics simulation for atmospheric pressure air plasmas in a streamer regime, *J Appl. Phys.* **97**(7), 073301 (2005).

[63] B. M. Annaratone, M. W. Allen, and J. E. Allen, Ion currents to cylindrical langmuir probes in RF plasmas, *J. Phys. D: Appl. Phys.* **25**(3), 417 (1992).

[64] F. F. Chen, Langmuir probe analysis for high density plasmas, *Phys. Plasmas.* **8**(6), 3029–3041 (2001).

[65] F. F. Chen, J. D. Evans, and D. Arnush, A floating potential method for measuring ion density, *Phys. Plasmas.* **9**(4), 1449–1455 (2002).

[66] F. F. Chen, Langmuir probes in RF plasma: Surprising validity of OML theory, *Plasma Sources Sci. Tech.* **18**(3), 035012 (2009).

[67] H. Timko, P. S. Crozier, M. M. Hopkins, K. Matyash, and R. Schneider, Why Perform Code-to-Code Comparisons: A Vacuum Arc Discharge Simulation Case Study, *Contrib. Plasma Phys.* **52**(4), 295–308 (2012).

[68] V. A. Godyak, R. B. Piejak, and B. M. Alexandrovich, Measurement of electron energy distribution in low-pressure RF discharges, *Plasma Sources Sci. Tech.* **1**(1), 36–58 (1992).

[69] I. D. Sudit and F. F. Chen, RF compensated probes for high-density discharges, *Plasma Sources Sci. Tech.* **3**(2), 162–168 (1994).

[70] C. Riccardi, R. Barni, and M. Fontanesi, Experimental study and simulations of electronegative discharges at low pressure, *J. Appl. Phys.* **90**(8), 3735–3742 (2001).

[71] R. A. Gottscho and V. M. Donnelly, Optical emission actinometry and spectral line shapes in RF glow discharges, *J. Appl. Phys.* **56**(2), 245–250 (1984).

[72] J. B. Fenn, Mass spectrometric implications of high-pressure ion sources, *Int. J. of Mass Spectrometry.* **200**(1–3), 459–478 (2000).

[73] H. R. Murphy and D. R. Miller, Effects of nozzle geometry on kinetics in free-jet expansions, *J. Phys. Chem.* **88**(20), 4474–4478 (1984).

[74] H. Ashkenas and F. S. Sherman, The structure and utilization of supersonic free jets in low density wind tunnels. In J. H. De Leeuw (ed.), *Rarefied Gas Dynamics — Volume 2. Proceedings of the Fourth International Symposium held at the Institute for Aerospace Studies, Toronto, 1964*, pp. 84–105. Academic Press, Toronto (1966).

[75] C. Huang, W. T. Nichols, D. T. O'Brien, M. F. Becker, D. Kovar, and J. W. Keto, Supersonic jet deposition of silver nanoparticle aerosols: Correlations of impact conditions and film morphologies, *J. Appl. Phys.* **101**(6), 064902 (2007).

[76] O. Abouali, S. Saadabadi, and H. Emdad, Numerical investigation of the flow field and cut-off characteristics of supersonic/hypersonic impactors, *J. Aerosol Science.* **42**(2), 65–77 (2011).

[77] R. Campargue, Aerodynamic separation effect on gas and isotope mixtures induced by invasion of the free jet shock wave structure, *J. Chem. Phys.* **52**(4), 1795–1982 (1970).

[78] V. H. Reis and J. B. Fenn, Separation of gas mixtures in supersonic jets, *J. Chem. Phys.* **39**(12), 3240–3250 (1963).

[79] P. C. Waterman and S. A. Stern, Separation of gas mixtures in a supersonic jet, *J. Chem. Phys.* **31**(2), 405–419 (1959).

[80] E. L. Knuth, Composition distortion in MBMS sampling, *Combust. Flame.* **103**(3), 171–180 (1995).

[81] P. Milani, P. Piseri, E. Barborini, A. Podesta, and C. Lenardi, Cluster beam synthesis of nanostructured thin films, *J. Vac. Sci. Technol. A: Vacuum, Surfaces, Films.* **19**(4), 2025 (2001).

[82] T. A. Witten Jr and L. M. Sander, Diffusion-limited aggregation, a kinetic critical phenomenon, *Phys. Rev. Lett.* **47**(19), 1400 (1981).

[83] X. Chen and S. S. Mao, Synthesis of titanium dioxide nanomaterials, *J. Nanosci. Nanotechnol.* **6**(4), 906–925 (2006).

[84] M. Paulose, O. K. Varghese, G. K. Mor, C. A. Grimes, and K. G. Ong, Unprecedented ultra-high hydrogen gas sensitivity in undoped titania nanotubes, *Nanotechnology.* **17**(2), 398 (2005).

[85] J. Burschka, N. Pellet, S. J. Moon, R. Humphry-Baker, P. Gao, M. K. Nazeeruddin, and M. Grätzel, Sequential deposition as a route to high-performance perovskite-sensitized solar cells, *Nature.* **499**(7458), 316–319 (2013).

[86] A. Robledo, C. N. Grabill, S. M. Kuebler, A. Dutta, H. Heinrich, and A. Bhattacharya, Morphologies from slippery ballistic deposition model: A bottom-up approach for nanofabrication, *Phys. Rev. E.* **83**(5), 051604 (2011).

[87] H. Shin, W. Zhu, L. Xu, V. M. Donnelly, and D. J. Economou, Control of ion energy distributions using a pulsed plasma with synchronous bias on a boundary electrode, *Plasma Sources Sci. Tech.* **20**(5), 055001 (2011).

[88] M. Fivaz, S. Brunner, W. Schwarzenbach, A. A. Howling, and C. Hollenstein, Reconstruction of the time-averaged sheath potential profile in an argon radiofrequency plasma using the ion energy distribution, *Plasma Sources Sci. Tech.* **4**(3), 373–378 (1995).

[89] S. Caldirola, H. E. Roman, and C. Riccardi, Ion dynamics in a supersonic jet: Experiments and simulations, *Phys. Rev. E.* **93**(3), 033202 (2016).

[90] D. Barton, J. W. Bradley, D. A. Steele, and R. D. Short, Investigating radio frequency plasmas used for the modification of polymer surfaces, *J. Phys. Chem. B.* **103**(21), 4423–4430 (1999).

[91] S. D. Tanner, Plasma temperature from ion kinetic energies and implications for the source of diatomic oxide ions in inductively coupled plasma mass spectrometry, *J. Anal. Atomic Spectrometry.* **8**(6), 891 (1993).

[92] J. K. Olthoff, R. J. Van Brunt, and S. B. Radovanov, Effect of electrode material on measured ion energy distributions in radio-frequency discharges, *Appl. Phys. Lett.* **67**(4), 473 (1995).

[93] J. B. Lee, H. Y. Chang, and S. H. Seo, Collisional effect on the time evolution of ion energy distributions outside the sheath during the afterglow of pulsed inductively coupled plasmas, *Plasma Sources Sci. Tech.* **22**(6), 065008 (2013).

[94] H. H. Michels, R. H. Hobbs, and L. A. Wright, Electronic structure of the noble gas dimer ions. I. potential energy curves and spectroscopic constants, *J. Chem. Phys.* **69**(11), 5151 (1978).

[95] J. Soler, J. Saenz, N. García, and O. Echt, The effect of ionization on magic numbers of rare-gas clusters, *Chem. Phys. Lett.* **109**(1), 71–75 (1984).

[96] S. A. Maiorov, Ion drift in a gas in an external electric field, *Plasma Phys. Rep.* **35**(9), 802–812 (2009).

[97] W. L. T. Chen, J. Heberlein, and E. Pfender, Critical analysis of viscosity data of thermal argon plasmas at atmospheric pressure, *Plasma Chem. Plasma Process.* **16**(4), 635–650 (1996).

[98] J. A. White, Lennard-Jones as a model for argon and test of extended renormalization group calculations, *J. Chem. Phys.* **111**(20), 9352 (1999).

[99] B. H. Bransden and M. R. C. McDowell, *Charge Exchange and the Theory of Ion-atom Collisions*. Clarendon Press, Oxford (1992).

[100] D. Rapp and W. E. Francis, Charge exchange between gaseous ions and atoms, *J. Chem. Phys.* **37**(11), 2631 (1962).

[101] A. V. Phelps, Cross sections and swarm coefficients for nitrogen ions and neutrals in N_2 and argon ions and neutrals in Ar for energies from 0.1 ev to 10 kev, *J. Phys. Chem. Ref. Data.* **20**(3), 557–573 (1991).

[102] R. Hippler, J. Kredl, and V. Vartolomei, Ion energy distribution of an inductively coupled radiofrequency discharge in argon and oxygen, *Vacuum.* **83**(4), 732–737 (2008).

[103] E. Kawamura, V. Vahedi, M. A. Lieberman, and C. K. Birdsall, Ion energy distributions in RF sheaths: Review, analysis and simulation, *Plasma Sources Sci. Tech.* **8**(3), R45–R64 (1999).

[104] T. Chevolleau and W. Fukarek, Ion flux, ion energy distribution and neutral density in an inductively coupled argon discharge, *Plasma Sources Sci. Tech.* **9**(4), 568–573 (2000).

[105] M. R. Alexander, F. R. Jones, and R. D. Short, Mass spectral investigation of the radio-frequency plasma deposition of hexamethyldisiloxane, *J. Phys. Chem. B.* **101**(18), 3614–3619 (1997).

Index